Environmental Engineering

Series Editors: U. Förstner, R. J. Murphy, W. H. Rulkens

L. Theodore · A. Buonicore (Eds.)

Air Pollution Control Equipment

Selection, Design, Operation and Maintenance

With 129 Figures

Springer-Verlag

Berlin Heidelberg NewYork
London Paris Tokyo
Hong Kong Barcelona Budapest

Series Editors

Prof. Dr. U. Förstner

Arbeitsbereich Umweltschutztechnik
Technische Universität Hamburg-Harburg
Eißendorfer Straße 40
D-21073 Hamburg, Germany

Prof. Robert J. Murphy

Dept. of Civil Engineering and Mechanics
College of Engineering
University of South Florida
4202 East Fowler Avenue, ENG 118
Tampa, FL 33620-5350, USA

Prof. Dr. ir. W.H. Rulkens

Wageningen Agricultural University
Dept. of Environmental Technology
Bomenweg 2, P.O. Box 8129
NL-6700 EV Wageningen, The Netherlands

Editors

Prof. Louis Theodore
Dept. of Chemical Engineering
Manhattan College
Riverdale, NY 10471, USA

Anthony J. Buonicore, P.E; President
Environmental Data Resources Inc.
3530 Post Road
Southport, CT 06490, USA

ISBN-13: 978-3-642-85146-9   e-ISBN-13: 978-3-642-85144-5
DOI: 10.1007/978-3-642-85144-5

Cip data applied for

Typesetting: Camera-ready by author
SPIN:10427335   61/3020-5 4 3 2 1 0 - Printed on acid -free paper

ETS PROFESSIONAL TRAINING INSTITUTE

For almost two decades the staff of ETS International, Inc. has trained thousands of air pollution control professionals from industry, government agencies, and utilities.

Annual seminars have been offered for Fabric Filters since 1980. The training staff of ETS has provided an air pollution control technologies course to key technical personnel from developing nations. In addition, senior staff of ETS has led courses in particulate emissions control, gaseous emissions control, regulatory and permitting policies, cost containment, operation and maintenance, stationary source testing, continuous emission monitoring and general air pollution control technology.

The latest educational and training aids are utilized in every classroom. Textbooks and other study aids prepared at ETS enable students to continue learning after completing their seminars.

We look forward to helping you grow as a professional in the challenging, exciting, and critically important field of air pollution monitoring and control.

Other training tools available through the ETS Professional Training Institute.

ETS/Theodore Tutorials

- A series of self-instructional problem workbooks for the environmental professional.

ETS Textbooks

- Air Pollution Control Equipment
- Fabric Filter Baghouses I
- Fabric Filter Baghouses II
- Fabric Filter Baghouses III

ETS Training Seminars

- Custom On-Site Seminars Geared To Clients Specific Needs
- Fabric Filter Baghouse
- Air Permitting
- VOC Control
- Incinerator Emissions Control

ETS Cassette Tapes

- Fabric Filter Baghouse Seminar

ETS Professional Training Institute

ETS International, Inc. • 1401 Municipal Road • Roanoke, VA 24012-1309

Phone (703) 265-0004 • FAX (703) 265-0131

Contents

Preface

With the advent of the Clean Air Act in 1970, the number of air pollution control equipment installations has increased at an accelerated pace. Although much has been written on attaining collection performance with the various control devices, a major void has occurred in the identification and transfer of information needed to help reduce maintenance costs and to prevent deterioration of collector performance.

Although design and selection information is presented, it is the primary intention of this book to discuss operation and maintenance topics and explore many of the repetitive problems that have plagued users of air pollution control equipment. The existence of these problems may be related to the complexity of the process or to a lack of well-defined operation techniques, among other reasons. In any event, this book intends to emphasize where and how these factors can have a major impact on the maintenance problems of control devices.

Operation and maintenance problems have plagued users for nearly 100 years. The number and complexity of these problems has increased at a nearly exponential rate since the early 1970s. The 1950s and 1960s produced installations designed for the low to medium collection efficiency range. During the 1970s, designs of high efficiencies were provided that often contained more component parts in more complex arrangements. The time normally required for the training of field technicians and engineers on the problems (and solutions) with these newer installations for both gas and particulate control was not available. Although there has been a concentrated effort in recent years to better understand air pollution control equipment, the result—unfortunately—is akin to placing the cart before the horse.

The air pollution control device can represent a major portion of the investment in an industrial plant. Control equipment life and performance are essential to the economic operation of the plant; the need for a sound, well-planned operation and maintenance program is obvious. The success or failure of a particular installation will often rest on the implementation of this program, or its equivalent. Normal control equipment operations should include proper procedures for startup and shutdown, as well as guidelines for troubleshooting and improving operation. Once onstream, equipment failures need to be analyzed to examine major causes. Obtaining backup information for preventive maintenance programs and spare-parts requirements is also necessary. Although specific maintenance program requirements may vary from process to process, from plant to plant, and from industry to industry, some basic steps and procedures are common to all. These include safety, inspection programs, and definition of inspection and maintenance responsibilities.

There are several reasons for an operation and maintenance program. The most important of which include:

1. Continuously meeting present emission regulatory control codes and improving air quality
2. Prolonging control equipment life
3. Maintaining productivity (reducing production interruptions) of the process unit served by the control device
4. Prolonging process life
5. Reducing operating costs
6. Promoting better public relations and avoiding community alienation
7. Promoting better relations with regulatory officials

Regarding its contents, the book begins by considering the overall strategy and general approach employed in the presentation of this subject. The second chapter separately considers selection procedures for air pollution control equipment. The remainder of the book is divided into 10 chapters, with each chapter devoted entirely to a specific control device or process. Starting with Chapter 3, the following air pollution control equipment are reviewed: absorbers, adsorbers, incinerators, condensers, cyclones, baghouses, scrubbers, and electrostatic precipitators. Flue gas desulfurization systems are considered in Chapter 11. The book concludes with a section (Chapter 12) on the very important subject of stacks. Each chapter considers the operation and maintenance characteristics of the particular control device in a somewhat analogous manner; topics such as description of control device, design procedure, installation, operation, maintenance, and methods of improving operation and performance are examined.

Knowledge of the information developed and presented in the various chapters is essential not only to the proper operation and maintenance but also to the design and selection of industrial control equipment for atmospheric pollutants. It will enable the reader to obtain a better understanding of both the system itself and those factors affecting system operation and performance. Such knowledge is essential in view of the stringency of existing and anticipated air pollution control regulations.

Louis Theodore
Anthony J. Buonicore

1

Introduction

Louis Theodore, D. Eng. Sc.

Professor of Chemical Engineering
Manhattan College
Bronx, New York

1.1 ABOUT THE BOOK

In the last two decades, the engineering profession has expanded its responsibilities to society to include the control of air pollution from industrial sources. Increasing numbers of engineers and applied scientists are being confronted with problems in this most important area. Although the design and construction of air pollution control equipment today is accomplished with some degree of success, operation and maintenance information on these control devices is sorely lacking. The environmental engineer and field technician of today and tomorrow must develop a proficiency and an improved understanding of not only the design and selection, but also the operation and maintenance of air pollution control equipment in order to cope with these problems and challenges. It was in this spirit that this book was undertaken.

The book begins by considering the overall strategy and general approach employed in the presentation of this subject, with particular emphasis on operation and maintenance programs. The second chapter separately considers selection procedures for air pollution control equipment. The remainder of the book is divided into 10 chapters, each chapter devoted entirely to a specific control device. Starting with Chapter 3, the following air pollution control equipment are reviewed: absorbers, adsorbers, incinerators, condensers, cyclones, baghouses, scrubbers, and electrostatic precipitators. Flue gas desulfurization systems are considered in Chapter 11. The book concludes with a chapter on the very important subject of stacks. Each chapter considers the operation and maintenance characteristics of the particular control device in a somewhat analogous manner; topics such as description of control device, design procedure, installation, operation, maintenance, and methods of improving operation and performance are examined.

1.2 THE NEED FOR AN OPERATION AND MAINTENANCE PROGRAM

Operation and maintenance problems have plagued users since the installation of the first pollution control device. However, because of the proliferation of equipment installed as a result of the Clean Air Act Amendments of 1970, the operation and maintenance problem has magnified intensely. The number and complexity of these problems has increased at a nearly exponential rate since the early 1970s. The 1950s and 1960s involved installations designed for the low to medium collection efficiency range. During the 1970s, designs of high efficiencies were provided that contained more and more component parts. The time normally required for the training of field technicians and engineers on the problems (and solutions) with these newer installations for both gas and particulate control was not available. Although there has been a concentrated effort in recent years to better understand air pollution control equipment, the result—unfortunately—is likened to placing the cart before the horse.

There are several reasons for an operation and maintenance program. The most important are:

1. Continuously meeting present emission regulatory control codes and improving air quality
2. Prolonging control equipment life
3. Maintaining productivity (reducing production interruptions) of the process unit served by the control device
4. Prolonging process life
5. Reducing operating costs
6. Promoting better public relations and avoiding community alienation
7. Promoting better relations with regulatory officials

Thus, the proper operation of the control device requires a planned program of operator's training, equipment know-how, preventive maintenance, and an adequate spare-parts inventory to maintain compliance with applicable air pollution emission codes, and it can prevent limitations of the plant's production schedule.

1.3 THE MAINTENANCE PROGRAM

The objective of this section is to provide a general discussion of the importance of maintenance to overall plant performance. The plant, including the air pollution control equipment, must be recognized as a highly specialized and complex facility producing an acceptable gaseous effluent. It is the plant management's responsibility to produce this atmospheric discharge at the lowest unit cost and at the highest quality possible. The key to fulfilling this responsibility is a sound maintenance management program.

This section outlines the basic features of a comprehensive maintenance management system; additional details are given later in the chapter. The following is a list of the basic features that should be included and should be applicable to all plants, regardless of type or size:

1. Equipment record system
2. Planning and scheduling
3. Storeroom and inventory system
4. Maintenance personnel
5. Cost and budgets for maintenance operations
6. Special tools and equipment
7. Lubrication
8. Major equipment information
9. Warranty and contract provisions

General details on each of these subsections are given below. Applications to specific control devices are treated in subsequent chapters. Specific details on system description, source sampling, records, personnel, and safety are discussed in later sections of this chapter.

Equipment Record System

Responsible individuals should prepare an equipment record system for the facility. The facility owner and/or administrator should be involved in selecting the record-keeping system to be used. The equipment record system should contain information on each item of equipment. This system may be only one card on each item of equipment, a number of cards for each item or for larger plants, a combination of information cards and data maintained on a computer.

The first step in establishing an equipment record system is to select an equipment numbering system that best satisfies the needs of the particular air pollution control facility. Each item of equipment in the plant requiring maintenance should be assigned a number for easy identification and to help ensure that all equipment receives proper attention. After each item of equipment has been assigned a number, a catalog should be prepared that lists equipment descriptions, locations, and equipment numbers. The catalog will provide a convenient reference for locating equipment and identifying equipment numbers. Various card files are available, such as single equipment cards, a three-card system, edge-punched cards, and card files which are set upright or are located in a horizontal position and have an edge exposed on which color tabs are placed. These color-coding tabs mark the month and week in which preventive maintenance work is to be performed. These card systems are readily available through most office supply agencies, and the systems can be adapted to plants based on their individual needs. The person recommending the equipment record system should ensure that the system contains the following information:

1. Description of equipment and equipment number with location in plant.
2. Supplier with address, representative, phone number, and date of purchase, with cost.
3. Size, model, type, and serial number.
4. Electrical and/or mechanical data.
5. Inventory of spare parts on hand.
6. Preventive maintenance (PM) items to be accomplished with their frequency. Space to note when PM was performed, by whom, and pertinent comments. Data on worker-hours, cost, and material or supplies consumed.
7. Information on corrective maintenance work should be maintained in a manner similar to that outlined above for preventive maintenance.
8. A system to compile this information for use in determining costs and for future use in budget development.

Planning and Scheduling

Air pollution control facilities often do not observe holidays and vacation shutdowns. The facilities do experience variations in flows and maintenance work loads. Under these conditions, it is imperative that maintenance be planned and scheduled so that there is no idle time or peak work load period. Maintenance scheduling will vary with the size and complexity of the facility and with the type of personnel available. However, proper management will provide a maintenance plan no matter what the facility size.

In planning and scheduling preventive maintenance work, the size and capabilities of the maintenance staff will affect the amount of maintenance work that can be accomplished. Corrective maintenance functions should be considered when setting up the preventive maintenance program. Corrective maintenance requirements can be estimated from past experience and maintenance history records. Maintenance labor standards will aid in determining the time required to accomplish specific maintenance tasks. Provisions should be made in planning and scheduling to allow time in the maintenance schedule to perform corrective maintenance. The extent to which the corrective maintenance can be planned will aid in determining the actual time that will be available to perform preventive maintenance tasks.

A schedule chart with priorities of subjects, personnel, and time is a convenient aid to reduce impulse searches for work for idle personnel. The schedule chart may be divided into daily, weekly, monthly, quarterly, semiannually, and yearly sections so that the entire range of maintenance functions can be observed. Color tabs and labels can be coded to account for all personnel and their duties at a specific point in time. The removal of the tag from the schedule chart board indicates that the work is under way or has been completed. The chart board provides a graphic indication of progress and manpower usage. The chart board also provides a graphic indication of tasks that are running behind. Chart boards are available from most office supply companies. The size, method of use, and detail of the schedule chart board depends upon the control facility management.

Indoor and outdoor maintenance should be scheduled to take advantage of open or inclement weather, low load (concentration) or flow-rate periods, and other variable conditions beyond the control of the operating staff. All maintenance work should be scheduled just as the operating routine has to be scheduled. Preventive maintenance should not be a haphazard procedure to be done if time permits. Some type of maintenance is often scheduled for the once-a-year opportunity when the plant load normally is at its lowest.

The manufacturer's maintenance manual is generally the best guide for preventive maintenance instructions for any item of equipment. Most equipment is mass-produced on a competitive basis and the cost of its maintenance should be consistent with its value, life expectancy, and replacement costs. Equipment should be rated as to its critical position in the plant operating system and its maintenance priority. Unnecessary or too frequent preventive maintenance can be as wasteful as improper maintenance procedures.

The plant maintenance personnel should be cautioned to continuously monitor the plant operations to determine other maintenance work to be accomplished. When emergencies and breakdowns occur, these tasks must be reviewed and work initiated to return the plant to its full capacity and efficiency.

Administrators should be guided to arrangements with contractors and repair services to aid the facility maintenance personnel in performing various maintenance tasks and emergency repairs. These tasks may include electrical and mechanical problems or other malfunctions. Other general maintenance tasks, such as snow removal, ground work, painting, or other minor functions, can be contracted out. A suggested list of contractors and contract jobs should be readily available. A work-order system should be established to initiate all corrective maintenance tasks.

The work-order system will aid in identifying work to be accomplished, procedure priority, and information on any special aspects of the job. A log of the work orders will provide a record of when the work order was initiated and completed.

Storeroom and Inventory System

Each control facility should have a storeroom and an inventory system. A central storeroom for spare parts, equipment, and supplies should be maintained. A review of the equipment and the manufacturer's recommendations will aid in determining what spare parts and miscellaneous supplies should be readily available. The spare parts and components should be listed in a central catalog and assigned a number. A minimum and maximum quantity to be stocked should be established. A card system to record information on quantity, item number, description, when last purchased, cost, date, vendor, and other information is helpful. A system for arranging items in the storeroom should be established to aid in locating items. When items are taken from this stock, the date and use should be noted on the card file. For items that may be long-lead-time items, a reorder point should be established to aid in resupply. A storeroom clerk could maintain accountability of the parts, keep records, initiate information for purchase orders, and handle a stores withdrawal system to maintain information on parts obtained by each division. A store ticket or withdrawal slip should be completed when any item is used. The ticket will be a record showing when the item was used and for what purpose. This information can be transferred to the card for the item to aid in determining when reorder is required. This system will provide an inventory of items currently in stock.

A purchase-order system should also be established. The system should provide a record of the date an item was ordered and when received, quantity, unit cost, total cost, supplier, and item destination (stock, preventive maintenance, or corrective maintenance.) Standing purchase orders can be used effectively for spreading out delivery of large quantities of supplies.

Maintenance Personnel

Only properly trained personnel should be expected to perform satisfactory inspections, repairs, and preventive maintenance tasks. Properly trained personnel should possess a thorough knowledge of the functions and operations of their equipment and the procedures for servicing it. A good maintenance management program must consider the limitations of plant operators and maintenance or factory representatives to perform certain required maintenance functions.

Job titles, job descriptions, and qualifications for maintenance personnel should be provided. Any general information regarding maintenance personnel or particular information on possible sources of maintenance help should be available. Additional details on personnel are given in Section 1.7.

Costs and Budgets
for Maintenance Operations

Plant information on maintenance cost and the development of a maintenance budget are very important for their incorporation into the plant's total operation and maintenance budget. Before an accurate estimate of maintenance cost can be made or a sound maintenance budget can be prepared, it is necessary to divide the maintenance operations into service categories, such as preventive maintenance, corrective maintenance, and major repairs or alterations. With the maintenance operations defined, the information in the equipment record system on work performed, work contracted out, items used from storeroom stock and purchased, and a breakdown of worker-hours provide information on maintenance cost. Using these costs and making allowances for equipment replacement, expansion, and information on maintenance history for the plant, a maintenance budget can be developed.

The needed information on contract maintenance and cost of items on purchase orders may require the plant to set up a filing system to retain this information. The information could be maintained by log books with file folders to maintain purchase orders, receipts, and other important papers.

For plants in which most personnel are performing certain maintenance duties, the hours worked for each employee should be broken into each half-hour of the work shift, with charge numbers established to separate operations and maintenance work. For the maintenance work, individual charge numbers should be established to cover each phase of maintenance work, such as preventive maintenance, corrective maintenance, and special charge numbers to handle major repairs and alterations.

In conjunction with the maintenance records previously discussed, logs are often kept of preventive and corrective maintenance work accomplished with description, worker-hours, cost, date, and maintenance personnel who accomplished work. History records should be maintained with maintenance cards for the previous years and breakdown reports for all major problems encountered. These records will aid in budget planning, scheduling, manpower, operating expenses, and other needed information as required. Such records will assist in determining the time at which it is more economical to replace an item of equipment instead of repairing it. For example, this information can assist in making decisions for baghouses as to whether to replace only the broken bags or to replace all the bags in a compartment (or the entire baghouse).

Housekeeping of buildings and grounds should receive the same attention as the operating equipment. A clean and neat appearance with unnecessary odors eliminated will promote public support for the facility. Routine housekeeping can be incorporated into shift standard operating procedures. Process equipment should be cleaned and painted as required for appearance and to minimize odor sources. The doors, windows, floors, walls, and other areas should be kept clean and in good repair. Outside maintenance work, such as mowing, snow removal, painting, and cleaning gutters and drains, should be scheduled as required.

Special Tools and Equipment

A review of the work to be performed by the maintenance personnel will aid in developing a list of the tools and equipment required for the facility. The tools and equipment should be maintained in good working order and be available when required. Where some items may be required in more than one location, this should be taken into account. It is suggested that maintenance tools and supplies not be placed any more than 100 to 200 feet from the point or points of use. Tool boards with specialized or frequently used tools should be located with appropriate equipment where required.

Control facilities large enough to warrant a central tool room should maintain a complete tool inventory. Tools should be issued under a tool check control system. All tools should be regularly checked for their condition in terms of personnel safety and equipment protection.

Lubrication

Equipment manufacturers' lubrication specifications should be readily available. An interchangeable lubricants chart should be provided. This should contain information on the use of color-coded lubrication tags for all equipment. Sample forms for recording quantities of lubricants consumed and in stock should be included. A sample lubrication route should be outlined to assist maintenance supervisors in developing lubrication routes for the control facility.

Those responsible for preventive maintenance should also be responsible for lubrication. Their duties should include the following:

1. Conduct lubrication studies
2. Prepare lubrication specifications
3. Establish schedules
4. Train lubricators
5. Standardize application methods
6. Maintain consumption and inventory records
7. Establish proper handling and storage
8. Investigate new lubricants; evaluate and revise specifications as necessary
9. Standardize lubricants whenever possible to eliminate stocks of identical material under various trade names

The most important step in establishing a lubrication system involves the gathering of basic lubrication data. Lubrication specifications can be developed from manufacturers' lubrication recommendations, and ASTM Standards. Lubrication routes should be established and every item of equipment given a route number. Lubrication points, types, and frequencies should be defined for each item of equipment. All data pertinent to lubrication of selected equipment should be assembled into a lubrication guidebook. Equipment can be color-coded with decals to indicate point of service, frequency of application, and type of lubricant. Specifying the highest-grade lubricant required for more than one application and specifying a single midrange viscosity oil to replace several within a certain viscosity range are two methods for the consolidation of lubricants.

The lubrication frequency is determined by many factors, but a lubrication schedule should be established and followed to ensure proper operation of the facility. The equipment card for each piece of equipment requiring lubrication should list the lubricant to be used and frequency advisable for efficient operation. For convenience and simplification of the lubrication process, a color-coded tag or decal label can be used to identify the part, frequency, and type of lubricant required for the moving part in question. The following is a suggested color code for various categories:

Color and Type	*Frequency*
Blue—gear lube	Daily
White—spindle oil	Weekly
Green—way oil	Monthly
Red—hydraulic oil	Semiannually
Brown—grease	Annually

The tag should also indicate the appropriate lubricant.

The plant size will determine the worker-hours required for the lubrication routine. The same lubricator should perform the service each time if possible in order to narrow the range or responsibility for lubrication activities. A fixed hour or day should be established for the routine when possible and a record filed on completion of the routine. A master card may be used to facilitate routing and recording operations. In large operations a computer printout for the various frequencies and lubricants may be of assistance in reducing time requirements. Machine number and data may be keypunched for the preventive maintenance supervisor and completed route records filed with him.

Oil-change schedules and operations must also be maintained and recorded with the maintenance supervisor. Oil may be sampled to determine when changes are required. Air-conditioning equipment, heat exchangers, and related services require coolants and antifreeze liquids for operations under wide ranges of temperature and service conditions. These liquids and their concentration should be checked and records made of routine procedures in a manner similar to those for lubrication oils.

Major Equipment Information

The basic maintenance considerations for each type of air pollution control equipment are usually found in the manufacturer's catalog. A responsible individual should determine if the manufacturer's catalog contains an adequate maintenance section that is presented in usable form. If the catalog information is adequate, it should be properly referenced. In referencing these catalogs a numerical system patterned after the equipment numbering system can be used. If the manufacturer's maintenance information is not considered adequate, the responsible individual mentioned above should prepare detailed maintenance guidelines for the item of equipment. These maintenance guidelines may be bound separately.

Information should be available that outlines the procedure for ordering parts/ components or new items of equipment. This information should be provided for each major item of equipment.

Warranty and Contract Provisions

Many air pollution control devices are constructed to include a 1-year guarantee period. During this period, the contractor has the responsibility to repair, correct, or replace any equipment or material that fails to perform in accordance with the terms and provisions of the contract. Workers should be cautioned that alteration of supplied equipment by control facility personnel, without the knowledge and consent of the contractor and manufacturer, may result in refusal of these parties to accept responsibility for any subsequent problems. Equipment that is not regularly in service during the guarantee period must be maintained and should be operated periodically to prevent problems that might arise due to the equipment not being operated. The fact that a guarantee is in effect should not be cause to allow improper maintenance or operation of equipment and thus reduce its useful life.

Because of plant size, complexity of equipment, and/or maintenance personnel qualifications, the facility may find it feasible to contract some maintenance work. Many equipment manufacturers will provide maintenance service for their equipment on a contract basis. However, contracting routine inspections, lubrication, and minor parts replacement can become very costly. Plant personnel should develop their maintenance capabilities in order to perform all but the most complex tasks.

1.4   SYSTEM DESCRIPTION

It is important that operating personnel be provided with detailed descriptions of the unit operations and processes for their systems. Guidance on operating and controlling the air pollution control device is provided later in this text. The purpose of this section is to assist the reader in understanding the construction drawings, the purpose and functions of the plant, and to simply state the engineer's concept of operation and control of the air pollution control process(es).

A detailed description of the operation, control, and relationship to other plant units of each control process and its components should be made available. Included should be photographs and/or schematic diagrams to supplement the verbal descriptions of routine, alternate, and emergency operation. The user should also be aware of all applicable manual, automatic, physical, chemical, analytical, and laboratory control measures. Discussions concerning applicable safety features; fail-safe features; operating problems, causes, and suggested cures; and startup guidance should also be available.

Suitable texts, reports, and references for further study by plant operators should be provided, with appropriate consideration given to the complexity of the control units/processes and the expertise of the plant personnel.

Process descriptions can be divided into two major divisions: a general division containing information generally applicable to all similar control processes, and a specific section containing operating information pertinent to the specific plant in question.

An example of the information required follows.

I. General
   A. Unit description
      1. Basic principles
      2. Operational features
      3. Design efficiency
      4. Performance test results
   B. Relationship to adjacent units
   C. Classification and control
   D. Major components
   E. Common operating problems
      1. Pumps
         a. Pump will not start
         b. Reduced rate of discharge
         c. High power requirements
         d. Excess noise
         e. Other
      2. Plugging
      3. Others
   F. Laboratory controls
   G. Startup
      1. Equipment inspection
      2. Startup procedures
      3. Controlling startup

II. Specific plant operation
   A. Normal operation
      1. General
      2. Schematics

    B.   Alternate operation
        1.   General description of alternate operation modes
        2.   Information for alternate operating conditions
        3.   References to construction drawings, equipment shop drawings, and construction specifications
    C.   Emergency operations and fail-safe features
        1.   Warning devices
        2.   Standby power
        3.   Pump capacities
        4.   Piping configuration

1.5 SOURCE SAMPLING

Although details on actual measurement methods for both gaseous and particulate pollutants are beyond the scope of this book, the presentation of some general information on sampling appears warranted. An accurate quantitative analysis of the discharge of pollutants from a process must be determined prior to the design and/or selection of control equipment. If the unit is properly engineered utilizing the emission data as input to the control device and the code requirements as maximum effluent limitations, most gaseous and particulate pollutants can be successfully controlled by one or a combination of the methods to be discussed later in the text.

Sampling is the keystone of source analysis. Sampling methods and tools vary in their complexity according to the specific task; therefore, a degree of both technical knowledge and common sense is needed to design a sampling function. Sampling is done to measure quantities or concentrations of pollutants in effluent gas streams, to measure the efficiency of a pollution abatement device, to guide the designer of pollution control equipment and facilities, and/or to appraise contamination from a process or a source. A complete measurement requires determination of the concentration and contaminant characteristics as well as the associated gas flow. Most statutory limitations require mass rates of emission; both concentration and volumetric flow-rate data are therefore required.

The selection of a sampling site and the number of sampling points required are based on attempts to get representative samples. To accomplish this, the sampling site should be at least 8 stack or duct diameters downstream and 2 diameters upstream from any bend, expansion, contraction, valve, fitting, or visible flame. Once these traverse points have been determined, velocity measurements are made to determine gas flow. The stack gas velocity is usually determined by means of a pitot tube and differential pressure gauge. Once a flow profile has been established, sampling strategy can be considered. Since sampling collection can be simplified and greatly reduced, depending on flow characteristics, it is best to complete the flow-profile measurement before sampling or measuring pollutant concentrations.

Specific details on source sampling procedures for both gaseous and particulate pollutants may be readily found in the air pollution literature.

An ongoing air pollution sampling program also provides the basis for process control and produces a record of how the control system is operating. This information keeps the operating personnel informed of plant efficiencies and helps in predicting problems that are developing in the system. Because sampling results are a record of plant performance, they are often also evaluated by governing and/ or regulatory bodies. For these reasons, it is essential that a control system's testing program produce complete and accurate results. A suggested schedule for performing the various sampling tests should be outlined. A list of reference materials for use should also be given. All references considered essential for proper laboratory operation should be provided. Samples of the laboratory worksheets recommended for use should be available.

1.6 RECORDS

An important factor in any efficient air pollution control system is the maintenance of accurate operational and financial records. Without a record of past operational performance, it is impossible to identify trends in any process. Operating cost records are also essential if meaningful budgets are to be prepared. Accurate records permit plant operating personnel and management to maintain control of their facility.

The objective of this section is to describe the records and reports that should be maintained. The importance of keeping neat and accurate records is stressed. The suggested types of records that should be maintained include:

1. Monthly operating report
2. Daily log, by shift, of process operations
3. Operating cost records

Other records, such as those pertaining to laboratory, maintenance, and safety, should also be kept.

Operators' worksheets should be maintained. These sheets are temporary until data are transferred to the daily operating log, or they may be kept if they apply to a remote location. A daily operating log should also be maintained. This log should be bound in notebooks to prevent the destruction or alteration of these important records. Information on this daily log can include the following:

1. Routine operational duties
2. Unusual conditions (operational and maintenance)
3. Accidents to personnel
4. Complaints (odor, etc.)
5. Power consumption
6. Plant visitors

Physical plant records should be available for reference at the plant and should include:

1.   Plant operation and maintenance manual
2.   As-built engineering drawings
3.   Copy of construction specifications
4.   Equipment suppliers' manuals
5.   Piping and wiring diagrams
6.   Data cards on all equipment
7.   Construction photographs

Operating costs should also be maintained. The major categories of operating costs are labor, utilities, chemicals, and supplies. Labor is usually broken down into operation, administration, and maintenance. Utilities include electricity, fuel oil, telephone, gas, and water. Chemicals should be limited to those used in the control processes (e.g., absorption with chemical reaction). Supplies include chemicals, cleaning materials, maintenance supplies, and other expendable items.

Finally, records that reflect such things as training individuals have received and employee turnover rate are valuable. A record of emergency conditions affecting the control system should also be maintained. A system for maintaining these records should be developed and available to management.

1.7 PERSONNEL

Regardless of the care that goes into the design and construction of an air pollution control facility, without qualified personnel in adequate numbers to operate the process(es), the full capabilities of the control facility cannot be realized. A well-thought-out staff requirement will assist the system's management as they seek funds for staffing their facility. The manpower requirements recommended should be compatible with existing federal and state guidelines, if required and applicable.

Up-to-date training for operators and maintenance personnel should be stressed as being of critical importance in the proper functioning of the control facility. The purpose is to protect the huge investment in plant equipment from damage or deterioration, and to reduce the quantity of pollutants discharged into the atmosphere.

To prepare manpower recommendations adequately, a task analysis of each job within the control system should be made. This analysis will provide details on the skills and qualifications required for each position. There are specialized methods and techniques and professionals available for the determination of manpower and training requirements. Persons responsible for this task should consider using personnel with specialized skills in manpower factors to determine personnel and training needs.

The qualifications for all types of personnel should be listed. Personnel types include superintendent, operators, maintenance personnel, chemists, and assistants for all levels. Job descriptions for all personnel should be given. This description might include the following for an operator: Follow shift standard procedures, keep daily operating records, check all meters and recorders for proper operation and check, and so on. The following is a list of some of the types of personnel commonly employed for the operation and maintenance of air pollution control systems.

Occupation Description
1. Superintendent
2. Assistant superintendent
3. Clerk-typist
4. Operations supervisor
5. Shift foreperson
6. Operator II
7. Operator I
8. Automotive equipment operator
9. Maintenance supervisor
10. Mechanical maintenance foreperson
11. Maintenance mechanic II
12. Maintenance mechanic I
13. Electrician II
14. Electrician I
15. Maintenance helper
16. Laborer
17. Painter
18. Storekeeper
19. Custodian
20. Chemist
21. Laboratory technician

The following outline provides a job description qualifications profile.

1. Formal education
2. General requirements
3. General educational development
   a. Reasoning
   b. Mathematical
   c. Language
4. Special vocational preparation

5. Aptitudes as relative to general working conditions
    a. Intelligence
    b. Verbal
    c. Numerical
    d. Form perception
    e. Spatial
    f. Clerical perception
    g. Motor coordination
    h. Finger dexterity
    i. Manual dexterity
    j. Eye–hand–foot coordination
    k. Color discrimination
6. Interests
7. Temperament
8. Physical demands
9. Working conditions

Complete and accurate job descriptions are sometimes difficult to prepare. Good job descriptions should include but are not limited to the following:

1. List items, equipment, or processes that an individual must operate
2. State if monitoring of gauges or meters is required
3. Discuss interpreting of any meter or gauge readings for process control actions
4. List any logs or records to be maintained
5. Outline any maintenance duties required
6. State any other title that the individual might carry
7. Discuss decision-making requirements
8. State responsibilities and authority given to individual in job being described
9. List any report or budget functions that must be performed
10. Discuss any supervisory or inspection functions
11. Discuss training requirements, if applicable

The pertinent details of any existing state certification program should be discussed, with emphasis on how they apply to the control system at hand. This discussion should include a copy of the rules and regulations of the state certification board.

1.8 SAFETY

The safety hazards associated with air pollution control processes are many and varied. Control equipment personnel should be made aware of all these hazards. They should be protected from these hazards to the greatest extent possible and should receive proper first aid training in case an accident does occur. A control system with a poor safety record will generally be providing only marginal air pollution control as well.

Hazards may be classified into the following broad categories:

1. Physical injuries
2. Bacterial infections
3. Mechanical equipment
4. Explosion and fire
5. Radiological hazards

Information on safety equipment (showers, goggles, etc.) and chemicals handling should be readily available to all personnel. The telephone numbers of several local physicians, the nearest hospital, police and fire departments, ambulance services, and rescue squad should be posted at each phone in the control system.

An important aspect of plant safety is the prompt reporting of personnel injuries to the control system's insurance company. Details on reporting accidents should be outlined and all other pertinent insurance information discussed with all personnel.

2

Considerations in the Selection of Air Pollution Control Equipment

Anthony J. Buonicore P.E.

President
Buonicore-Cashman Associates, Inc.
Bridgeport, Connecticut

and

Louis Theodore

Professor of Chemical Engineering
Manhattan College
Bronx, New York

2.1 INTRODUCTION

In the last decade the engineering profession has been heavily influenced by its responsibility to society. This responsibility has directed itself in particular toward the protection of public health and welfare, and is guided by a host of environmental regulations. The extent to which the engineering profession responds to this challenge depends largely upon the limits imposed by three principal considerations:

1. Legal limitations imposed for the protection of public health and welfare
2. Social limitations imposed by the community in which the pollution source is or will be located
3. Economic limitations imposed by marketplace constraints

Careful evaluation within the framework of all three limitations is now essential and often integral with corporate strategic planning processes.

The control strategy for environmental impact assessment often focuses on five alternatives whose purpose would be the reduction and/or elimination of pollutant emissions:

1. Elimination of the operation entirely or in part
2. Modification of the operation
3. Relocation of the operation
4. Application of appropriate control technology
5. Combination thereof

In view of the relatively high costs often associated with pollution control systems, engineers today are directing considerable effort toward process modification to eliminate as much of the pollution problem as possible *at the source*. This includes evaluating alternative manufacturing and production techniques, substitute raw materials, improved process control methods, and so on. Unfortunately, if there is no alternative, the application of pollution control equipment must be considered. In view of the relatively high costs, proper selection of this equipment is essential. The equipment must be designed to comply with regulatory emission limitations on a continual basis, with interruptions subject to severe penalty depending upon the circumstances. The requirement for design performance on a continual basis places very heavy emphasis on operation and maintenance practices. In fact, it is not unusual that favorable operation and maintenance requirements associated with a particular piece of equipment can strongly influence its selection, despite the fact that its capital cost may be higher. The rapidly escalating costs of energy, labor, and materials can make operation and maintenance considerations even more important than original cost.

2.2 FACTORS IN CONTROL EQUIPMENT SELECTION

There are a number of factors to be considered prior to selecting a particular piece of air pollution control hardware [1]. In general, they can be grouped into three categories: environmental, engineering, and economic.

Environmental
1. Equipment location
2. Available space
3. Ambient conditions
4. Availability of adequate utilities (i.e., power, water, etc.) and ancillary system facilities (i.e., waste treatment and disposal, etc.)
5. Maximum allowable emission (air pollution codes)
6. Aesthetic considerations (i.e., visible steam or water vapor plume, etc.)

7. Contribution of air pollution control system to wastewater and land pollution
8. Contribution of air pollution control system to plant noise levels

Engineering
1. Contaminant characteristics (i.e., physical and chemical properties, concentration, particulate shape and size distribution—in the case of particulates, chemical reactivity, corrosivity, abrasiveness, toxicity, etc.)
2. Gas stream characteristics (i.e., volume flow rate, temperature, pressure, humidity, composition, viscosity, density, reactivity, combustibility, corrosivity, toxicity, etc.)
3. Design and performance characteristics of the particular control system (i.e., size and weight, fractional efficiency curves—in the case of particulates, mass transfer and/or contaminant destruction capability—in the case of gases or vapors, pressure drop, reliability and dependability, turndown capability, power requirements, utility requirements, temperature limitations, maintenance requirements, flexibility toward complying with more stringent air pollution codes, etc.)

Economic
1. Capital cost (equipment, installation, engineering, etc.)
2. Operating cost (utilities, maintenance, etc.)
3. Expected equipment lifetime and salvage value

Prior to the purchase of control equipment, experience has shown that the following points should be emphasized.

1. Refrain from purchasing any control equipment without reviewing *certified independent test data* on its performance under a similar application. Request the manufacturer to provide performance information and design specifications.
2. In the event that sufficient performance data are unavailable, request that the equipment supplier provide a small pilot model for evaluation under existing conditions.
3. Request participation of the local control authorities in the decision-making process.
4. Prepare a good set of specifications. Include a *strong performance guarantee* from the manufacturer to ensure that the control equipment will meet all applicable local, state, and federal codes at specific process conditions.
5. Closely review your process and economic fundamentals. Assess the possibility for emission trade-offs (offsets) and/or applying the "bubble concept." The bubble concept permits a plant to find the most efficient way to control its emissions as a whole, rather than the Environmental Protection Agency regulating the emissions from individual sources. Reduc-

tions at a source where emissions can be lessened for the least cost can offset emissions of the same pollutant from another source in the plant.

6. Make a careful material balance study before authorizing an emission test or purchasing control equipment.

7. Refrain from purchasing any equipment until *firm* installation cost estimates have been added to the equipment cost. *Escalating installation costs are the rule rather than the exception.*

8. Give operation and maintenance costs high priority on the list of equipment selection factors.

9. Refrain from purchasing any equipment until a solid commitment from the utility supplier(s) is obtained. Make every effort to ensure that the new system will utilize fuel, controllers, filters, motors, and so on, that are compatible with those already available at the plant.

10. The specification should include written assurance of *prompt* technical assistance from the equipment supplier. This, together with a complete operating manual (with parts list and full schematics), is essential and is too often forgotten in the rush to get the equipment operating.

11. Schedules, particularly on projects being completed under a court order or consent judgment, can be critical. In such cases, delivery guarantees should be obtained from the manufacturers and penalties identified.

12. The air pollution equipment should be of fail-safe design with built-in indicators to show when performance is deteriorating.

13. Withhold 10 to 15% of the purchase price until compliance is clearly demonstrated.

The usual design/procurement/construction/startup problems can be further compounded by any one or combination of the following:

1. Unfamiliarity of process engineers with air pollution engineering
2. New and changing air pollution codes
3. New suppliers with frequently unproven equipment
4. Lack of industry standards in some key areas
5. Interpretations of control agency field personnel
6. Compliance schedules that are too tight
7. Vague specifications
8. Weak guarantees for the new control equipment
9. Unreliable delivery schedules
10. Process unreliability problems

Proper selection of a particular system for a specific application can be extremely difficult and complicated. In view of the multitude of complex and often ambiguous pollution control regulations, it is in the best interest of the prospective user to work closely with regulatory officials as early in the process as possible. Finally, previous experience on a similar application cannot be overemphasized. This is treated in more detail in Section 2.5.

2.3 GENERALIZED DESIGN REVIEW PROCEDURE

Design reviews for air pollution control equipment are performed for a variety of reasons, including:

1. To anticipate compliance with applicable air pollution codes
2. To estimate performance of existing control equipment
3. To evaluate the feasibility of a proposed equipment design
4. To assess the effect on control equipment of process modification

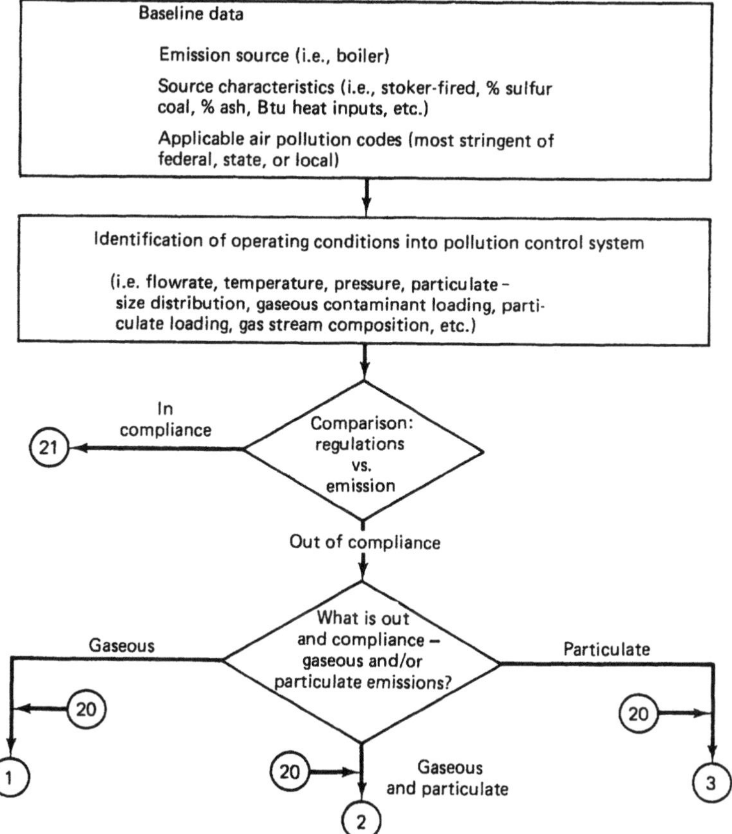

FIGURE 2-1 Typical generalized design review approach.

1
Has control device been proposed?
Yes
No
Select from possible control alternatives:
Absorption systems (packed columns, plate columns, etc.)
Adsorbers
Combustion systems (thermal, catalytic, etc.)
Condensers
Combinations thereof
Absorption system design
Adsorption system design
Combustion system design
Condensation system design
Combination systems design
5
6
7
8
9
3
Has control device been proposed?
Yes
No
Select from possible control alternatives:
Mechanical collectors
Wet scrubbers
Electrostatic precipitators
Fabric filters
Combinations thereof
Mechanical collector system design
Wet scrubber system design
Electrostatic precipitator system design
Fabric filter system design
Combination systems design
15
16
17
18
19

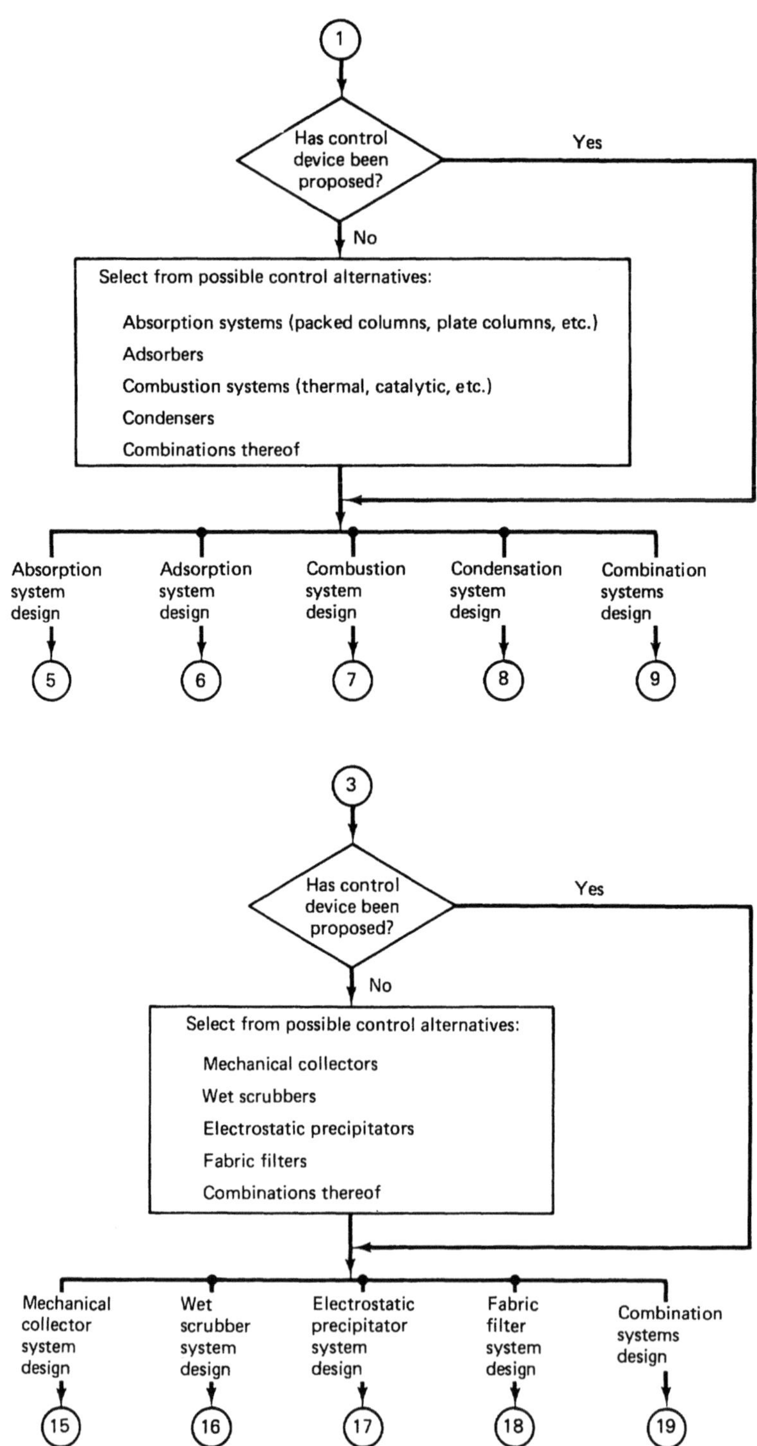

24

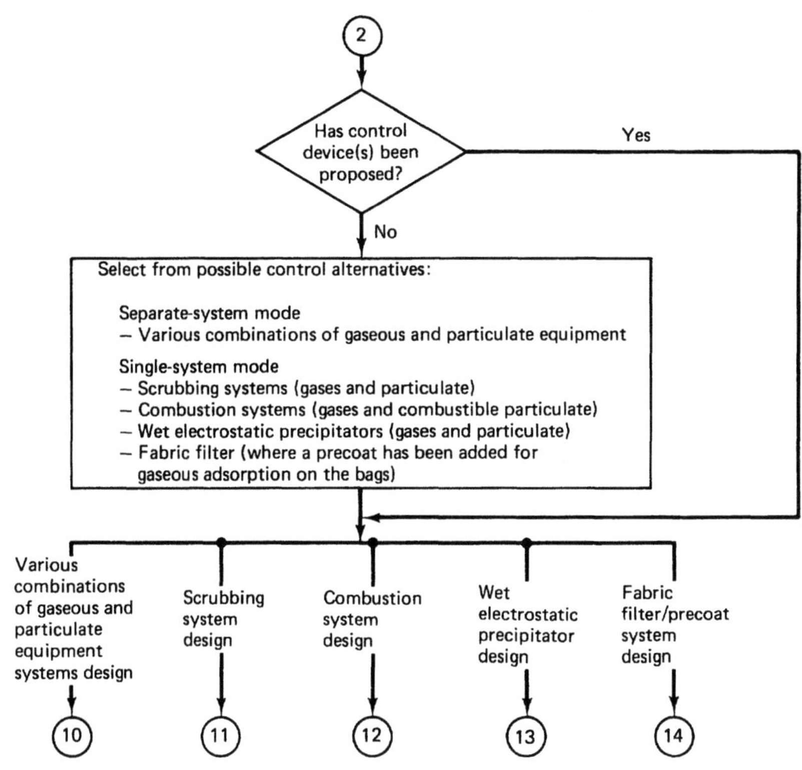

2
Has control device(s) been proposed?
Yes
No
Select from possible control alternatives:
Separate-system mode
— Various combinations of gaseous and particulate equipment
Single-system mode
— Scrubbing systems (gases and particulate)
— Combustion systems (gases and combustible particulate)
— Wet electrostatic precipitators (gases and particulate)
— Fabric filter (where a precoat has been added for gaseous adsorption on the bags)
Various combinations of gaseous and particulate equipment systems design
Scrubbing system design
Combustion system design
Wet electrostatic precipitator design
Fabric filter/precoat system design
10
11
12
13
14

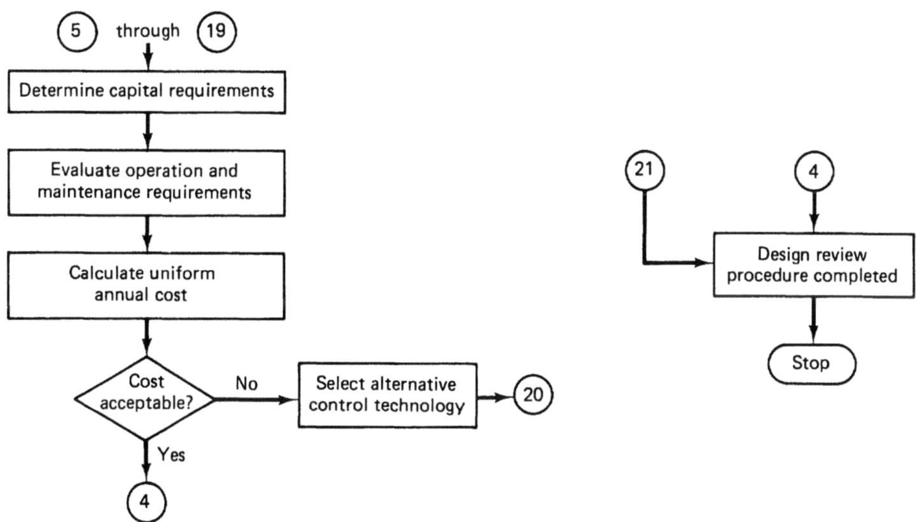

5
through
19
Determine capital requirements
Evaluate operation and maintenance requirements
Calculate uniform annual cost
Cost acceptable?
No
Select alternative control technology
20
Yes
4
21
4
Design review procedure completed
Stop

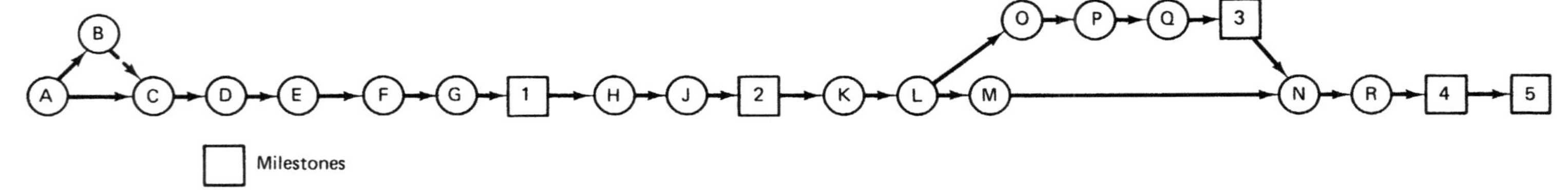

Milestones

1 Date of submittal of final control plan to appropriate agency.
2 Date of award of control device contract.
3 Date of initiation of onsite construction or installation of emission control equipment.
4 Date by which onsite construction or installation of emission control equipment is completed.
5 Date by which final compliance is achieved.

Activities

Designation

A–C	Preliminary investigation
A–B	Source tests, if necessary
C–D	Evaluate control alternatives
D–E	Commit funds for total program
E–F	Prepare preliminary control plan and compliance schedule for agency
F–G	Agency review and approval
G–1	Finalize plans and specifications
1–H	Procure control device bids
H–J	Evaluate control device bids
J–2	Award control device contract
2–K	Vendor prepares assembly drawings

Designation

K–L	Review and approval of assembly drawings
L–M	Vendor prepares fabrication drawings
M–N	Fabricate control device
L–O	Prepare engineering drawings
O–P	Procure construction bids
P–Q	Evaluate construction bids
Q–3	Award construction contract
3–N	Onsite construction
N–R	Install control device
R–4	Complete construction (system tie-in)
4–5	Startup, shakedown, source test

FIGURE 2-2 Compliance activity and schedule chart.

A typical generalized design review approach is presented in Figure 2-1. The design review investigation is an activity performed early in the evaluation process. Other activities that must be accomplished before final compliance is achieved are presented in Figure 2-2.

2.4 COMPARING CONTROL EQUIPMENT ALTERNATIVES

The final choice in equipment selection is usually dictated by that equipment capable of achieving compliance with regulatory codes at the lowest uniform annual cost (amortized capital investment plus operation and maintenance costs). In order to compare specific control equipment alternatives, knowledge of the particular application and site is essential. A preliminary screening, however, may be performed by reviewing the advantages and disadvantages of each type of air pollution control equipment. For example, if water or a waste treatment system is not available at the site, this may preclude use of a wet scrubber system and instead focus particulate removal on dry systems, such as cyclones, baghouses, and/or electrostatic precipitators. If auxiliary fuel is unavailable on a continuous basis, it may not be possible to combust organic pollutant vapors in an incineration system. If the particulate-size distribution in the gas stream is relatively fine, cyclone collectors most probably would not be considered. If the pollutant vapors can be reused in the process, control efforts may be directed to adsorption systems. There are many more situations where knowledge of the capabilities of the various control options, combined with common sense, will simplify the selection procedure. General advantages and disadvantages of the most popular types of air pollution control equipment for gases and particulates are presented in Tables 2-1 to 2-9.

TABLE 2-1   Advantages and disadvantages of cyclone collectors.

Advantages
1. Low cost of construction.
2. Relatively simple equipment with few maintenance problems.
3. Relatively low operating pressure drops (for degree of particulate removal obtained) in the range of approximately 2 to 6 in. water column.
4. Temperature and pressure limitations imposed only by the materials of construction used.
5. Dry collection and disposal.
6. Relatively small space requirements.

Disadvantages
1. Relatively low overall particulate collection efficiencies, especially on particulates below 10 μm in size.
2. Inability to handle tacky materials.

TABLE 2-2 Advantages and disadvantages of wet scrubbers.

Advantages
1. No secondary dust sources.
2. Relatively small space requirements.
3. Ability to collect gases as well as particulates (especially "sticky" ones).
4. Ability to handle high-temperature, high-humidity gas streams.
5. Capital cost is low (if wastewater treatment system not required).
6. For some processes, the gas stream is already at high pressures (so pressure drop considerations may not be significant).
7. Ability to achieve high collection efficiencies on fine particulates (however, at the expense of pressure drop).

Disadvantages
1. May create water disposal problem.
2. Product is collected wet.
3. Corrosion problems are more severe than with dry systems.
4. Steam plume opacity and/or droplet entrainment may be objectionable.
5. Pressure drop and horsepower requirements may be high.
6. Solids buildup at the wet–dry interface may be a problem.
7. Relatively high maintenance costs.

TABLE 2-3 Advantages and disadvantages of electrostatic precipitators.

Advantages
1. Extremely high particulate (coarse and fine) collection efficiencies can be attained (at a relatively low expenditure of energy).
2. Dry collection and disposal.
3. Low pressure drop (typically less than 0.5 in. water column).
4. Designed for continuous operation with minimum maintenance requirements.
5. Relatively low operating costs.
6. Capable of operation under high pressure (to 150 psi) or vacuum conditions.
7. Capable of operation at high temperatures (to 1300°F).
8. Relatively large gas flow rates can be effectively handled.

Disadvantages
1. High capital cost.
2. Very sensitive to fluctuations in gas stream conditions (in particular, flows, temperatures, particulate and gas composition, and particulate loadings).
3. Certain particulates are difficult to collect due to extremely high or low resistivity characteristics.
4. Relatively large space requirements required for installation.
5. Explosion hazard when treating combustible gases and/or collecting combustible particulates.
6. Special precautions are required to safeguard personnel from the high voltage.
7. Ozone is produced by the negatively charged discharge electrode during gas ionization.
8. Relatively sophisticated maintenance personnel required.

28

TABLE 2-4 Advantages and disadvantages of fabric filter systems.

Advantages

1. Extremely high collection efficiency on both coarse and fine (submicron) particulates.
2. Relatively insensitive to gas stream fluctuation. Efficiency and pressure drop are relatively unaffected by large changes in inlet dust loadings for continuously cleaned filters.
3. Filter outlet air may be recirculated within the plant in many cases (for energy conservation).
4. Collected material is recovered dry for subsequent processing or disposal.
5. No problems with liquid waste disposal, water pollution, or liquid freezing.
6. Corrosion and rusting of components are usually not problems.
7. There is no hazard of high voltage, simplifying maintenance and repair and permitting collection of flammable dusts.
8. Use of selected fibrous or granular filter aids (precoating) permits the high-efficiency collection of submicron smokes and gaseous contaminants.
9. Filter collectors are available in a large number of configurations, resulting in a range of dimensions and inlet and outlet flange locations to suit installation requirements.
10. Relatively simple operation.

Disadvantages

1. Temperatures much in excess of 550°F require special refractory mineral or metallic fabrics that are still in the developmental stage and can be very expensive.
2. Certain dusts may require fabric treatments to reduce dust seeping or, in other cases, assist in the removal of the collected dust.
3. Concentrations of some dusts in the collector ($\sim$50 g/m^3) may represent a fire or explosion hazard if a spark or flame is admitted by accident. Fabrics can burn if readily oxidizable dust is being collected.
4. Relatively high maintenance requirements (bag replacement, etc.).
5. Fabric life may be shortened at elevated temperatures and in the presence of acid or alkaline particulate or gas constituents.
6. Hygroscopic materials, condensation of moisture, or tarry adhesive components may cause crusty caking or plugging of the fabric or require special additives.
7. Replacement of fabric may require respiratory protection for maintenance personnel.
8. Medium pressure-drop requirements, typically in the range 4 to 10 in. water column.

TABLE 2-5 Advantages and disadvantages of absorption systems
(packed and plate columns).

Advantages

1. Relatively low pressure drop.
2. Standardization in fiberglass-reinforced plastic (FRP) construction permits operation in highly corrosive atmospheres.
3. Capable of achieving relatively high mass-transfer efficiencies.
4. Increasing the height and/or type of packing or number of plates can improve mass transfer without purchasing a new piece of equipment.
5. Relatively low capital cost.
6. Relatively small space requirements.
7. Ability to collect particulates as well as gases.

Disadvantages

1. May create water (or liquid) disposal problem.
2. Product collected wet.
3. Particulates deposition may cause plugging of the bed or plates.
4. When FRP construction used, it is sensitive to temperature.
5. Relatively high maintenance costs.

TABLE 2-6 Comparison of plate and packed columns.

Packed column
1. Lower pressure drop.
2. Simpler and cheaper to construct.
3. Preferable for liquids, with high foaming tendencies.

Plate column
1. Less susceptible to plugging.
2. Less weight.
3. Less of a problem with channeling.
4. Temperature surge will result in less damage.

TABLE 2-7 Advantages and disadvantages of adsorption systems.

Advantages
1. Product recovery may be possible.
2. Excellent control and response to process changes.
3. No chemical disposal problem when pollutant (product) recovered and returned to process.
4. Capability of systems for fully automatic, unattended operation.
5. Capability to remove gaseous or vapor contaminants from process streams to extremely low levels.

Disadvantages
1. Product recovery may require an exotic, expensive distillation (or extraction) scheme.
2. Adsorbent progressively deteriorates in capacity as the number of cycles increases.
3. Adsorbent regeneration requires a steam or vacuum source.
4. Relatively high capital cost.
5. Prefiltering of gas stream may be required to remove any particulate capable of plugging the adsorbent bed.
6. Cooling of the gas stream may be required to get to the usual range of operation (less than 120°F).
7. Relatively high steam requirements to desorb high-molecular-weight hydrocarbons.

TABLE 2-8 Advantages and disadvantages of combustion systems.

Advantages
1. Simplicity of operation.
2. Capability of steam generation or heat recovery in other forms.
3. Capability for virtually complete destruction of organic contaminants.

Disadvantages
1. Relatively high operating costs (particularly associated with fuel requirements).
2. Potential for flashback and subsequent explosion hazard.
3. Catalyst poisoning (in the case of catalytic incineration).
4. Incomplete combustion can create potentially worse pollution problems.

30

TABLE 2-9   Advantages and disadvantages of condensers.

Advantages
1. Pure product recovery (in the case of indirect-contact condensers).
2. Water used as the coolant in an indirect contact condenser (i.e., shell-and-tube heat exchanger) does not contact the contaminated gas stream and can be reused after cooling.

Disadvantages
1. Relatively low removal efficiency for gaseous contaminants (at concentrations typical of pollution control applications).
2. Coolant requirements may be extremely expensive.

2.5 SELECTING CONTROL EQUIPMENT FOR SPECIFIC INDUSTRIES

The basic types of emission control devices are mechanical collectors, wet scrubbers, baghouses, electrostatic precipitators, combustion systems, condensers, absorbers, and adsorbers. All of these have been used to some extent to control emissions from a nearly infinite variety of processes, with the selection procedure almost always dictated by experience. This section provides information on suggested control measures for a small number of industries. These limited applications are presented in Table 2-10. Information on petroleum refineries, chemical plants, mining, and so on, are also available in the literature [2].

2.6 REGULATORY CONSIDERATIONS

The federal government's involvement in air pollution control began in 1955 with Public Law 159. This law authorized funding for the U.S. Public Health Service to initiate research into the nature and extent of the nation's air pollution problem. With the passage of the Clean Air Act of 1963, grants were authorized to state and local agencies to assist them in their own control programs. It also provided some limited authority to the federal government to take action to relieve interstate pollution problems. The basic federal control authority was expanded and strengthened by the Air Quality Act of 1967. One of the more significant measures gave citizens, for the first time, a statutory right to participate in the control process (through public hearings). However, it was not until the Clean Air Act Amendments of 1970 that the regulatory effort first had any real bite.

The Clean Air Act Amendments of 1970

The historic Clean Air Act Amendments of 1970 broadened and accelerated the nation's earlier air pollution control programs and is the basis for most of the present efforts toward abating air pollution. The Amendments reaffirmed that state and local

TABLE 2-10 Control techniques applicable to unit processes at important emission sources.

Industry	Process of Operation	Air Contaminants Emitted	Control Techniques
Aluminum reduction plants	Materials handling: buckets and belt conveyor or pneumatic conveyor	Particulates (dust)	Exhaust systems and baghouse
	Anode and cathode electrode preparation		Exhaust systems and mechanical collectors
	Cathode: baking	Hydrocarbon emissions from binder	
	Anode: grinding and blending	Particulates (dust)	
	Baking	Particulates (dust), CO, SO_2, hydrocarbons, and fluorides	High-efficiency cyclone, electrostatic precipitators, scrubbers, catalytic combustion or incinerators, flares, baghouse
	Pot charging	Particulates (dust), CO, HF, SO_2, CF_4, and hydrocarbons	High-efficiency cyclone, baghouse, spray towers, floating-bed scrubber, electrostatic precipitators, chemisorption, wet electrostatic precipitators
	Metal casting	Cl_2, HCl, CO, and particulates (dust)	Exhaust systems and scrubbers
Asphalt batch plants	Materials handling, storage and classifiers: elevators, chutes, vibrating screens	Particulates (dust)	Local exhaust systems with a cyclone precleaner and a scrubber or baghouse
	Drying: rotary oil- or gas-fired	Particulates and smoke	Proper combustion controls, fuel-oil preheating where required; local exhaust system, cyclone and a scrubber or baghouse

Source	Operation	Emissions	Control methods
Cement plants	Truck traffic	Dust	Wetting down truck routes
	Quarrying: primary crusher, secondary crusher, conveying, storage	Particulates (dust)	Wetting, exhaust systems with mechanical collectors
	Dry processes: materials handling, air separator (hot-air furnace)	Particulates (dust)	Local exhaust system and mechanical collectors and baghouse
	Grinding	Particulates (dust)	Local exhaust system with cyclones and baghouse
	Pneumatic, conveying and storage	Particulates (dust)	
	Wet process: materials handling, grinding, storage	Wet materials, no dust	
	Kiln operations Rotary kiln	Particulates (dust), CO, SO_x, NO_x, hydrocarbons, aldehydes, ketones	Electrostatic precipitators and baghouses, scrubber, flare
	Clinker cooling: materials handling	Particulates (dust)	Local exhaust system and mechanical collectors
	Grinding and packaging: air separator, grinding, pneumatic conveying, materials handling, packaging	Particulates (dust)	Local exhaust systems and mechanical collectors
Coal preparation plants	Materials handling: conveyors, elevators, chutes	Particulates (dust)	Local exhaust systems and cyclones
	Sizing: crushing, screening, classifying	Particulates (dust)	Local exhaust systems and cyclones
	De-dusting	Particulates (dust)	Local exhaust system, cyclone precleaners, and baghouse
	Storing coal in piles	Blowing particulates (dust)	Wetting, plastic spray covering

33

TABLE 2-10 *(continued)*

Industry	Process of Operation	Air Contaminants Emitted	Control Techniques
	Refuse piles	H_2S, particulates, and smoke from burning storage piles	Digging out fire, pumping water onto fire area, blanket with incombustible material
	Coal drying: rotary, screen, suspension, fluid bed, cascade	Dust, smoke, particulates, sulfur oxides, H_2S	Exhaust systems with cyclones and venturi scrubbers
Coke plants	By-product ovens charging	Smoke, particulates (dust)	Pipeline charging, careful charging techniques; portable hooding and scrubber or baghouses
	Pushing	Smoke, particulates (dust), SO_2	Minimize green coke pushing, scrubbers and baghouses
	Quenching	Smoke, particulates (dust and mists), phenols, and ammonia	Baffles and spray tower
	By-product processing	CO, H_2S, methane, ammonia, H_2, phenols, hydrogen cyanide, N_2, benzene, xylene, etc.	Electrostatic precipitator, scrubber, flaring
	Material storage (coal and coke)	Particulates (dust)	Wetting, plastic spray, fire-prevention techniques
Fertilizer industry (chemical)	Phosphate fertilizers: crushing, grinding, and calcining	Particulates (dust)	Exhaust system, scrubber, cyclone, baghouse
	Hydrolysis of P_2O_5	PH_3, P_2O_5, H_3PO_4 mist	Scrubbers, flare
	Acidulation and curing	HF, SiF_4	Scrubbers
	Granulation	Particulates (dust) (product recovery)	Exhaust system, scrubber, or baghouse
	Ammoniation	NH_3, NH_4Cl, SiF_4, HF	Cyclone, electrostatic precipitator, baghouse, high-energy scrubber
	Nitric acid acidulation	NO_x, gaseous fluoride compounds	Scrubber, addition of urea
	Superphosphate storage and shipping	Particulates (dust)	Exhaust system, cyclone, or baghouse

	Ammonium nitrate reactor	NH_3, NO_x	Scrubber
	Prilling tower	NH_4, NO_3	Proper operation control, scrubbers
Foundries	Melting (cupola):		
Iron	Charging	Smoke and particulates	Closed top with exhaust system, CO afterburner, gas-cooling device and scrubbers, baghouse or electrostatic precipitator, wetting to extinguish fire
	Melting	Smoke and particulates, fume	
	Pouring	Oil, mist, CO	
	Bottom drop	Smoke and particulates	
Brass bronze	Melting		
	Charging	Smoke particulates, oil mist	Low-zinc-content red brass: use good combustion controls and slag cover; high-zinc-content brass: use good combustion controls, local exhaust system, and baghouse or scrubber
	Melting	Zinc oxide fume, particulates, smoke zinc oxide	
	Pouring	Fume, lead oxide fume	
Aluminum	Melting: charging, melting, pouring	Smoke and particulates	Charge clean material (no paint or grease); proper operation should be required; no air pollution control equipment if no fluxes are used and degassing is not required; dirty charge requires exhaust system with scrubbers and baghouses
Zinc	Melting		
	charging	Smoke and particulates	Exhaust system with cyclone and baghouse; charge clean material (no paint or grease)
	Melting	Zinc oxide fume	Careful skimming of dross
	pouring	Oil mist and hydrocarbons from die-casting machines	Use low-smoking die-casting lubricants
	Sand handling shake-out	Particulates (dust), smoke, organic vapors	
	Magnetic pulley, conveyors and elevators, rotary cooler, screening, crusher mixer	Particulates (dust)	Exhaust system, cyclone, and baghouse

TABLE 2-10 (continued)

Industry	Process of Operation	Air Contaminants Emitted	Control Techniques
	Coke-making ovens	Organic acids, aldehydes, smoke, hydrocarbons	Use of binders that will allow ovens to operate at less than 400°F or exhaust systems and afterburners
Galvanizing operations	Hot-dip galvanizing tank kettle: dipping material into the molten zinc; dusting flux onto the surface of the molten zinc	Fumes, particulates (liquid), vapors: NH_4Cl, ZnO, $ZnCl_2$, Zn, NH_3, oil, and carbon	Close-fitting hoods with high indraft velocities (in some cases the hood may not be able to be close to the kettle, so that the in-draft velocity must be very high), baghouses, electrostatic precipitators
Kraft pulp mills	Digesters: batch and continuous	Mercaptans, methanol (odors)	Condensers and use of lime kiln, hog fuel boiler, or furnaces as afterburners
	Multiple-effect evaporators	H_2S, other odors	Caustic-scrubbing and thermal oxidation of noncondensibles
	Recovery furnace	H_2S, mercaptans, organic sulfides, and disulfides	Paper combustion controls for fluctuating load and unrestricted primary and secondary air flow to furnace and scrubber or electrostatic precipitator
	Weak and strong black liquor oxidation	H_2S	Packed tower and cyclone
	Smelt tanks	Particulates (mist or dust)	Demisters, venturi, packed tower, or impingement-type scrubbers
	Lime kiln	Particulates (dust), H_2S	Venturi scrubbers
Municipal and industrial incinerators	Single-chamber incinerators	Particulates, smoke, volatiles, CO, SO_x, ammonia, organic acids, aldehydes, NO_x, hydrocarbons, odors, HCl	
	Flue-fed		Settling chambers, scrubbers, afterburner, bypass flue, ash cleanout

Source		Air contaminants	Control methods
	Multiple-chamber incinerators: retort, inline	Particulates, smoke, and combustion contaminants	Operating at rated capacity, using auxiliary fuel as specified and good maintenance, including timely cleanout of ash
	Flue-fed	Particulates, smoke, and combustion contaminants	Use of charging gates and automatic controls for draft
	Wood waste	Particulates, smoke, and combustion contaminants	Continuous-feed systems; operate at design load and excess air; limit charging of oily material
	Municipal incinerators: 50–100 tons/day	Particulates, smoke, volatiles, CO, ammonia, organic acids, aldehydes, NO_x, hydrocarbons, SO_x, hydrogen chloride, odors	Preparation of materials, including weighing, grinding, shredding; control of tipping area, furnace design with proper automatic controls; proper startup techniques; maintenance of design operating temperatures; use of electrostatic precipitators, scrubbers and baghouses; proper ash cleanout
	Pathological incinerators	Odors, hydrocarbons	Proper charging
	Wood waste and industrial waste	Particulates, smoke, and combustion contaminants	Modified fuel feed, auxiliary fuel and dryer systems
	Box type	Particulates, smoke, and combustion contaminants	Allow proper startup, charge material slowly, don't overload
Nonferrous Smelters, primary Copper	Roasting	SO_2, particulates, fume	Exhaust system, settling chambers, cyclones or scrubbers and electrostatic precipitators for dust and fumes and sulfuric acid plant for SO_2
	Reverberatory furnace	Smoke, particulates, fume, SO_2	Exhaust system, settling chambers, cyclones or scrubbers and electrostatic precipitators for dust and fumes and sulfuric acid plant for SO_2

TABLE 2-10 *(continued)*

Industry	Process of Operation	Air Contaminants Emitted	Control Techniques
	Converters: charging, slag skim, pouring, air or oxygen blow	Smoke, fume, SO_2	Exhaust system, settling chambers, cyclones or scrubbers, electrostatic precipitators for dust and fumes and sulfuric acid plant for SO_2
Lead	Sintering	SO_2, particulates, smoke	Exhaust system, cyclones and baghouse or precipitators for dust and fumes, sulfuric acid plant for SO_2
	Blast furnace	SO_2, CO, particulates, lead oxide, zinc oxide	Exhaust system, settling chambers, afterburner and cooling device, cyclone, and baghouse
	Dross reverberatory furnace	SO_2, particulates, fume	Exhaust system, settling chambers, cyclone and cooling device, baghouse
	Refining kettles	SO_2, particulates	Local exhaust system, cooling device, baghouse or precipitator
Cadmium	Roasters, slag, fuming furnaces, deleading kilns	Particulates	Local exhaust system, baghouse or precipitator
Zinc	Roasting	Particulates (dust) and SO_2	Exhaust system, humidifier, cyclone scrubber, electrostatic precipitator, and acid plant
	Sintering	Particulates (dust) and SO_2	Exhaust system, humidifier, electrostatic precipitator and acid plant
	Calcining	Zinc oxide fume, particulates, SO_2, CO	Exhaust system, baghouse
	Retorts: electric arc		

Nonferrous smelters, secondary	Blast furnaces and cupolas—recover metal from scrap and slag	Dust, fumes, particulates, oil vapor, smoke, CO	Exhaust systems, cooling devices, CO burners and baghouses or precipitators
	Reverberatory furnaces	Dust, fumes, particulates, smoke, gaseous fluxing materials	Exhaust systems and baghouses, or precipitators or venturi scrubbers
	Crucible furnaces	*See* Nonferrous smelters	
	Sweat furnaces	Smoke, particulates, fumes	Precleaning metal and exhaust systems with afterburner and baghouse
	Wire reclamation and autobody burning	Smoke, particulates	Scrubbers and afterburners
Paint and varnish manufacturing	Resin manufacturing: closed reaction vessel	Acrolein, other aldehydes and fatty acids (odors); phthalic anhydride (subl.)	Exhaust systems with scrubbers and fume burners
	Varnish cooking—open or closed vessels	Ketones, fatty acids, formic acids, acetic acid, glycerine, acrolein, other aldehydes, phenols and terpenes; from tall oils, hydrogen sulfide, alkyl sulfide, butyl mercaptan, and thiofene (odors)	Exhaust system with scrubbers and fume burners—close-fitting hoods are required for open kettles
	Solvent thinning	Olefins, branched chain aromatics and ketones (odors), solvents	Exhaust system with fume burners
Rendering plants	Feedstock storage and housekeeping	Odors	Quick processing, washdown of all concrete surfaces, pave dirt roads, proper sewer maintenance, packed towers
	Cookers and percolators	SO_2, mercaptans, ammonia, odors	Exhaust system, condenser, scrubber, or incinerator
	Grinding	Particulates (dust)	Exhaust system and scrubber

TABLE 2-10 (continued)

Industry	Process of Operation	Air Contaminants Emitted	Control Techniques
Roofing plants (asphalt saturators)	Felt or paper saturators: spray section, asphalt tank, wet looper	Asphalt vapors and particulates (liquid)	Exhaust system with high inlet velocity at hoods (>200 ft/min) with either spray scrubbers, baghouses, or two-stage low-voltage electrostatic precipitators
	Crushed rock or other minerals handling	Particulates (dust)	Local exhaust system, cyclone or multiple cyclones
Steel mills	Blast furnaces: charging, pouring	CO, fumes, smoke, Particulates (dust)	Good maintenance, seal leaks; use of higher ratio of pelletized or sintered ore; CO burned in waste heat boilers, stoves or coke ovens; cyclone, scrubber, electrostatic precipitator, or venturi scrubber
	Electric steel furnaces: charging, pouring, oxygen blow	Fumes, smoke, particulates (dust), CO	Segregate dirty scrap; proper hooding, baghouses, venturi scrubbers, or electrostatic precipitator
	Open-hearth furnaces: oxygen blow, pouring	Fumes, smoke, SO_x, particulates, (dust), CO, NO_x	Proper hooding, settling chambers, waste-heat boiler, baghouse, electrostatic precipitator, or venturi scrubber
	Basic oxygen furnaces: oxygen blowing	Fumes, smoke, CO, particulates, (dust)	Proper hooding (capture emissions and dilute CO), scrubbers, or electrostatic precipitator
	Raw material storage	Particulates (dust)	Wetting or application of plastic spray
	Pelletizing	Particulates (dust)	Proper hooding, cyclone, baghouse
	Sintering	Smoke, particulates (dust), SO_2, NO_x	Proper hooding, cyclones, venturi scrubbers, baghouse, or precipitator

governments had the primary responsibility to control air pollution, but strengthened the federal government's role in air pollution control. This legislation empowered the Environmental Protection Agency to establish ambient air quality standards to protect the public health and welfare and then ensure that they were enforced. The Environmental Protection Agency was given authority to grant permits to construct or modify an emission source in any of the categories of significant stationary sources. Emission standards were promulgated together with specific sampling methodologies and reporting requirements. An ongoing procedure to update the list of significant stationary source categories was included.

Emissions standard development for existing stationary sources (except sources of hazardous pollutants) was required of the states. The states were also responsible for enforcing the emission standards, as well as attaining and maintaining the federal ambient air quality standards within their respective boundaries. The State Implementation Plan (SIP) was to be the means by which the states were to comply with the Clean Air Act.

Criteria pollutants   Criteria pollutants were identified pursuant to Sections 108 and 109 of the 1970 Clean Air Act. Particulate matter, sulfur oxides, nitrogen oxides, hydrocarbons, and carbon monoxide were defined as criteria pollutants when the 1970 Clean Air Act was promulgated. (On March 31, 1976, lead was added to the list of criteria pollutants.) Each state was given responsibility for promulgating regulations to control these pollutants such that the National Ambient Air Quality Standards (NAAQS) would be achieved. These regulations, part of each State Implementation Plan (SIP), could vary from state to state. The regulations were to cover both existing and new sources.

Hazardous pollutants   The Clean Air Act differentiated between nonhazardous and hazardous pollutants. The U.S. EPA Administrator was charged with promulgation of standards for hazardous air pollutants regardless of whether they emanated from new or existing sources. Standards have since been promulgated for a number of hazardous materials, including asbestos, beryllium, mercury, vinyl chloride, arsenic, and benzene. Enforcement of these standards is the responsibility of the EPA; however, at the states' option, implementation plans submitted to EPA may include regulatory control procedures for these materials. After EPA approval of the implementation plan, states would then be authorized to enforce the hazardous pollutant standards within their jurisdiction.

Source test methods were identified with the hazardous pollutant standards. These test methods were the only approved means by which compliance could be determined. The regulations also specify reporting requirements for the source. Not only must operational data be maintained, but application must also be made to EPA prior to any modification of existing sources. No new or modified source of a hazardous pollutant may start operation without prior notification of EPA.

State permit systems   Since the early 1970s, and in many jurisdictions well before that, operating permits have been required for various processes that are vented to the atmosphere. A thorough review of regulations applicable to each plant site is required to determine precisely the processes for which permits are required.

In some jurisdictions, the law requires that permits be obtained before a new source is constructed or an existing source is modified. Permit forms for these operations allow the control agency to evaluate the planned emission control equipment and assess potential compliance with applicable regulations. If the agency judges that the source as planned will not operate in compliance with regulations, agency officials may require changes in the design of the process or installation. Only processes believed to operate in compliance with the applicable regulations can receive permits.

Prevention of Significant Deterioration (PSD) Areas

PSD requirements were developed for those areas of the country already in compliance with National Ambient Air Quality Standards. The requirements made use of an area classification scheme whose basic premise was that a moderate amount of industrial development should be routinely permitted in all areas but that industrialization should not be allowed to degrade air quality to the point where it barely complies with air quality standards. In addition, an opportunity should be provided for states to designate certain areas where pristine air quality was especially valued and any growth generating significant emissions of pollutants should be tightly curtailed. Three classes were subsequently designated. The Class I category was to include the pristine areas subject to tightest control. Class II covered areas of moderate growth and Class III was for areas of major industrialization. The EPA regulations also established another critical concept known as the increment. This would be the numerical definition of the amount of additional pollution that would be allowed through the combined effects of all new growth in a particular area.

EPA did impose one major additional requirement to PSD areas to assure that the increments would not be used hastily. It specified that each major new plant must install the best available control technology (BACT) to limit its emissions. Where new source performance standards had been promulgated, they would control determinations of BACT. Where such standards had not been promulgated, an ad hoc determination would be called for in each case. To implement these controls, EPA imposed a requirement that each new source undergo a preconstruction review.

Nonattainment Areas

Nonattainment requirements were developed for those areas of the country not yet in compliance with the National Ambient Air Quality Standards. In any non-attainment area (where any ambient air quality standard was being violated), no major new source could be constructed without a permit. The permit would impose stringent control requirements and require sufficient "offsets" to assure progress toward compliance. Approval for a new source in a nonattainment area would require the new source to be equipped with pollution controls to assure the lowest achievable emission rate (LAER), all existing sources owned by an applicant in the same region to be in compliance or under an approved schedule to achieve compliance, the applicant to come up with sufficient "offsets" to more than make up for the emissions to be generated by the new source (after application of LAER), and the emission offset must provide a positive net air quality benefit in the affected area.

Clean Air Act Amendments of 1977

The 1977 amendments further strengthened air pollution control standards and enforcement. The basic control strategy remained the same; however, additional enforcement power was given to EPA. The significance of the PSD program established by EPA in 1974 and the Offsets Policy interpretive ruling (for nonattainment areas) issued by EPA in 1976 was that when Congress in the 1977 amendments provided the statutory foundations for PSD and nonattainment areas, it adopted in toto the basic concepts of the EPA programs. Congress did, however, make numerous changes in critical elements, and in virtually every case the effect was to broaden the program and tighten its requirements.

Regulatory Direction

The current direction of present governmental air pollution control efforts is toward continued strengthening of the regulatory impact.[3] Such efforts indicate concern for:

1. Fine particulate (defined as particulates less than 2 μm in size) emissions
2. Hazardous (e.g., certain heavy metals) emissions
3. Acid rain (resulting from sulfur and nitrogen oxide emissions)
4. Volatile organic compound emissions
5. Carcinogenic organic emissions

Fine-particle emission control may take any number of strategies:

1. Direct control via a fine-particle standard. For example, mass emission limitations and/or efficiencies may be specified for specific particle-size ranges.
2. Indirect control via inhalable particulate regulations and/or visibility restrictions.

The control of hazardous heavy-metal emissions might also be related to any fine-particle emission standards. Most heavy-metal emissions are in the extremely fine particle size range.

The emphasis on carcinogenic emissions is expected to continue as one of the highest priorities at EPA, with standards promulgation anticipated once a satisfactory supporting data base is generated.

Although it is not possible to predict the future, it is possible by monitoring the direction of current research to prepare for it. It is highly recommended that maximum flexibility be designed into the final equipment selected. For example, if the choice is between an electrostatic precipitator or fabric filter for a particular application, selection in favor of the fabric filter may be the wisest choice in view of the fact that fabric filters quite often can more than adequately comply with particulate removal requirements and demonstrate the highest collection efficiencies in the fine-particle-size range. Assuming that the capital costs are comparable, the higher operating costs usually associated with the fabric filter may be worth paying for the greater flexibility. Such trade-offs must be carefully evaluated and incorporated into the overall selection process.

2.7 REFERENCES

[1] THEODORE, L., AND BUONICORE, A. J., *Industrial Air Pollution Control Equipment for Particulates*, Chap. 6. CRC Press, Inc., West Palm Beach, Flda., 1976.

[2] *Field Operations and Enforcement Manual for Air Pollution Control*, Vol. 2: *Control Technology and General Source Inspection.* EPA APTD-1101, 1972.

[3] BUONICORE, A. J., "Air Pollution Control," *Chemical Engineering*, Vol. 87, No. 13, 81–101, June 30, 1980.

3

Absorbers

James W. Macdonald

Senior Process Design Engineer
The Ceilcote Company
A Unit of General Signal Corp.
Berea, Ohio

3.1 DESCRIPTION OF CONTROL DEVICE

Air pollution control equipment design has had a reputation as being a mystical art rather than theoretically predictable designed devices. As air pollution equipment has grown out of its infancy, much of the mysticism has been replaced by the acceptance of specific design parameters for sizing and predicting performance. As these design concepts have become more widely accepted, greater emphasis is now being placed on methods of operating cost reduction and improved performance. Cost reduction and improved performance are directly related to the proper operation and maintenance of the device.

In this chapter it is the intent to provide a guide to the operation of absorber equipment and its maintenance. Absorption of gaseous pollutants occurs in any device that provides contact between the contaminated air and a liquid in which the gas is soluble or with which it will chemically react. The degree of removal depends on several factors: solubility of the gas, gas and liquid throughput rates, contact time, mechanism of contact, and the type of wet collector. Many types of wet collectors are used, not only for the removal of gaseous pollutants, but also for removal of solid and mist particulates.

In industrial applications, the packed scrubber (absorber) is used predominately for control of gaseous pollutants where efficiencies in excess of 90 to 95% are required. There are three modes of operation of packed scrubbers: countercurrent, cocurrent, and cross-flow [1].*

The countercurrent model (Figure 3-1) is the most popular design for removing gas contaminants. In this mode, the airstream flows vertically upward while the liquid flows downward in direct opposition. It is useful in handling the more difficult gases because the packing height is not restricted and can achieve very high removal. Although the countercurrent packed scrubber provides the greatest removal efficiency, it also has the highest pressure drop and the least solids-handling capacity of any of the three modes [1].

In the cocurrent design (Figure 3-2), the airstream and liquid flow in the same direction through the scrubber. Most commonly, this is vertically downward through the packing. This is the least used of the basic modes because it has a finite limit on absorption efficiency. It is useful where high solids loadings are present because

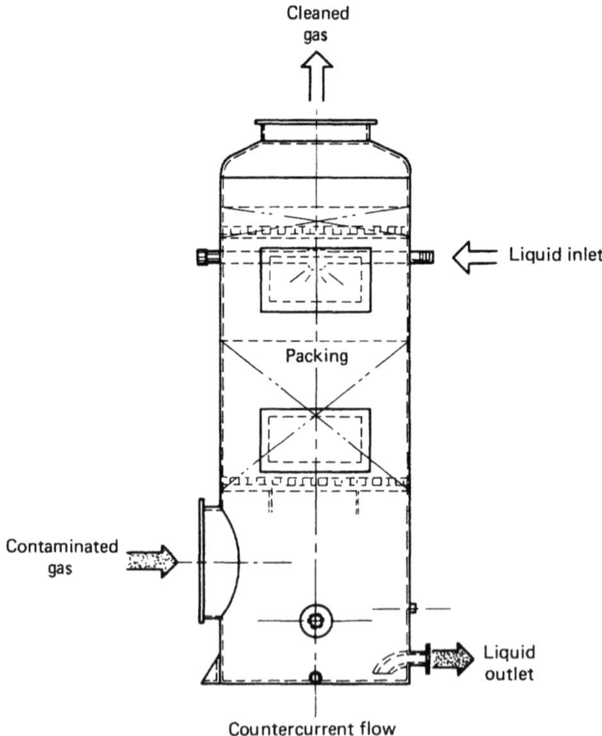

FIGURE 3-1 Countercurrent model.

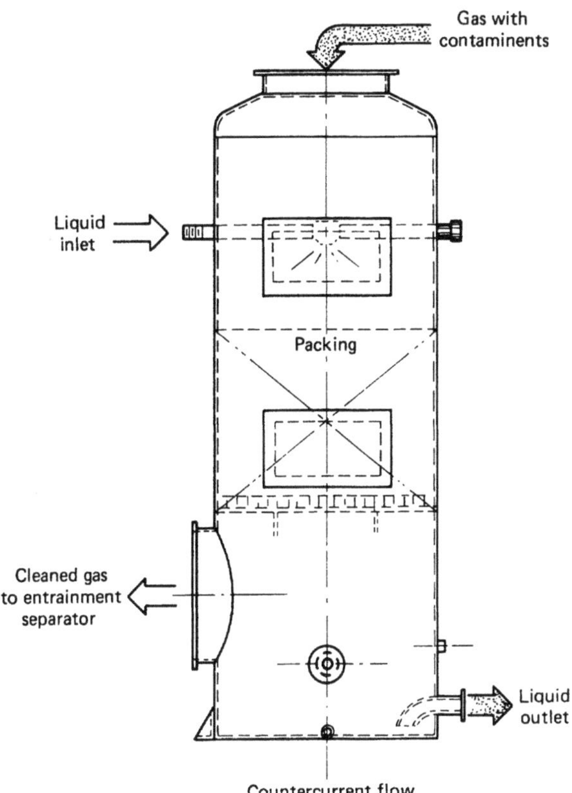

FIGURE 3-2 Cocurrent design.

it can operate at much higher liquid rates to prevent plugging. The pressure drop can be kept low or the tower area can be reduced for space considerations. Caution must be taken with the cocurrent flow to prevent liquid entrainment from being emitted to the atmosphere. To prevent this, an entrainment separator must be provided either in the bottom gas outlet or external of the absorber [1].

The cross-flow design (Figure 3-3) is the most unique and has become a commonly used mode for the removal of highly or moderately soluble gases, or where rapid chemical reactions will occur. In this mode the airstream flows horizontally through the packing while the liquid flows vertically downward. The cross-flow mode has some attractive features over the countercurrent design, while affording lower pressure drop. Lower pressure drop results in the cross-flow where the liquid is at right angles to the air flow resulting in less pressure drop when compared to the countercurrent mode, where the gas and liquid flows oppose each other. In addition, the cross-flow has greater solids-handling capacity without causing fouling. This can be achieved by increasing the liquid flow over the first section of packing to flush off any collected solids to prevent buildup, and by adding a front

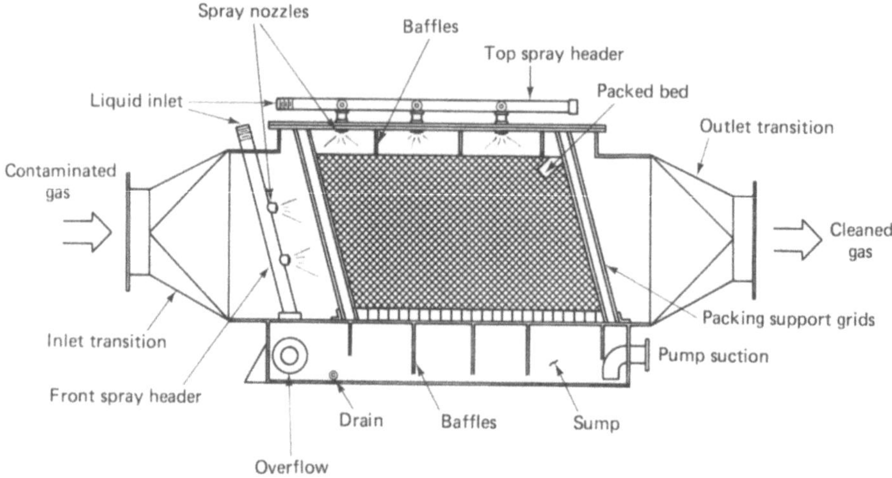

FIGURE 3-3 Cross-flow design.

cocurrent spray to prevent blinding at the front packing support plate. Care must be taken to prevent gas from bypassing the packed bed. This bypassing occurs primarily in the liquid distribution zone and in the sump. By the proper use of internal baffles, this can be eliminated [1].

3.2 DESIGN PROCEDURES

Basic gas absorption principles are common to all packed scrubbers and wet collectors used for gas absorption, regardless of their operating modes. The gas to be absorbed into the liquid must be soluble or be capable of reacting chemically with the scrubbing liquid. The degree of solubility depends on the type of gas, its concentration (mole fraction or vapor pressure), the system pressure, the type of absorbing liquid, and the liquid temperature. Any liquid has a limited amount of gas which it can absorb before it becomes saturated. This is known as *equilibrium*. If the liquid reaches equilibrium within a packed scrubber, it cannot absorb additional gas. In most packed scrubbers the liquid is recirculated from a reservoir (sump) to the liquid distributor. This liquid cannot be recycled indefinitely or it will reach saturation. Equilibrium or saturation will be reached first at the gas outlet of the scrubber because it is here that the contaminant has the lowest concentration in the air stream. To prevent equilibrium from occurring and drastically reducing the removal efficiency, fresh liquid is introduced and contaminated liquid is withdrawn from the system continuously, or by introducing chemicals to react with the absorbed gaseous pollutant.

In addition to equilibrium considerations, one must consider how the gas can be absorbed into the liquid most effectively. Since it is the liquid that is absorbing the gas, a means of extending the liquid surface must be provided. This is the premise for the packed absorber itself. By providing a packing medium over which the liquid is spread, a greater area of contact can be achieved. One could interpret, therefore, that the area of gas–liquid contact is the area of the packing. Were this true, the relative efficiency of two packings would be proportional to the ratio of their areas. Yet ½-in. Raschig rings having twice the surface area are only 70% more efficient than 1-in. Raschig rings. Therefore, there must be more than the area of the packing relating to performance. This additional factor, liquid surface renewal, has a greater effect with the use of thermoplastic packing than with ceramic packings.

Liquid surface renewal is related to the number of agglomeration points (sometimes referred to as interstitial holdup points). At an agglomeration point, liquid that has flowed down a section of packing will be momentarily held at this point until the weight of the liquid exceeds the surface tension forces and then flow will continue. Liquid flowing down an adjacent section of packing can also flow into this retention zone and mix with the resident liquid. By falling onto the next packing section, the droplet breaks up and presents fresh liquid surface for gas contact. This retention (agglomeration) and continued flow (dispersion) mechanism causes a surface renewal. This agglomeration–dispersion mechanism is also one of the reasons for the effectiveness of packed columns; it is not the surface area of the packing alone, especially where thermoplastic packings are considered [2].

The choice of type, size, and material of construction of the packing medium therefore becomes extremely important. The specific packing that is chosen will affect the initial size and subsequent cost of the equipment. It will also relate to pressure drop, type of support for the packing, type of liquid distribution, fouling, and packing replacement. Packings are available in three basic materials of construction: metal, ceramic, or thermoplastics. Metal packings are rarely used for gas absorption because most gaseous pollutants are so highly corrosive that the type of exotic metal required becomes cost-prohibitive. Ceramic materials are used, but in limited applications, primarily where high system temperatures exist. Ceramic packings are very brittle and extreme care must be taken in filling the absorber to prevent or reduce breakage. This is often accomplished by filling the column with water and allowing the packing to slowly settle into position, or by physically stacking the packing in a dry column. During operation, ceramics can erode or may be subjected to thermal shock, causing packing wear or breakage. Occasionally, the columns should be inspected and additional packing added, if required. Generally, because of the rough surfaces of ceramic packings, fouling or plugging will occur more readily than with thermoplastic packings. If fouling is observed, periodic scheduled maintenance is recommended to prevent eventual total blinding of the packing media. The packing can usually be cleaned by water flushing, chemical cleaning, or steam cleaning, depending on the deposited material. Other factors that should be considered in choosing ceramic materials are its higher pressure drop and greater weight compared to thermoplastic materials.

Thermoplastic materials of construction for packings are the most widely used for gas absorption in air pollution control (Figure 3-4). Care must be taken to choose

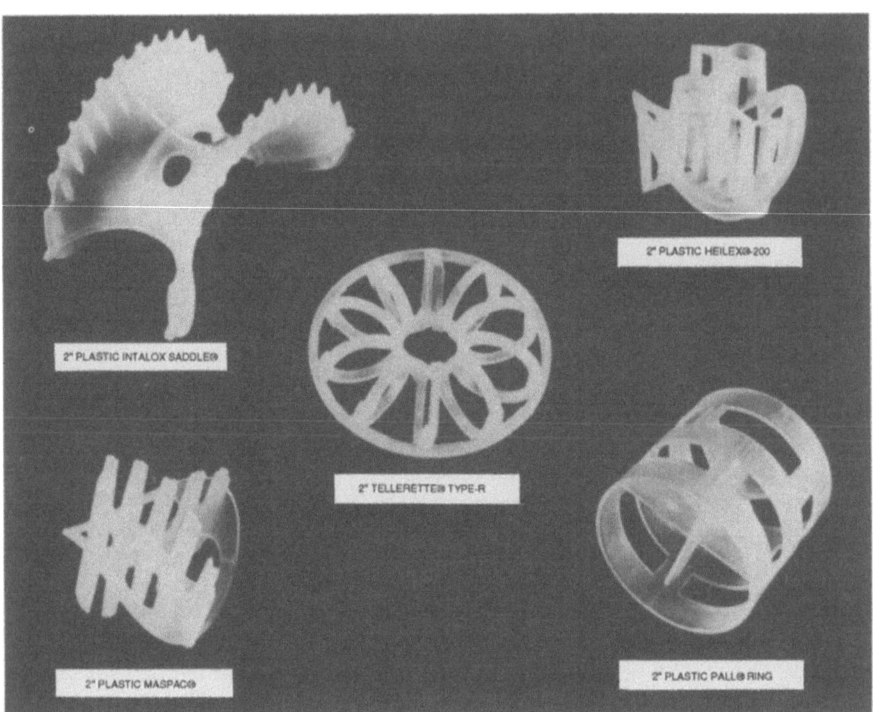

FIGURE 3-4 Typical absorber plastic packings: (a) 2-in plastic Intalox saddle (Intalox, registered trademark of Norton Co.); (b) 2-in. plastic Heilex 200 (Heilex, registered trademark of Heil Process Equipment Corp.); (c) 2-in. Tellerette Type R (Tellerette, registered trademark of the Ceilcote Co.); (d) 2-in. plastic Maspac (Maspac, registered trademark of the Dow Chemical Co.); (d) 2-in. plastic Pall ring (Pall, registered trademark of Norton Co.).

the proper thermoplastic material for the corrosive environment and temperature. Thermoplastic packings are available in polypropylene, polyethylene, polyvinyl chloride (PVC), CPVC, Noryl,[†] Kynar,[‡] Udel[§] polysulfone, and Tefzel,[‖] to name the most common. If the proper material is chosen, it can be virtually inert to a corrosive environment. The smooth surface of plastic packings reduces fouling since any solids that may be deposited do not cling, as they would on a rough surface, and are more easily washed away. Lightweight plastic allows a more open support plate to be used, reducing pressure drop and solids deposition. This lighter weight also reduces the shell cost and is less expensive to install and support. Plastic packings are easily installed by simply dumping the packing into a dry column. When installing the packing, care should be taken to prevent bridging of the packing.

[†] Noryl is a registered trademark of General Electric Co.

[‡] Kynar is a registered trademark of Pennwalt Corp.

[§] Udel is a registered trademark of Union Carbide Corp.

[‖] Tefzel is a registered trademark of E. I. du Pont de Nemours and Co.

If bridging occurs, this area will be void of packing, causing what is known as channeling. This will reduce the efficiency of the absorber. Channeling is a condition providing nonuniform cross-sectional distribution of a liquid or a gas flowing through the packing media. Most plastic packings will settle slightly during the first two weeks of operation. It is recommended that the absorber packing height be checked after this initial 2-week period, and if it has settled, that additional packing be added to bring the packing level to its proper height.

The most common maintenance problem associated with packing media is fouling or plugging. Plugging can occur due to deposition of undissolved solids in the liquid, deposition of insoluble dust in the inlet airstream, precipitation of dissolved salts in the liquid which exceed their solubility, and precipitation of an insoluble salt which is formed by the reaction of a soluble salt in solution with an absorbed gas. Of those mentioned, the most prevalent and the most difficult to prevent is the precipitation of an insoluble salt formed by reaction. Typically, this will form due to the Ca^{2+} or Mg^{2+} in hard water reacting with the absorbed CO_2 from the air to form $CaCO_3$ and $MgCO_3$ scaling. In most cases this will occur when using an alkaline scrubbing liquid. This scaling can be removed by periodic flushing with an acidic solution such as HCl. Scaling can be reduced by using a packing having a large free volume. The other plugging problems mentioned can be easily reduced by increasing the liquid flow through the column, adding a strainer in the recycled liquid piping, or by increasing the water-makeup and overflow rates.

Although the type and size of packing are important in designing the physical absorber size and packing depth to achieve a specified removal efficiency of the gaseous pollutants, an equally important factor is the method of liquid distribution over the packing. If the liquid is not distributed evenly across the packing, or if portions of the packing are not wetted, channeling will occur and reduce performance. Again, bear in mind that it is the liquid that absorbs the gases, not the packing. Also, packing that is not wetted will tend to plug much faster because there is no liquid continually washing away any solids deposition. There are four basic types of liquid distributors: spray, weir, perforated pipe, and open splash plate. The spray type using spray nozzles provides the most effective means of evenly distributing the liquid over the packing. It also provides an area above the packed section for additional contact between the air stream and the spray droplets for gas absorption. A properly designed spray distributor should provide overlapping sprays, and the top of the packing should be at a specified distance below the sprays for optimum effectiveness. If the packing level is too high, some areas near the tower wall will not be wetted; conversely, if the packing level is too low, too great a portion of the liquid is sprayed on the vessel wall and becomes ineffective. Full cone-type spray nozzles are suggested and should be as nonclogging as possible. Spray nozzle plugging is the major disadvantage; however, with the proper nozzle selection and by installing a strainer in the recycle piping which is capable of removing solids smaller than the nozzle orifice size, nozzle plugging can be drastically reduced, if not eliminated.

Weir-type distributors are also used frequently, primarily to eliminate the plugging associated with spray types. Liquid is fed into the open weir boxes, which contain V notches from which the liquid overflows onto the packing. These V notches (points of liquid overflow) are typically spaced at 6-in. increments throughout the absorber cross-sectional area. With weir distributors the distance between the packing and the point of liquid feed is not as critical. This is normally 4 to 12 in. below the weir box. When these liquid streams contact the packing, they splatter and spread out over the packing surfaces. The spreading of the liquid occurs quickly on rough-surfaced packing such as ceramic material. On plastic packing having extended surfaces this occurs less rapidly, and on plastic packings having no extended surfaces the liquid may not spread out evenly, if at all, prior to leaving the packed section. Uneven liquid distribution will also occur if the weir box is not level. This liquid channeling will drastically reduce tower performance.

Perforated pipe distributors are not recommended. Because of the extremely small hole size and number of holes required to provide even distribution, plugging becomes a major operational problem. These holes (orifices) can be as small as 1/16 in. and can be very difficult to clean. If the holes are made larger and spaced less frequently to reduce clogging, the liquid distribution becomes uneven [3].

Open-splash-type distributors are used very infrequently. This type is simply an open pipe with a plate suspended below it to provide a means of distributing the liquid over a larger area. They are used when the liquid being introduced is known to have large quantities of suspended solids and filtration is not practical, such as the use of river water where silt would be present periodically. Should this type of distributor be required, it is suggested that greater liquid flows be used to compensate for uneven liquid distribution [3].

Another major consideration in absorption equipment is the entrainment separation section (sometimes referred to as the mist eliminator). In most packed absorbers the entrainment separator is incorporated as an integral part of the absorber. It is placed above the liquid distributor in the countercurrent absorber, in the gas outlet, or externally in the cocurrent unit or after the wetted packed section in the cross-flow mode. Some manufacturers provide this as a dry section integral with the wetted bed in the cross-flow mode. The purpose of an entrainment separator is sometimes misunderstood. Its specific function is to remove entrained liquid particles which the air stream captures as it passes through the packed section and liquid distributor. It is not designed for the specific purpose of removing liquid mists entering the inlet air stream. The entrained liquid particles are generally large in size and can be removed easily, by impingement or interception mechanisms, in excess of 99.9% by weight. Three types of entrainment separators are used: mesh pads, packing media, and chevron baffles. Because of their fine filamentous construction, mesh pads provide the most effective removal. However, their fine filamentous wire construction can cause rapid plugging and require frequent cleaning. It is because of this potential plugging that a shallow bed of the more open packing media is often used. Although not quite as efficient, it does provide a much larger open area, reducing the frequency of cleaning if plugging or scaling occurs. Where large plugging or scaling potentials exist, chevron-type separators are used. These are less effective than mesh or packing, but do provide maximum protection against solids plugging.

Our discussion has centered around the function of an absorber and the internals of a packed scrubber; however, we have neglected one of the most important considerations for any absorber—the material of construction of the shell. If the proper material is not chosen, repair and maintenance costs can be extremely high because of corrosion. The material chosen for the absorber shell should be dictated by the types of gaseous pollutants present in the airstream, the liquid used for absorption, the operating temperature, and the reaction products formed in solution (if a reaction is required for removal). Typical gaseous pollutants include HCl, HF, Cl_2, SO_2, H_2S, NH_3, NO_x, amines, alcohols, and organics. Often, caustic or acid solutions are used, rather than water, to react with the absorbed gases to form soluble salts. As can be seen, most of these are extremely corrosive and require special care in selecting the proper materials of construction. Absorbers are constructed of stainless steel, lined steel, exotic metals, PVC, and fiberglass-reinforced plastic (FRP). In most applications, economics dictate the use of either lined steel or fiberglass. If fiberglass is chosen, care must be taken to select the proper resin material since not all resins have equal protection against corrosion. If lined steel is selected, care must be exercised not only in selecting the appropriate lining material, but also on the surface preparation of the steel to which the lining is to be applied. In addition, it is recommended that steel shells be externally coated or protected against atmospheric corrosion.

As can be seen, gas absorbers within themselves are very simple devices with no moving parts. With the selection of proper materials of construction to protect against corrosion, there is very little maintenance required other than to guard against scaling or plugging.

It becomes apparent that the absorber cannot function without equipment to provide a means of motivating the air and liquid streams to and from the device. In order to provide an operational system for gas absorption (Figure 3-5), we must also consider the operation and maintenance of the duct, fan, pumps, piping, valves, chemical feed system, liquid reservoir (sump), and instrumentation.

The motivating force for the liquid is a pump, which in most cases is used for recycling the liquid from the absorber sump to the liquid distributor. This is done to conserve liquid consumption and reduce treatment of the contaminated liquid. This pump must be carefully chosen because it provides the means by which the absorber functions. Care must also be taken in choosing the proper material of construction not only against corrosion, but also against abrasion if solids are present. Normally, a base-mounted centrifugal pump with a coupling connection to the motor is used. A pump with a packed stuffing box is used in many cases for easy maintenance; however, it should be recognized that leakage will and should occur. This drippage can be a nuisance and an external corrosion problem. It can be eliminated by the use of a properly chosen mechanical seal or by piping away this leakage. On pumps that handle solids, it is generally recommended that a continuous freshwater flush be provided to prevent scoring of the shaft. To reduce

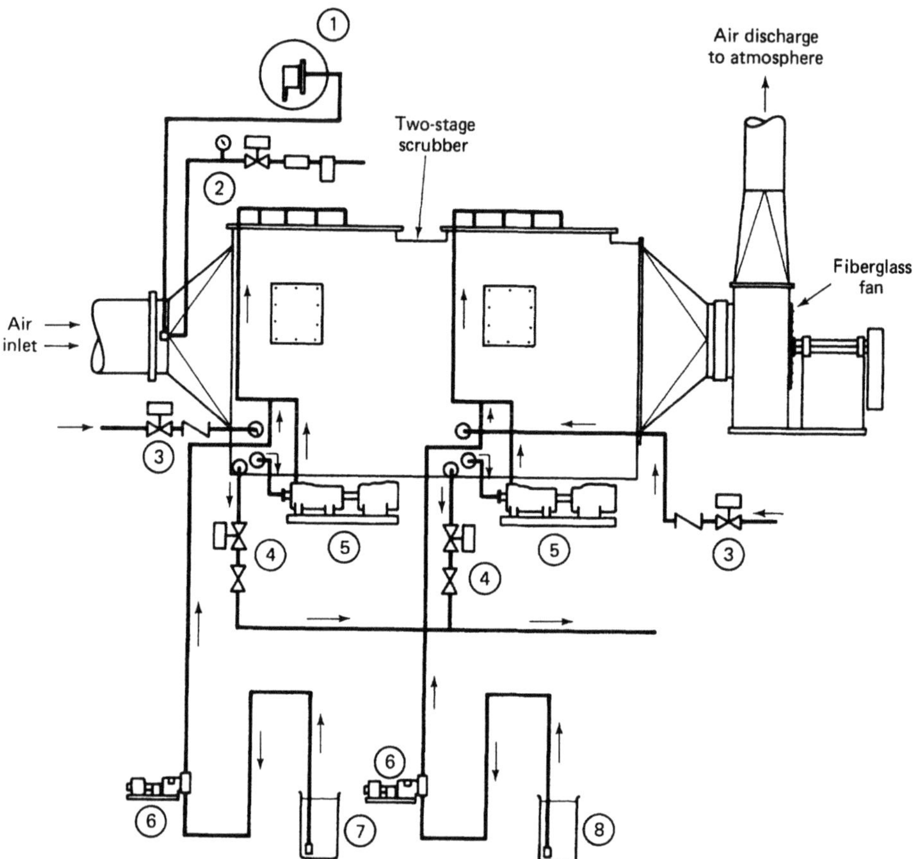

FIGURE 3-5 Schematic of a two-stage cross-flow absorber: 1, chlorine cylinder and regulator; 2, compressed air supply; 3, freshwater-makeup solenoid valves; 4, drain solenoid valves; 5, chemical solution recirculation pumps; 6, chemical makeup metering pumps; 7, concentrated sulfuric acid storage; 8, concentrated sodium hydroxide storage.

impeller wear and maintenance where solids are present, a 1750- or 1150-rpm pump is recommended. If the recycle pump is to be used outdoors in a cold climate where winter freezing problems may exist, it is recommended that a drain be provided in the casing or, if this cannot be provided, that the pump be thermally protected against the liquid freezing in the bottom of the casing when the pump is not operating.

Many problems associated with pumps are not necessarily due to the pump itself but to the piping connections to and from the pump and the physical placement of the pump with respect to the absorber sump. The pump should be placed so that the pump suction line is completely filled with liquid (flooded suction) and the liquid level in the sump at a level above the pump suction to provide the Net Positive Suction Head Required (NPSHR) for the pump. In addition, the pump suction connection in the absorber should be designed to prevent cavitation or vortexing from occurring, especially in shallow sump reserviors. Cavitation causes air to be drawn into the pump with the liquid and will cause the pump to operate erratically. This will produce low liquid pumping rates and excessive wear. The recycle piping should be of a proper material of construction to withstand the corrosive environment, temperature, and pressure conditions. Typical materials used are PVC, CPVC, polypropylene, fiberglass (FRP), lined steel, and stainless steel. The suction-line piping should be at least one pipe size larger than the pump suction connection on the recycle pump. It is also suggested that a large coarse screen or filter be used on the suction side of the pump to prevent broken pieces of packing, rags, or other large foreign material from entering the pump and damaging the impeller or casing. This screen or filter should be large enough to prevent restriction of flow to the pump. The discharge-line piping should be at least one pipe size larger than the pump discharge size. A throttling valve should be provided in the discharge line of the pump to allow for adjustment of the proper liquid flow. Where spray-type liquid distributors are used in the absorber, it is recommended that a strainer or filter be provided in the pump discharge piping. The strainer should be designed to filter out particles equal to or smaller than the spray nozzle orifice or particle-free passage size. This will reduce or prevent plugging of the spray nozzles. Normally, a shutoff valve is installed in both the suction and discharge piping to allow for isolation of the pump for maintenance. In cold climates, consideration must also be given for insulation of piping systems that are not self-draining when the system is shut down.

Another essential piece of equipment is the fan or blower to circulate the air through the absorber and duct system. Most absorber systems locate the fan to induce the draft (vacuum) on the system, to prevent out-leakage of the contaminated air. There are occasions, however, where a forced-draft system, using the fan on the inlet side of the absorber, is necessary, and more practical, because of the nature of the corrosion environment. In the selection of the fan material of construction, the location of the fan with respect to the absorber is an important factor to consider, as well as the maintenance required for operational reliability. On systems where hot unsaturated airstreams are entering the absorber, with low or little particulate solids present, it may be advantageous to provide a forced-draft system since, at high temperatures with corrosive gases, corrosion is not as large a problem as saturated airstreams containing corrosives, and the fan material of construction could be of lower-grade alloy materials. If, however, the inlet air is at low temperature and highly saturated with water vapor, or contains a high, or relatively high, solids loading, it is suggested that the fan be located after the absorber to reduce solids buildup on the fan blades. It should be recognized that systems using induced-draft fans after the absorber will still be exposed to highly corrosive gases, because no absorber is 100% effective. In addition, one should be cognizant of the fact that the airstream coming from any absorber is saturated with water vapor. This in turn will cause a certain amount of condensation to take place in the fan housing at the walls, which will absorb any remaining corrosive pollutants and become very corrosive. Therefore, particular attention should be given to the material of construction of the fan and housing. Materials generally used are PVC, fiberglass (FRP), or lined steel housings with exotic-alloy wheels and shafts. In certain cases, lined impellers are used, provided that the fan tip speed is low enough to prevent erosion of the lining. Corrosion and erosion of the fan wheel and blades is a major consideration, and a fan should be selected at minimal rpm and tip speed to reduce the abrasion that may occur and reduce maintenance. Corrosion of the fan blades can be caused not only from solids abrasion, but also from water droplets or condensate, which can be very abrasive. Fans used on absorber systems are normally the centrifugal type and should be equipped with access openings and drains for inspection of the wheel. The drains on the fans should be U-trapped or barometrically sealed to allow any condensate formed in the fan housing to drain away. In designing the static pressure for the fan, system designers usually add safety factors to the estimated system pressure drop. Although this has advantages in overdesign of the fan, it can cause problems resulting in more air being delivered to the system than is designed. This increased capacity will be detrimental to the absorber performance, and also require more horsepower, and thus increased energy consumption for the system. To prevent this from happening, many system designers employ a damper in the duct ahead of the fan for volume adjustment. On fans with adjustable sheaves or with belt drives, if large increases in capacity are encountered, it may be advantageous in reducing horespower consumption simply to change the sheave to lower the fan speed to the desired system capacity. All blowers or fans in a system should be located for ease of accessibility for maintenance, and all drives should be adequately protected for safety reasons. The fans or blowers should be adequately isolated by use of flexible connectors to the duct on the inlet and outlet side to prevent vibration from being transmitted to the duct system. Fans can be equipped with vibration switches to shut down the fan automatically should vibration be excessive. This is suggested on systems where solids or excessive buildup may be expected. The system ductwork should be designed of an appropriate material of construction against the corrosive environment. If the system is handling solids or any amount of particulate matter that may be deposited in the duct, it is recommended that periodic cleanout ports be designed into the ductwork for inspection and/or cleaning. The duct should also contain a slope in its installation, so that any condensate formed will run to a low point and can be drained off so that no pools are created in the ductwork. In extremely corrosive applications, the duct is generally constructed of materials such as PVC, polypropylene, fiberglass (FRP), or lined steel.

Auxiliaries

Chemical feed systems   In many absorber (scrubber) applications, a chemical is added in the recycle liquid stream which will react with the absorbed contaminated gases. If the absorber is designed utilizing this chemical feed, it is imperative that the chemical feed be maintained and controlled at the proper feed rate for continuous effective removal efficiency in the absorber. The types of chemicals typically used are acids, such as sulfuric acid, and caustics, such as sodium hydroxide or sodium carbonate solutions. Some applications also require oxidizing agent additives, such as potassium permanganate, sodium hypochloride, or hydrogen peroxide. There are two approaches to using a chemical feed to a system. One is continuous control, the other is a batch-type system. The batch system will typically be used for intermittent-type operation. It consists of filling the absorber sump with a fairly high concentration of chemical solution, which is recycled until a minimum chemical concentration is reached, such that the amount of chemical in the recycle solution will stoichiometrically react with the absorbed pollutants. With this approach there is no continuous overflow or makeup addition to the scrubber sump; however, water must be added to account for evaporative losses in the system. Once the minimum-strength chemical solution is reached, the system must be shut down and the sump drained and recharged with the chemical solution. The disadvantage of this approach is that the chemical cannot be totally utilized, because of the minimum-strength requirement for maintaining absorption efficiency. This increases the operating cost by the value of the unused chemical solution. The other approach consists of continuous addition of chemical usually utilizing a pH control system, with either on/off or proportioning control using a metering pump or control valve. The chemical feed to the metering-pump or proportioning valve is fed from a chemical storage tank of sufficient size to allow long operation without loss of chemical solution. The pH control is normally inserted either in the scrubber sump, the overflow, or the recycle piping, to the specified pH control set point. This allows maximum utilization of the chemical feed and reduces operating costs. In most air pollution control systems, an on/off type of control is used, which allows periodic input of chemical into the recycle piping or above the packed zone for treating with the absorbed gases. Where chemical feeds are used, it is important that a continuous overflow or bleed stream from the absorber be maintained to preclude the buildup of dissolved solids in the recycle system to prevent crystallization of salts on the packing medium, ultimately causing plugging. It is recommended that the chemical solution feed tank be equipped with a low-level alarm to preclude the system from running out of chemical feed solution. It is also recommended that weekly inspection of the pH probes be provided and that the probes be cleaned or replaced when required to maintain an operable system.

Liquid-level controls   In many absorber applications, it is good practice to add a liquid-level control to the absorber sump. Liquid-level controls are typically bubbler types, float types, capacitance types, sonic types, and so on. All serve the same function, that is to minimize the water makeup in the system, and to protect the system against low liquid level in the sump. Should the continuous freshwater makeup be inadvertently shut off, or the liquid makeup be at such a low rate that a continuous overflow is not achieved, the automatic level control will protect the absorber sump from running dry. This will also prevent pump cavitation from occurring, protecting the recycle pump from damage. Since most absorbers contain an integral sump section, and turbulence of the liquid in the sump creates a problem in maintaining an accurate liquid level, it is suggested that an external stilling well for control be used to allow greater accuracy in maintaining the correct liquid level. In addition, external mounting of the liquid-level control allows periodic maintenance for the liquid-level control without shutting down the system.

Freeze protection   For absorbers installed in climates where winter conditions create a freezing problem as regards the liquid in the sump or the recycle piping, precautions should be taken to eliminate any freezing when the system is not in use. In most absorber applications, the liquid will not freeze as long as the system is in operation because the inlet air to the absorber is generally warm and will prevent the liquid from freezing. However, during a system shutdown or over a weekend when the system is not in use, the scrubber sump, if exposed to external conditions, should either be drained completely or some type of immersion heater should be used in the sump to prevent freeze-up of the liquid. The type of material used for the immersion heater for this freeze protection depend on the corrosive nature of the solutions and gases. To avoid corrosion in this area, exotic alloy materials of construction or graphite material are generally used. If an immersion heater is selected, care must be taken to prevent the liquid level from being too low, thus exposing the immersion heater to the air above the liquid, resulting in heater burnout.

Steam is often used directly into the sump to heat the solution and thus protect against freeze. Caution should be used, however, because steam can cause deterioration and failure if not properly controlled. The materials used to construct most air pollution absorbers—thermoplastic internals with absorber shells of thermoplastic, fiberglass, or lined steel—will fail if exposed to high-temperature steam for prolonged periods of time.

In many cases it is recommended that the recycle piping be designed to be self-draining when the recycle pump is not in operation. This will eliminate water being held up in the lines and protect against freezing. Certain portions of the piping will not be self-draining, and these should be heat-traced or insulated to protect against freezing.

Instrumentation   There is a wide variety of instrumentation that can be used with an absorber system. This instrumentation should be provided for safety, automatic operation, and operating information of the equipment. Instrumentation used for safety considerations should protect the equipment and the operator against hazards during upset or emergency conditions. The type of instrumentation chosen depends largely on the specific system involved. Generally, high-temperature alarms, low-liquid-flow-rate alarms, low-liquid-makeup alarms, and chemical alarms may be required. Instrumentation used for operational information is useful in determining any problems or misoperation of the equipment during normal operation. Instrumentation used in this regard can include the following:

1. Inlet and outlet temperature indication of the gas and liquid streams
2. Pressure-drop indication across the absorber, the wetted packed section, and the entrainment separator
3. Continuous monitoring of the recycle liquid flow rate(s) to the absorber
4. Indication of pump discharge pressure for both high and low conditions; available with set points for alarms if desired
5. Continuous monitoring or control of the fresh liquid makeup rate to the absorber
6. Monitoring the rate of chemical feed to the absorber
7. pH control of the chemical feed rate
8. Low-liquid-level alarm for the chemical solution feed tank
9. Monitoring or alarming high or low airflow rates to the absorber
10. Continuous indication of fan and pump motor amperage
11. Low-liquid-level alarm for the absorber sump

These, if recorded and reviewed on a daily basis, can indicate changes in operating conditions that could signal problems with the system operation. Immediate or scheduled maintenance can then be provided to prevent possible extensive equipment failure if left undetected.

3.3 INSTALLATION PROCEDURES

Most equipment is shipped by truck or rail and should be inspected immediately upon receipt for any signs of damage to the exterior and interior of the gas absorber or other equipment. Specific attention should be given to checking any protruding parts, points that come in contact with the shipping media, or areas where lifting lugs, tie-down lugs, ladder clips, or other exterior components are attached, and around all internal components. If damage is observed, report the damage to the carrier and the supplier promptly. Any damage should be repaired before placing the equipment into service. When unloading large scrubbers and setting on foundations, always use proper rigging and handling procedures as recommended by the supplier. Some helpful suggestions in handling are:

1. Exercise care to prevent dropping, abrasion, and impact.
2. Always use padding where necessary to prevent abrasion and impact.
3. Always provide sufficient support to prevent undue deflection and distribute pressure over a large surface area.
4. Use lifting lugs, keeping pull on the lugs' centerline, in a radial direction only.
5. Always use guidelines to control the absorber when moving to prevent striking objects.

6. Never place concentrated stresses at any one point.
7. Do not roll or drag the absorber over anything that will cause damage.
8. Never roll over a fitting, lug, or other component.
9. Do not lift or pull the absorber by fittings or other components except lifting lugs.

Scrubbers with lifting lugs can be lifted into vertical position using a crane. When setting the absorber in place, make certain that the area has been swept clean. Set the unit down gently to avoid damage. Most absorbers must have full bottom support. The foundation must have sufficient strength to support the absorber when full without sagging or deflecting. The absorber should be well anchored by securing with all the hold-down lugs before proceeding with the piping. On scrubbers with bottom drain nozzles, a drain recess must be provided in the installation pad which does not allow the weight of the vessel to rest on the drain nozzle. After the absorber has been placed in position, it should be inspected again to ensure that no damage has been incurred during installation.

Once the absorber is positioned and anchored, the absorber internals must be installed. The packing support plates and entrainment separator support plates are often shipped separately to prevent breakage. These plates should be carefully installed through access doors and oriented properly. If the support plates were factory-installed, check to ensure that they are positioned properly and not damaged. The packing material, which is normally shipped separately, can now be placed in the absorber. If thermoplastic packing is used, it can be randomly dumped into the absorber through the gas outlet or the upper access door, making sure that bridging of the packing does not occur. Do not tamp down the packing, because this will cause a higher pressure drop. After filling to the proper height, the packing should be leveled. This can be easily done by hand or with a long-handled tool such as a rake. If ceramic packing is used, the tower should be filled with water and the packing material carefully and slowly added to prevent breakage. After filling to the proper packing height, the water should be emptied from the absorber and the packing leveled.

The liquid distributor can now be installed through the access provided. If a weir-type distributor is provided, it will normally be shipped loose and must be placed in the absorber on the internal supports provided. These weir boxes must be placed apart at the specified distances and must be leveled to prevent maldistribution of liquid. If spray-type distributors are provided, these are normally assembled and installed in the scrubber prior to shipment Often, the spray nozzles are shipped loose and must be installed in the field. If the spray nozzles are not installed, it is recommended that the piping and fittings be inspected to ensure they are clean and are at least hand-tight. If possible, blow out the lines with air. Install the spray nozzles by screwing them in hand-tight, then, using a strap wrench, tighten them one or two turns more. On thermoplastic threads it is suggested that a single layer of Teflon ribbon thread seal be used. Some spray distributors are designed so that they can be completely removed from the scrubber. If this type is provided, care must be taken when reinstalling the spray distributors that the spray nozzles are pointed in the proper direction. The entrainment separator should now be installed above the liquid distributor as recommended by the equipment manufacturer. Once the absorber internals are properly installed, the recycle piping (if required) and all piping and instrument connections to the absorber can be made. All connections should be properly aligned and supported to prevent abnormal stresses on the shell nozzles.

The overflow and drain connections should be barometrically sealed or a U-trap provided to allow the liquid to flow out properly through these connections. Inlet and outlet duct connections should be made, taking care not to force the flanges to match up with the absorber. If a fan or blower is connected to the absorber or duct, it is suggested that a flexible rubber joint be used to prevent vibration from affecting the scrubber's operation.

Pre-Startup Check

After installation is completed, it is recommended that each component of the system be checked prior to actual startup of the system.

1. The scrubber sump should be filled with fresh water and then drained to remove any dirt or foreign material.
2. Remove all foreign matter from the duct system and fan housing.
3. Turn the fan impeller by hand to assure that it runs freely and does not contact the fan inlet or housing.
4. Check the fan belt tension for tightness and adjust if necessary.
5. Bump the fan start switch to ensure proper rotation of the fan impeller.
6. Check the recycle pump and motor alignment.
7. If the pump is provided with a packed stuffing box, check to ensure that the packing and seal cage have been installed.
8. Turn the pump shaft by hand to be sure that all rotating parts run freely.
9. Disconnect the pump coupling and bump the pump start switch to ensure proper rotation of the impeller. Some pumps will be severely damaged if operated in the wrong rotation. After proper rotation is achieved, reconnect the coupling.
10. Check to ensure that oil has been added for lubrication of the pump bearings.
11. If a mechanical seal rather than a packed stuffing box is provided on the pump, it may require lubrication or continuous freshwater flushing. If so, adjust the water rate as recommended by the pump and seal manufacturer.
12. Close the absorber drain valve and fill the sump with fresh water until it begins to flow out of the overflow connection.

13. Adjust the continuous water-makeup rate to the absorber, if required, as recommended by the manufacturer.

14. Open all valves in the recycle piping and start the recycle pump. The liquid level in the sump will fall slightly, due to the liquid being held up in the packing and piping. This is normal, and the liquid level will again rise to the overflow position after a few minutes of operation.

15. Set the recycle liquid flow rate by using the throttling valve in the recycle piping.

16. Check for leaks in the recycle piping and tighten flanges where leaks occur.

17. Visually observe the spray pattern or liquid distribution in the absorber to make certain that the liquid is evenly distributed. Adjust if necessary to achieve even distribution.

18. Close all absorber access doors, and open all dampers in the system except for the inlet damper on the fan suction, which should be closed. Start the fan and slowly open the inlet damper to the design volume.

19. The system volume should be adjusted to its designed capacity. This can be accomplished by adjusting the dampers in the system.

20. Check the fan for excessive noise or vibration, which may indicate misalignment. This should be adjusted if required.

21. Check all instrumentation and record all system functions under these clean conditions.

22. After allowing the recycle pump and fan to operate for approximately 2 hours, shut down the system and drain the sump.

23. Recheck the spray nozzles and strainers for plugging, and clean if required.

Startup

Once the pre-startup check has been completed, a routine startup of the system can be initiated.

1. Close the sump drain valve and fill the sump with water to the overflow position.

2. Open the makeup water valve. The rate of makeup has been preset during the initial system check.

3. Start the pump; the throttling valve has already been preset for the proper flow rate. Allow the pump to run for 2 to 3 minutes, then start the fan. Always start the pump before initiating the fan. On system shutdown, always stop the fan before the pump.

4. Initiate the chemical feed system, making sure that the chemical feed tank is filled.

5. Check all liquid flow rates, gas flow rates, pressures, and the pressure drop across the absorber.

3.4 OPERATION

After startup, the system liquid and gas flow rates should again be checked to ensure that they are operating within the design parameters of the system. After approximately 2 weeks of operation, the system should be shut down and inspected for any possible nozzle pluggage or settling of the packing, which frequently occurs during the first 2 weeks of operation. If the packing has settled, the packing should be topped off to the appropriate designed level. During normal operation, daily checks should be made of the recycle liquid flow to the liquid distributor and the liquid makeup rate. Also, if a chemical feed system is employed, daily checks of the chemical solution should be made to ensure an adequate supply of chemical. Daily logs should also be kept of any instrumentation readings, such as pressure indicators, temperature indicators, flow indicators, and all other operations pertinent to the specific system. Should any drastic variance be noted in these readings, a close investigation should be conducted to indicate the cause of these abnormal readings. The following list is a guide for any abnormalities that may be encountered during operation of the equipment [4].

1. If the static pressure drop across the scrubber continually increases over a long period of time, this could indicate one of the following:
    a. The liquid flow rate to the liquid distributor has increased and should be checked.
    b. The packing in the irrigated bed could be partially plugged due to solids deposition, and may require cleaning.
    c. The entrainment separator could be partially plugged and may require cleaning.
    d. The packing support plate at the bottom of the packed section could be blinding and causing increased pressure drop, and will require cleaning.
    e. The packing could be settling due to corrosion or to solids deposition, again requiring cleaning or additional packing.
    f. The airflow rate through the absorber could have been increased by a change in the damper setting, which may need readjustment.
2. The pressure drop across the absorber begins to decrease either slowly or rapidly. This could be caused by the following possibilities:
    a. The liquid flow rate to the distributor has decreased and should be adjusted accordingly.
    b. The airflow rate to the scrubber has decreased due to a change in the fan characteristics or due to a change in the system damper settings.

    c.   Partial plugging of the spray or liquid distributor, causing channeling through the scrubber, could be occurring. The liquid distributor should be inspected to ensure that it is totally operable.

    d.   The packing support plate could have been damaged and fallen into the bottom of the absorber, allowing the packing to fall to the bottom and produce a lower pressure drop. This should be checked.

3.   A pressure or flow change in the recycle liquid, indicating a lower liquid flow rate. This may be caused by the following:

    a.   A plugged strainer or filter in the recycle piping, which may require cleaning.

    b.   Plugged spray nozzles, which may require cleaning.

    c.   The piping may be becoming partially plugged with solids and need cleaning.

    d.   The liquid level in the sump could have decreased, causing pump cavitation.

    e.   The pump impeller could have been worn excessively.

    f.   A valve in either the suction or discharge side of the pump could have been inadvertently closed.

4.   A high liquid flow indicates the following:

    a.   A break in the internal distributor piping.

    b.   A spray nozzle that has been inadvertently "uninstalled."

    c.   A spray nozzle that may have come loose or eroded away, creating a low-pressure drop.

    d.   A change in the throttling valve setting on the discharge side of the pump, allowing larger liquid flow; reset to the proper conditions.

5.   Excessive entrainment carryover from the outlet of the absorber. This could be caused by the following:

    a.   A partially plugged entrainment separator, causing channeling and reentrainment of the collected liquid droplets.

    b.   The airflow rate to the absorber could have increased above the design capability, causing reentrainment.

    c.   If a packed-type entrainment separator is used, the packing may not be level, causing channeling and reentrainment of moisture.

    d.   If a packed entrainment separator is used, and a sudden surge of air through the absorber occurred, this could have caused the packing to be carried out of the absorber or the packing being blown aside, creating an open area "hole" through which the air passes.

    e.   The entrainment support plate could be damaged and have dropped, causing channeling through the separator.

    f.   The velocity through the absorber has decreased to a point that absorption does not effectively take place, and low removal is achieved.

   g. Conceivably during wintertime operation, some moisture due to condensation may be observed and considered a problem with the entrainment separator. However, it should be recognized that this is not uncommon, since the air off an absorber is saturated with water vapor, and any difference or lowering of the temperature will cause condensation to occur.

6. If the reading indicates a low airflow or no airflow across the absorber, the following may be the cause:

   a. The packing in the absorber may be plugged, causing a restriction to the airflow.

   b. The liquid flow rate to the absorber could have been increased inadvertently, again causing greater restriction and pressure drop, creating a lower gas flow rate.

   c. The fan belts have worn or loosened, reducing the airflow to the equipment.

   d. The fan impeller could be partially corroded, reducing fan efficiency.

   e. The ductwork to or from the absorber could be partially plugged with solids and may need cleaning.

   f. A damper in the system has been inadvertently closed or the setting changed.

   g. A break or a leak in the duct could have occurred due to corrosion.

7. Should the airflow through the absorber be increasing or has suddenly increased, it may be due to the following:

   a. A sudden opening of a damper in the system.

   b. Low liquid flow rate to the absorber.

   c. The packing has suddenly been damaged and has fallen to the bottom of the absorber.

8. A sudden decrease in the absorber efficiency could indicate the following:

   a. The liquid makeup rate to the absorber has been inadvertently shut off or throttled to a low level, decreasing the absorber efficiency.

   b. If a chemical feed system is employed, the system may have run out of the chemical required.

   c. This (b) could indicate a malfunction of pH probes if employed, requiring replacement.

   d. The set point on the pH control may have to be adjusted to allow more chemical feed.

   e. A problem may exist with the chemical metering pump, control valve, or line pluggage.

   f. The liquid flow rate to the scrubber may be too low for effective removal.

  g. Pluggage or solids buildup in the absorber liquid distributor or packing may have caused channeling [5].

The foregoing list indicates only some of the common operational problems encountered with absorber systems. There are a number of related problems. Excessive moisture may be blown from either the fan or the blower if it is on the suction side of the system. Excessive moisture droplets being emitted could be caused by condensation in the stack or in the fan. Improper condensate drainage and trapping or line pluggage from the fan could be the problem. Scrubber liquid levels can also be too high. This can occur by having too large an inlet continuous freshwater makeup with the overflow piping being too restrictive or with the overflow and drain piping trap or barometric seal being broken. Generally, an absorber system is overdesigned for the capacity it is to handle. On occasion, lower volumes are introduced through the system below the designed condition. These lower volumes usually increase gas absorption. However, if too low a gas input is used, creating very low static pressure drops through the absorber, channeling can occur and can lower the absorber's performance and entrainment separator performance. Caution should be taken to use the manufacturer's minimum gas-flow-rate recommendation to prevent this problem from occurring [4].

3.5 MAINTENANCE

Normal preventative maintenance requires only periodic checks of the fan, pumps, chemical feed system, piping, duct, and absorber liquid distributor. The normally irrigated packing may never have to be cleaned during the life of the absorber as long as the absorber has been properly designed and operated. However, should the absorber be run dry or the entering air stream contain unexpectedly high solids loadings, a heavy formation of solids, crystallized salts, or other foreign matter may accumulate on the packing, and this must be removed. In most cases, removal can be accomplished by recirculating, for a short period of time, a chemical solution into which the solids will dissolve or react. Sometimes, chemical removal will not remove the buildup and it may be necessary to use high-pressure water, hot water, or atmospheric steam. Prior to using any chemical, hot water, or steam, the absorber manufacturer should be consulted to verify the resistance of the internals and shell. In a very few instances, such as CaF_2 deposition or extremely heavy solids deposition, it may be necessary to remove the packing medium from the absorber for cleaning.

  The normally unirrigated entrainment separator must be periodically flushed with sprays to prevent buildup and eventual plugging. The intervals between routine washings must be determined by experience, as the collection of solid materials is a function of the specific operating conditions. To ensure proper operation and unexpected problems with the absorber system the following is a suggested maintenance checklist.

1. *Pump Maintenance*
   a. Check the bearing lubrication to make sure that the proper amount of oil is present. (weekly)
   b. If a packed stuffing box is provided, make sure that the drip rate is not excessive or that the packing is lubricated. (weekly)
   c. Check pump and motor bearings for unusual noise or heat. If this is noticed, it could indicate excessive wear on the bearings, which may result in excessive shaft runout, requiring frequent repacking of the stuffing box. (weekly)
   d. Inspect gasketing at the suction and discharge connections for leaks. (weekly)
   e. Check shaft alignment and levelness of the baseplate. (monthly)
   f. Look for crystallization of solution or solids in the packing, which could cause scoring of the shaft. (monthly)
   g. Check for water flush to the mechanical seal (if provided), and adjust if necessary. (weekly)

2. *Fan Maintenance*
   a. Check the fan motor and fan bearings for unusual vibration noise or heat. (weekly)
   b. Check the bearing lubrication to ensure that the proper amount of oil is present. (weekly)
   c. Inspect the belts for wear and replace if necessary. Also check the belts for proper tension. Adjust if necessary. (monthly)
   d. Check all fan bolts for looseness and tighten if required. (monthly)
   e. Inspect the fan impeller and blades for any solids buildup or erosion. Clean or replace the impeller if excessive buildup or erosion has occurred. Such buildup can unbalance the impeller and cause premature failure. (monthly)
   f. Check the fan housing drain to make sure it is not plugged so that it will permit proper condensate drainage. (monthly)

3. *Absorber Maintenance*
   a. Check the pressure drop across the absorber. Increased pressure drop will usually indicate that plugging of the packing is occurring. (weekly)
   b. Open the absorber access door provided for inspection of the liquid distributor. (Access doors should never be opened while the fan is in operation.) Observe the liquid distribution. Clean if required. (monthly to quarterly)
   c. Inspect the absorber sump for solids buildup and flush if necessary. (semiannually)

   d.   During the inspection of the liquid distributor, check the packing level for possible settling. Add packing to the proper level if necessary. (quarterly to semiannually)

   e.   Check the entrainment separator for solids or crystallization buildup and clean with sprays if required. (quarterly to semiannually)

   f.   Inspect the absorber packing for solids buildup and clean if required. (semiannually)

   g.   Open all access openings and inspect the tower internals for corrosion or breakage. (semiannally)

   h.   During the inspection of the tower internals, also inspect the absorber shell for any corrosion and repair if necessary. (semiannually)

4.   *System Maintenance*

   a.   Check and clean all strainers or filters in the piping system. (weekly to monthly)

   b.   Inspect the recycle piping, duct and absorber flanged connections, access doors, and flexible connections for leaks and tighten or replace gaskets if necessary. (monthly to quarterly)

   c.   Check all system bypass or control dampers and cycle each one a few times to ensure proper operation. (weekly)

   d.   Check all piping to and from the absorber, including the recycle piping, for erosion or plugging and clean or replace if required. (quarterly)

   e.   If a chemical feed system is employed, check the metering pump or valves and the pH probes weekly for proper operation. It is not unusual to replace the pH probe elements frequently if solids buildup occurs. (weekly)

   f.   Inspect the duct for solids buildup due to settling. This may require periodic flushing to prevent excessive restriction to air flow or weight. (semiannually)

   g.   Check the duct, fan, and absorber for any external corrosive attack or obvious deterioration. (semiannually)

   h.   Inspect structural supports, duct, and pipe hangers for any deterioration due to corrosion. (semiannually)

   i.   Frequent checks of any heat tracing or freeze protection during the winter season, if applicable, should be conducted on a weekly basis.

   j.   All instrumentation should be inspected, checked, and calibrated to ensure proper function and control. (monthly)

The periodic time intervals indicated for maintenance will vary depending on the specific system's operating conditions and the equipment manufacturer's recommendation.

3.6 IMPROVING OPERATION AND PERFORMANCE [3]

We have been discussing the basic design parameters associated with absorbers, primarily packed absorbers, and the installation, operation, and maintenance of this equipment. Much of the information presented is also applicable to other absorber types, such as spray towers, baffle towers, tray towers, wet cyclones, and so on. Information relating to operation and maintenance of these types is presented in Chapter 9.

There are options available that should be considered when purchasing a packed absorber or any absorber, to reduce maintenance costs and allow the equipment to be more operationally reliable between scheduled maintenance shutdown. Any options generally increase the initial cost of the equipment, and for this reason are often not considered. However, some of these items can pay for themselves in a short time by reducing maintenance cost.

One such item is the provision of adequate access openings to all internal portions of the scrubber. The areas that should be considered are the sump section of the absorber; the lower packed section for removal of the packing and internal packing support plate, if required; the liquid distributor for spray nozzle cleaning or removal of the internal liquid distributor; and, access to the entrainment separator for inspection and removal of the packing and support plates, if necessary. The access openings should be large enough for accessibility and total removal of the internal components of the absorber. To reduce the initial cost of equipment, some manufacturers provide only two or three access doors. In many cases it may be advantageous to consider clear, see-through access openings, especially in the liquid distributor region, for external observation of the absorber of the liquid distributor to ensure its proper operation without equipment shutdown.

The liquid distributor, especially spray-type distributors, are one area where maintenance is required on a periodic basis. Where spray-type distributors are used, many manufacturers provide these as an integral part of the absorber shell construction. Should the spray nozzles need cleaning or removal with this type of spray distributor, a worker must physically enter the absorber for removal of the spray nozzles for cleaning. This can be a continual source of irritation. Since most absorbers are handling corrosive-type gases, the absorber must be shut down and totally purged with air to allow entry. In addition, in most plants, it is necessary that a worker wear a rubber suit and mask for safety when entering a vessel. The need for entry into the absorber for maintenance of a spray-type distributor can be totally eliminated by designing the distributor so that it can be totally removed from the absorber.

The entrainment separator is another source requiring continuous maintenance. Because it is generally an unirrigated section, solids and/or salt deposition, which is not uncommon, needs to be eliminated periodically by washing. To clean the entrainment separator it is necessary to shut down the absorber, open the equipment, and physically flush down the separator media with liquid. Equipment shutdown for this maintenance can be eliminated by installation of a spray wash-down flush header above the entrainment separator, which can provide intermittent flushing of the separator media. This eliminates equipment shutdown and maintenance cost. If this is provided, it should be recognized that if the fan is operating, liquid will be carried over through the outlet of the system during the flush cycle. To prevent this, the fan should be shut down during this flushing cycle.

The absorber can be designed to provide future additional packing-height requirements should greater future gas absorption capability be required. This can be accomplished in two ways: the shell of the absorber can be flanged to allow a future stub section to be inserted, or the tower can be initially designed with a void space above the packing between the liquid distributors so that future packing heights can be added without shell modification. Although the initial cost of this may seem high, it could preclude the need to field-modify the absorber or to purchase a new absorber for greater efficiency, when the air pollution laws are tightened.

Another source of periodic maintenance relates to the strainers or filters in the recycle piping, and the recycle pumps. To prevent total system shutdown during cleaning or maintenance of these items, dual strainers or pumps may be considered, valved separately, so that the system can be maintained in a fully operational mode during strainer cleaning or pump maintenance checks.

As discussed earlier in the chapter, instrumentation used for the purpose of providing operational data is useful in determining areas where problems or maintenance may be required. The type of instrumentation will depend totally on the system requirements. Areas of specific interest would be pressure drop across the absorber, both the wetted bed and entrainment separator; low liquid recycle flow and/or pressure; and, liquid makeup rates to the system.

3.7 CONCLUSIONS

We have tried to present a practical rather than a theoretical approach to absorber design and maintenance. It should be recognized that an absorber system used for air pollution control is an operating system, as is any production operating equipment producing product. High maintenance costs associated with absorber systems are often due to misapplication of the equipment for the specific purpose. Air pollution control systems and absorber systems are considered nuisance areas required to prevent air pollution emissions to the atmosphere. They are a nonproductive item required of the manufacturer of an item. It is an added cost to produce this item. It is for this reason that maintenance is often neglected on air pollution equipment in preference to operating equipment producing a product. If the air pollution control system is considered as a process system needed to produce a product and preventive scheduled maintenance is provided, in the long run, operating and replacement cost will be lowered.

Air pollution control regulatory agencies are becoming more cognizant of the fact that preventive maintenance on air pollution control equipment has been neglected. Because of this, many states are requiring continuous stack emissions monitoring to prevent excessive gaseous pollutants from being emitted to the atmosphere due to the absorber or air pollution system becoming ineffective due to lack of maintenance. With the greater emphasis being placed on preventative maintenance of absorbers, it is hoped that the information presented in this chapter will be helpful to the user in obtaining a better understanding of absorber operation and maintenance for greater equipment reliability and lower operating costs.

3.8 REFERENCES

[1] HANF, E. B., "A Guide to Scrubber Selection," *Environ. Sci. Technol.*, *4*(2), 110–115, 1970.

[2] CEILCOTE CO., INC., *Tellerette Manual*. Berea, Ohio, 1970.

[3] Personal notes: J. Macdonald.

[4] CEILCOTE CO., INC., *Installation, Operation and Maintenance Manual*. Berea, Ohio, 1975.

[5] CROSS, JR., FRANK L. and HOWARD E. HESKETH, Ph.D., eds., "Handbook for the Operation and Maintenance of Air Pollution Control Equipment," pp. 175–176. Westport, Conn.: Technomic Publishing Co., Inc., 1975.

4

Adsorbers

Ron G. Wynne, P.E.

Western Electric Company, Inc.
Oklahoma City, Oklahoma

and

Lori P. Spencer

Environmental Consultant
Spencer Environmental Consultants
Birmingham, Alabama

4.1 DESCRIPTION OF CONTROL DEVICES

Introduction

Adsorption is a process wherein molecules of a fluid or gas contact and adhere to the surface of a solid. This phenomenon is believed to occur generally in nature, regardless of the particular solids and gases involved. The forces present in this intermolecular activity have been identified as being of two major types: van der Waals forces and chemical or ionic forces. The bonding force playing the major role in the adsorption process identifies either *physical adsorption,* in which the van der Waals molecular interaction force predominates, or *chemisorption,* where chemical reaction forces provide the predominant intermolecular bonds. Thus, if in the process of adsorption the individuality of the adsorbed molecule (adsorbate) and of the solid surface (adsorbent) is preserved, we have physical adsorption. If a chemical reaction occurs and the nature of the adsorbate or the adsorbent is altered through a valence electron exchange representing chemical bonds, then we have chemisorption.

Both physical adsorption and chemical adsorption are exothermic reactions and both liberate heat during the adsorption phase. The heat of physical adsorption, with its somewhat weak physical bonds, is of the same order of magnitude as the heat of liquefaction; the heat of chemisorption, possessing stronger bonding forces, is in the same order of magnitude as its associated chemical reaction. Regeneration therefore requires a corresponding equivalent amount of energy to free the gas molecule from the solid–gas bonding force. The strong bonding forces characterized in chemisorption make this particular method undesirable for use in regenerative applications because of the excessive amount of energy required for desorption. However, regeneration of adsorbents possessing predominately physical bonds can be accomplished quite easily utilizing heat energy available in steam, hot gas, or electricity. The solid–gas interface may also be broken by lowering the vapor pressure above the adsorbent, providing the liquefied gas molecule that is attached to the solid surface a sufficient amount of energy to break the weak van der Waals forces and again move into the gaseous state.

Although it is probable that all solids adsorb gases to some extent, adsorption as a rule is not very pronounced unless an adsorbent possesses a large surface area for a given mass. For this reason such adsorbents as charcoals, activated alumina, silica gel, and molecular sieves are particularly effective as adsorbing agents. These substances have a very porous structure and with their large exposed surfaces can take up appreciable volumes of various gases. The first three of these adsorbents are amorphous in nature with a nonuniform internal structure. Molecular sieves, however, are crystalline and possess an internal structure of regularly spaced cavities with interconnecting pores of a definite size.

Charcoal

Charcoal is an inefficient form of activated carbon. The activation process involves a high-temperature distillation of hydrocarbon impurities from a charcoal and leads to exposure of a larger free surface for possible adsorption of the gaseous molecule on the solid surface.

A large majority of the total surface area of an individual adsorbent is in the molecular-sized pores, as shown in the carbon granule (Figure 4-1). Obviously, the surface of the carbon cannot be "wetted" by the vapor if the gas molecule is too large to penetrate the pore structure. Logically, then, an adsorbent must be selected that will present to the particular gas molecule the largest surface area within the molecular-sized pore structure of the granule. [1]

Each individual activated carbon granule offers an abundance of surface area to the exposed gas molecule. Electron scanning microscope photographs (Figure 4-2) [2] clearly show the structure of the carbon surface that provides the massive adsorption capacity of activated carbon. Figure 4-2a shows a new individual activated carbon granule that has been precleaned in an ultrasonic cleaner to remove

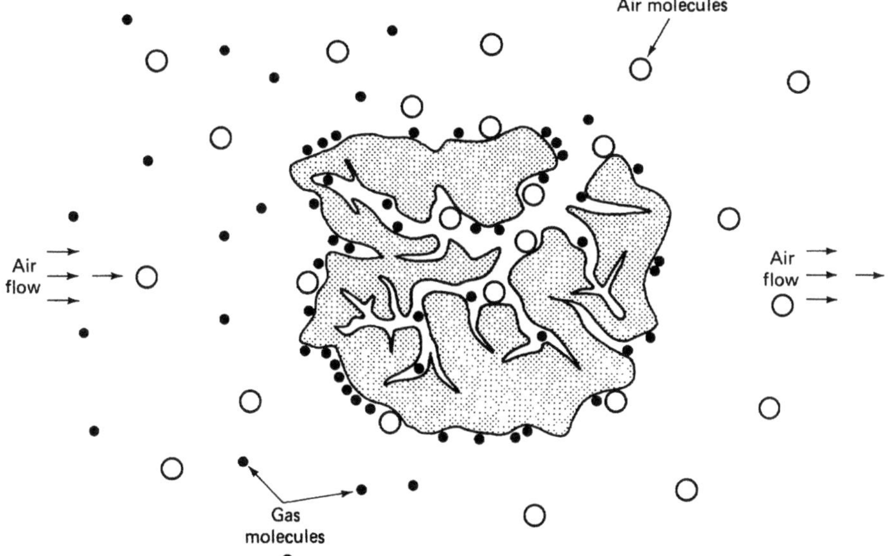

FIGURE 4-1 Carbon granule.

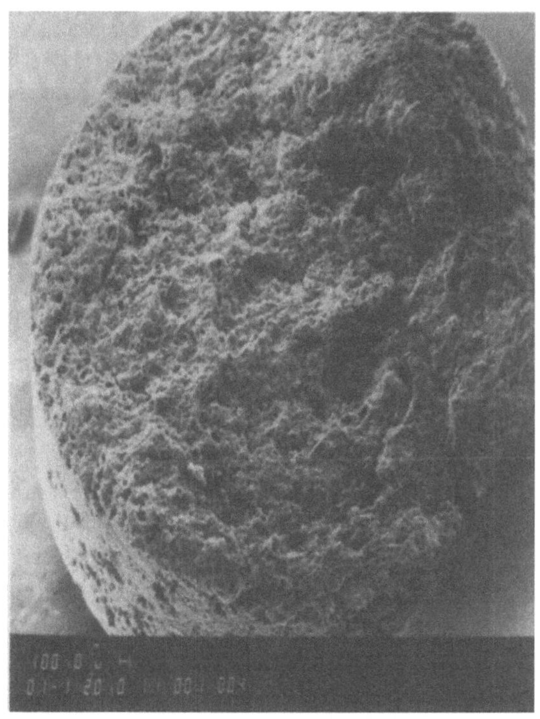

FIGURE 4-2(a) New individual activated carbon granule.

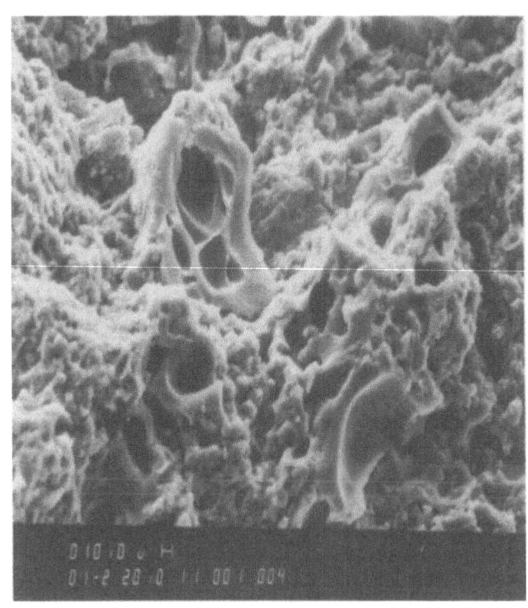

FIGURE 4-2(b) Irregular granule surface.

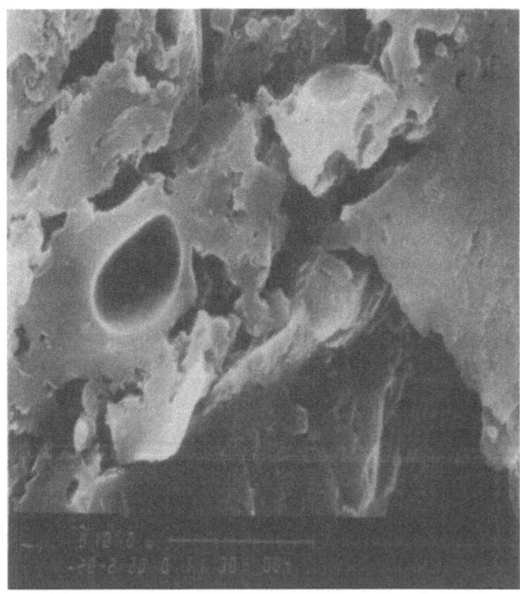

FIGURE 4-2(c) Interstices leading into granule.

carbon dust residues on the surface after manufacturing. The magnitude of the magnification of the 4 × 10 mesh granule is indicated by the 100-micrometer (μm) scale beneath the photograph. The surface of the granule is grossly irregular and is shown in Figure 4-2b. Notice the extremely porous nature of the particle. A good view of the interstices or pathways leading into the granule is shown in Figure 4-2c. Gaseous molecules travel through these macropores into the interior of the carbon, where they will be adsorbed and held fast by van der Waals forces until they are liberated by a sufficient amount of energy. Activated carbon with micropore structures ranging from 15 to 20 angstroms (Å) in diameter, coupled with a larger system of macropores of approximately 1000 Å provides an effective adsorbent for many organic compounds used in industry today, and has found extensive use in many air pollution control applications [3].

The following materials are used in the industrial manufacture of activated carbon: bituminus coal, bones, coconut shells, lignite, nutshells (walnut, pecan, etc.), peatmoss, petroleum sludges, pulpmill ash, wastewater treatment sludges, and wood. These gas-adsorbent carbons find primary application in solvent recovery, odor elimination, and gas purification.

Activated Alumina

Activated alumina (hydrated aluminum oxide) is produced by special heat treatment which is precipitated from native aluminas or bauxite. It is available in either granule or pellet form and is used principally for drying of a wet gas stream.

Silica Gel

The manufacture of silica gel consists of the neutralization of sodium silicate by mixing with dilute mineral acid, washing the gel formed to rid it of salts produced during the neutralization reaction, followed by drying, roasting, and grading processes. The name "gel" arises from the jellylike form of the material present during one stage of its production. It is generally used in granular form, although bead forms are also available. Silica gel also finds primary use in gas drying, although it has other applications (e.g., gas desulfurization and purification).

Molecular Sieves

Molecular sieves are dehydrated zeolites (i.e., aluminosilicates) in which atoms are arranged in a definite pattern. The fundamental building block of the molecular sieve is a tetrahedron of four oxygen anions surrounding a smaller silicon or aluminum cation. The cations serve to make up the positive charge deficit in the alumina tetrahedra. Each of the four oxygen anions is shared with another silica or alumina tetrahedron to extend the crystal lattice in three dimensions. The resulting crystal is unusual in that it is honeycombed with relatively large cavities, each

TABLE 4-1 Basic types of properties of molecular sieves.

Basic Type	Nominal Pore Diameter (Å)	Available Form	Bulk Density (lb/ft^3)	Heat of Adsorption (max.) ($Btu/lb\ H_2O$)	Equilibrium H_2O Capacity (wt %)[1]	Molecules Adsorbed[2]	Molecules Excluded	Applications
3A	3	Powder $^1/_{16}$-in. pellets $^1/_8$-in. pellets	30 47 47	1800	23 20 20	Molecules with an effective diameter <3, Å including H_2O and NH_3	Molecules with an effective diameter >3 Å (e.g., ethane)	The preferred molecular sieve adsorbent for the commercial dehydration of unsaturated hydrocarbon streams, such as cracked gas, propylene, butadiene, and acetylene
4A	4	Powder $^1/_{16}$-in. pellets $^1/_8$-in. pellets 8 × 12 beads 4 × 8 beads 14 × 30 mesh	30 45 45 45 45 45	1800	28.5 22 22 22 22 22	Molecules with an effective diameter <4 Å, including ethanol, H_2S, CO_2, SO_2, C_2H_4, C_2H_6	Molecules with an effective diameter >4 Å (e.g., propane)	The preferred molecular sieve adsorbent for static dehydration in a closed gas or liquid system; used as a static desiccant in household refrigeration systems; in packaging of drugs, electronic components, and perishable chemicals; and as a water scavenger in paint and plastic systems; also used commercially in drying saturated hydrocarbon streams

5A	5	Powder	30	1800	28	Molecules with an effective diameter <5 Å, including n-C_4H_9OH,[2] n-C_4H_{10},[2] C_3H_8 to $C_{22}H_{46}$,[1]	Molecules with an effective diameter >5 Å (e.g., iso compounds and all four carbon rings)	Separates normal paraffins from branched-chain and cyclic hydrocarbons through a selective adsorption process
		$^1/_{16}$-in. pellets	43		21.5			
		$^1/_8$-in. pellets	43		21.5			
10X	8	Powder	30	1800	36	Iso paraffins in olefins; C_6H_6 molecules with an effective diameter <8 Å	Di-n-butylamine and larger	Aromatic hydrocarbon separation
		$^1/_{16}$-in. pellets	36		28			
		$^1/_8$-in. pellets	36		28			
13X	10	Powder	30	1800	36	Molecules with an effective diameter <10 Å	Molecules with an effective diameter >10 Å (e.g., C_4F_9,)	Used commercially for general gas drying air plant feed purification (simultaneous removal of H_2O and CO_2) and liquid hydrocarbon and natural gas sweetening H_2S and mercaptan removal
		$^1/_{16}$-in. pellets	38		28.5			
		$^1/_8$-in. pellets	38		28.5			

[1] Lb H_2O/100 lb of activated adsorbent at 17.5 mm Hg and 25°C.
[2] Each type adsorbs listed molecules plus those of preceding type.

cavity connected with six adjacent ones through apertures or pores. These pores serve as the adsorption surface within a molecular sieve [4].

Molecular sieves will effectively remove acetylene from olefins, and ethylene or propylene from saturated hydrocarbons [4]. Some of the basic types and properties of molecular sieves are presented in Table 4-1 together with their applications.

Applications of Adsorbents

Although these four adsorbents are the most commonly used, others are available [5]. These adsorbents and others are listed with their applications in Table 4-2.

Industrial adsorbents are usually capable of adsorbing both organic and inorganic gases. However, their preferential adsorption characteristics and other physical properties make each one more or less specific for a particular application. For organic vapor adsorption, activated carbon has superior qualities, possessing hydrocarbon selective properties and high adsorption capacity for such materials. The three other common adsorbents (activated alumina, silica gel, and molecular sieves) are both hydrophilic (affinity for water molecules) and organic-adsorbing by nature. The hydrophilic properties of the adsorbent create a considerable drawback when considering solvent recovery by steam regeneration since the water molecule also attaches itself to the solid surface by adsorption and is not easily removed by conventional desorption methods.

TABLE 4-2   Common adsorbents and their applications.

Adsorbent	Application
Activated carbon	Solvent recovery, elimination of odors, purification of gases
Alumina	Drying of gases, air, and liquids
Bauxite	Treatment of petroleum fractions; drying of gases and liquids
Bone char	Decolorizing of sugar solutions
Decolorizing carbons	Decolorizing of oils, fats, and waxes; deodorizing of domestic water
Fuller's earth	Refining of lube oils and vegetable and animal oils, fats, and waxes
Magnesia	Treatment of gasoline and solvents; removal of metallic impurities from caustic solutions
Molecular sieves	Selective removal of contaminants from hydrocarbons
Silica gel	Drying and purification of gases
Strontium sulfate	Removal of iron from caustic solutions

Activated alumina and silica gel are capable of adsorbing organic vapors; however, regeneration is difficult because of strong bonding forces between the organic and adsorbent. Activated carbon adsorbent minimizes regeneration difficulties because of the compatibility of steam with activated carbon. Use of steam in silica gel or activated alumina is impossible because of the risk of breakdown of these materials when contacted with water.

The adsorbing action of molecular sieves is slightly different from those of other adsorbents, in that selectivity is determined more by the pore-size limitations of molecular sieves. It is important that the contaminant molecule to be removed be smaller than the available pore size, while the carrier gas molecule (usually air) is larger and thus not adsorbed by the molecular sieve. In any situation where the contaminant and the carrier molecule are of approximately the same size, molecular sieve adsorption is not feasible, since both molecules will enter the crystal structure and become adsorbed by the solid.

Regulatory agency analysis has established the effectiveness of carbon adsorption applications for air pollution control of volatile organic compounds. The benefits of carbon adsorption are considered to be solvent recovery for reuse and emission control for compliance and odor removal. However, to obtain the optimum results available from the control process, the facility must be properly designed for the manufacturing process to be controlled, perform within the operating parameters of the adsorption and desorption process, and receive routine maintenance inspections to ensure the operational integrity of the process.

Adsorbers

After determining the particular adsorbent necessary to control the emissions, an applicable adsorber system can be selected. Two major categories of adsorbers that are presently manufactured for industrial and commercial use are air purification devices (nonregenerative) and solvent recovery devices (regenerative).

Air purification applications are effective where the pollutants are emitted at very low concentrations (below 100 ppm) but need to be controlled because of their highly malodorous or toxic nature. These systems are cost-effective only at low concentrations because of their nonregenerative design. In solvent recovery applications, activated carbon may be used to economically recover and reclaim solvents from process exhaust streams at concentrations in the range of 100 to 1000 ppm. Activated-carbon regenerative systems, whether for pollution control or solvent recovery, are designated as carbon-resorb systems. The air pollution control device that is required for a particular application will depend upon the concentration range of the volatile organic compound to be controlled [6].

The basic function of control devices applicable in the low-concentration range is to purify both recirculated indoor air and ventilation air brought in from outdoors so that it will be odorless and safe for breathing. The control devices used for this purpose are designated as air purification systems. Air purification is generally carried out in a partially closed system; that is, the carbon-purified air is recirculated into the occupied spaces for reuse by the occupants. There are several configurations of air purification devices, such as radial flow canisters, panels, and pleated carbon beds; all are designed to treat contaminated air in low concentrations.

To attain the most effective and economical service from air purification systems utilizing carbon adsorbent, various parameters need to be optimized. These parameters are flow, velocity, carbon mesh size, and bed thickness. They determine the resistance to flow, the adsorptive capacity, and the carbon bed area. Another important factor is the determination of the optimum service time at which the carbon bed should be replaced. Pollutant concentration is also an important factor that should be recognized, because of its effect on capacity and service time.

To attain maximum but uninterrupted service from a carbon filter, it would be desirable to monitor the progress of the adsorbed-phase profile through the carbon bed so that the service time of the filter could be determined. The low capital cost of the cell does not warrant high expenditures for monitoring instruments that can determine this service time precisely. However, a determination can be made of the service time by means of a test element attached to the influent face of one of the cells. Prior to the estimated service time, the test element is removed from the cell and sent to the system's manufacturer for analysis and determination of the remaining service time of the cell.

The spent panels, cells, or canisters are shipped to the system's manufacturer, where they are emptied and reloaded with reactivated or virgin material. Special procedures are used in the reloading process to ensure tight packing of the carbon and thereby avoid later settling during transit or usage. Poorly packed beds can develop open areas through which impure air can pass. Some handling and activation losses occur on each cycle, and can be as high as 5%. Over a number of cycles, the used adsorbent is completely replaced with virgin makeup carbon or other suitable adsorbing material.

Solvent Recovery Systems

Because of the high cost of maintenance of air purification systems in applications of high concentrations of organic vapors, scientists and engineers have been forced into research and design of systems for solvent recovery. The result has been the development of three types of systems, differentiated by the manner in which the adsorbent bed is maintained or handled during both phases of the adsorption–regeneration cycle: (1) fixed or stationary bed, (2) moving bed, and (3) fluidized bed [6].

Figure 4-3 presents a flow diagram of a dual stationary-bed solvent recovery system with auxiliaries for collecting the vapor–air mixture from various point sources, then transporting through the particulate filter and into the on-stream carbon adsorber, in this case bed 1. The effluent air, which is virtually free of vapors, is usually vented outdoors. The lower carbon adsorber (number 2) is regenerated during the service time of bed 1. A steam generator or other source of steam is required. The effluent steam–solvent mixture from the adsorber is directed through

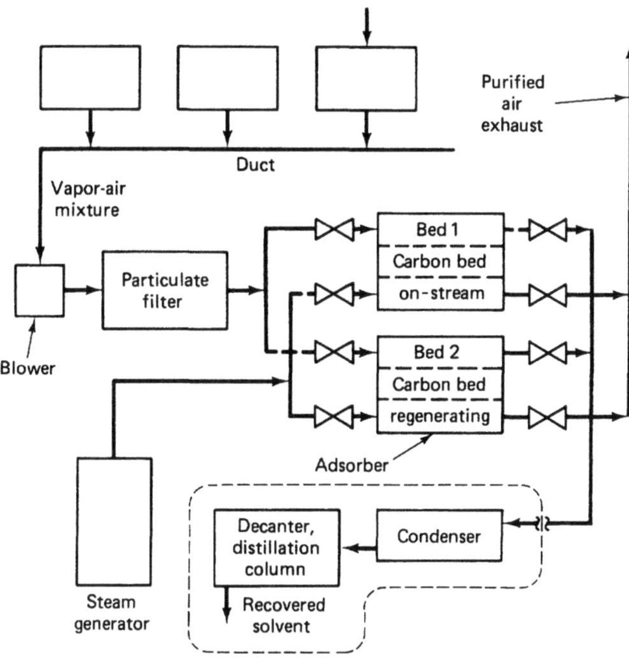

FIGURE 4-3 Stationary-bed carbon system with auxiliaries for vapor collection and solvent separation from steam condensate.

the condenser and the liquified mixture then run into the decanter and/or distillation column for separation of the solvent from the steam condensate [6].

Figures 4-4 and 4-5 [6] show two designs of stationary-bed solvent recovery systems. The type shown employs vertical cylindrical beds wherein the solvent-laden air flows axially down through the bed. This particular design is advertised for use in recovery of solvents used in degreasing and dry cleaning, although equally well suited for recovery of solvents from other industries. Solvents mentioned are trichloroethylene, tetrachloroethylene, toluene, Freon TF, and dichloromethane. Regeneration is accomplished with steam flowing upward through the bed and, since the above-mentioned solvents are immiscible in water, decantation is used to separate the condensed solvent from the steam condensate. The valves are of disk type, opened and closed by air-driven pistons. Water is used as coolant in the condenser. Steam, electric power to drive the blower, and cooling water are three operating-cost items. The cost of steam and cooling water increases with frequency of regeneration.

Figure 4-4 shows the features and arrangement of the component parts of a two-adsorber system. This particular unit is a Vic Model 572 with two adsorber beds used alternately (i.e., while one unit is on-stream adsorbing, the other is regenerating) [7].

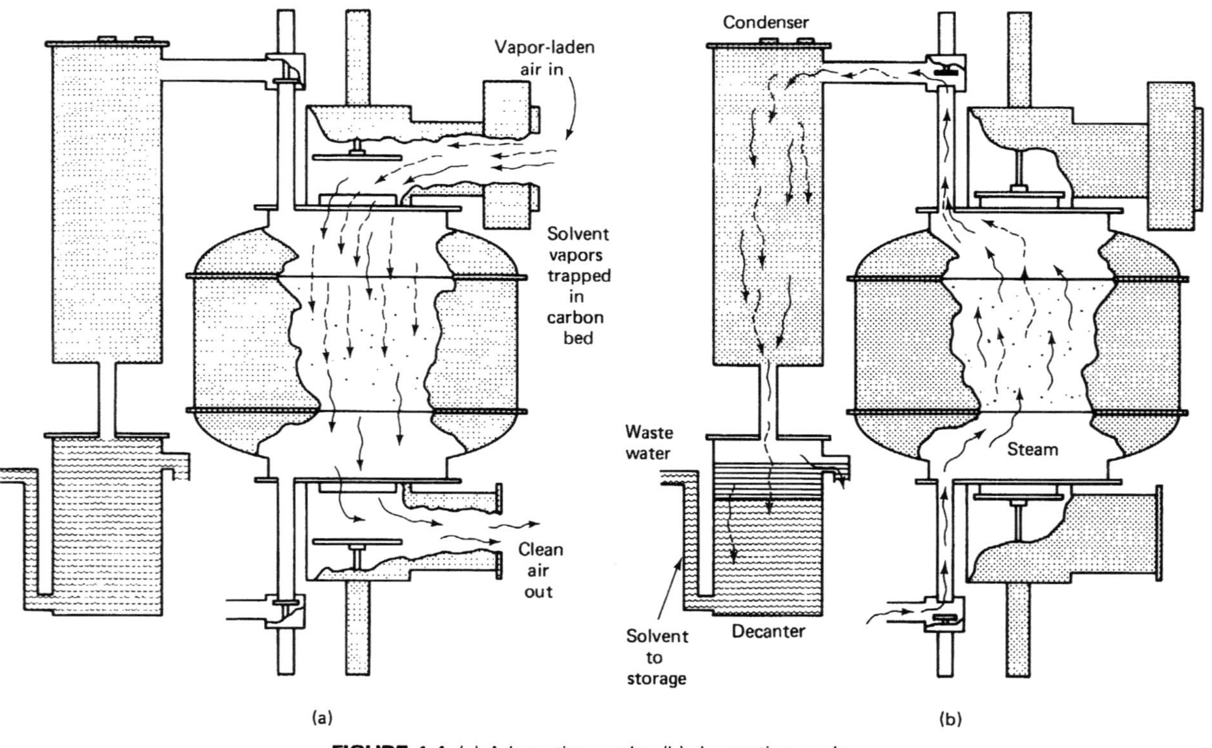

FIGURE 4-4 (a) Adsorption cycle; (b) desorption cycle.

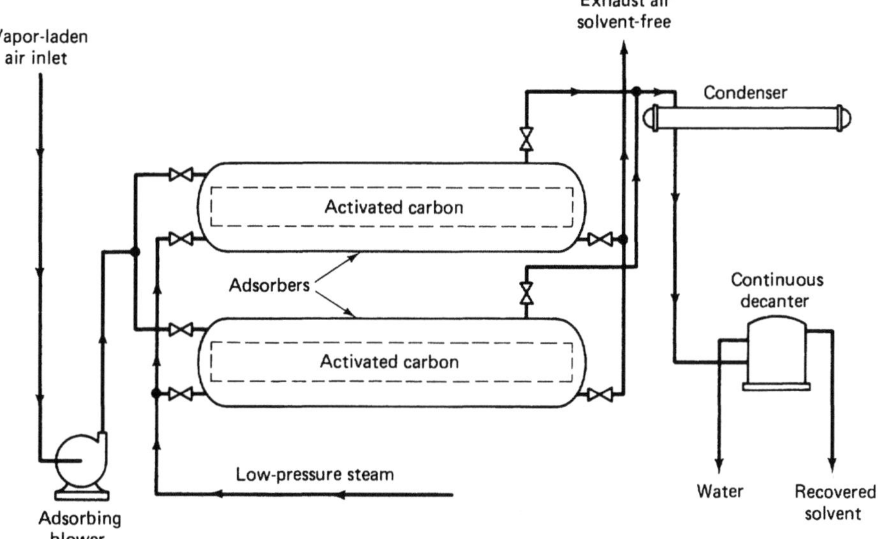

FIGURE 4-5 Flow diagram of a solvent recovery system.

Dual adsorber systems can also be operated with both beds on-stream simultaneously, especially when solvent concentrations are low. In this situation, regeneration is less frequent than a full work shift and may be accomplished during off-work hours. Operation in parallel almost doubles the air-handling capacity of the adsorption unit and may be an advantage in terms of operating cost.

The concept of the moving-bed system is illustrated in Figure 4-6. The rotary component of the system consists of four coaxial cylinders. The outer cylinder is impervious to gas flow except at the slots near the left end. They serve as impure air inlet ports where they are shown uncovered and as steam–vapor outlet ports as shown at the lower left end of the rotary. The carbon bed is retained between two cylinders made of screen or perforated metal and also segmented by partitions placed radially between the two cylinders. The inner cylinder is again impervious to gas flow except at the slots near the right end of the rotary. These slots serve as outlet ports for the purified air and as inlet ports for the regenerating steam. On each rotation of the rotary, each segment of the bed undergoes adsorption and regeneration. The desorbed solvent can then be separated from the steam by decantation or distillation [6].

Because of the continuous regeneration capability of the rotary bed, more efficient utilization of the carbon is possible than with stationary-bed systems. In most solvent recovery operations the adsorption zone and the saturated bed behind it are idle but add to the bulk of the system and increase airflow resistance through the bed. In deep beds of 12 to 36 in., as required in the stationary-bed systems, a large portion of the bed is idle at any one time [6]. By continuous regeneration, the regeneration time for each segment of the bed is shortened, and thereby shorter bed

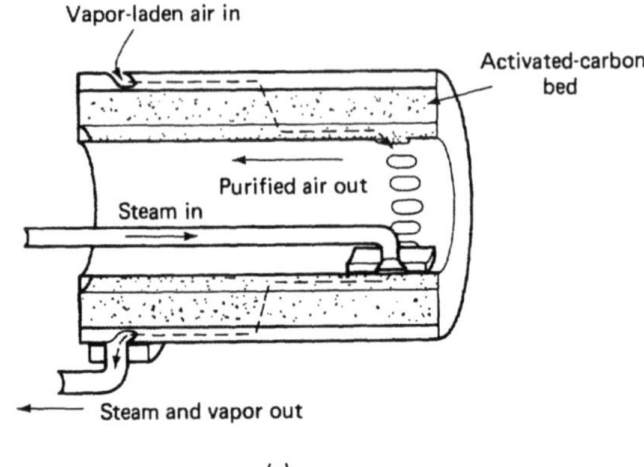

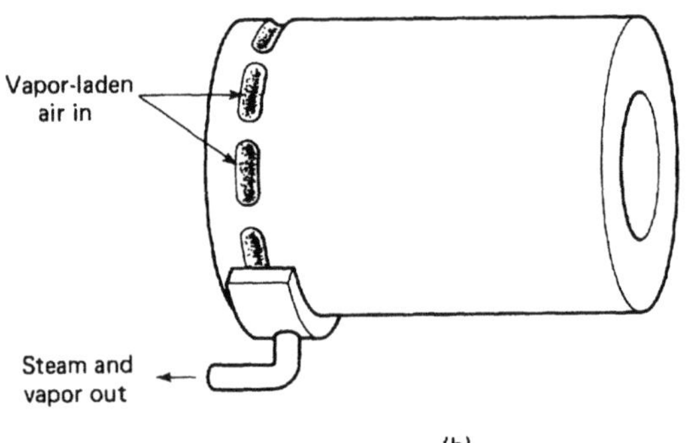

FIGURE 4-6 Continuous rotary bed: (a) cross-sectional view; (b) horizontal exterior view.

lengths can be used. This leads to two advantages: a more compact system and reduced pressure drop. The disadvantages are those associated with wear on moving parts and maintaining seals in contact with moving parts. The use of shorter beds also decreases the steam utilization efficiency.

Figure 4-7 shows a flow diagram for a fluidized-bed solvent recovery system [6]. The carbon is recirculated continuously through the adsorption–regeneration cycle. The spent carbon, saturated with solvent, is elevated to the surge bin and then passed down into the regeneration bed, where it is contacted to an upward flow of steam. The regenerated carbon is then metered into the adsorber, where the carbon traverses nine beds while fluidized with the upward flow of the vapor-laden air. An air velocity of approximately 240 ft/min is required to cause fluidization of the bed. In both the regeneration and adsorption phases, the carbon is moved countercurrent to the gas or vapor.

The countercurrent movement increases the efficiency of regeneration, that is, the steam/solvent ratio is less than for a stationary bed under otherwise comparable conditions. In addition to the beneficial effects of the countercurrent movement, the bed length can be increased to further improve steam utilization.

The countercurrent movement also increases the effective use of the carbon; more solvent can be recovered with less carbon than with stationary- or rotary-bed systems. By adjustments or balance of the carbon and solvent input rates, the total carbon in the adsorber can be made part of the adsorption zone reaching saturation in the lowest bed just before it is discharged into the elevator. Very little of the carbon is then idle; hence, maximum utilization is made of the carbon.

The fact that the carbon has reached saturation when delivered to the regeneration phase is another factor in the reduced steam requirement; in this respect, the fluidized-bed system is most favorable. In rotary-bed operations, the carbon moved

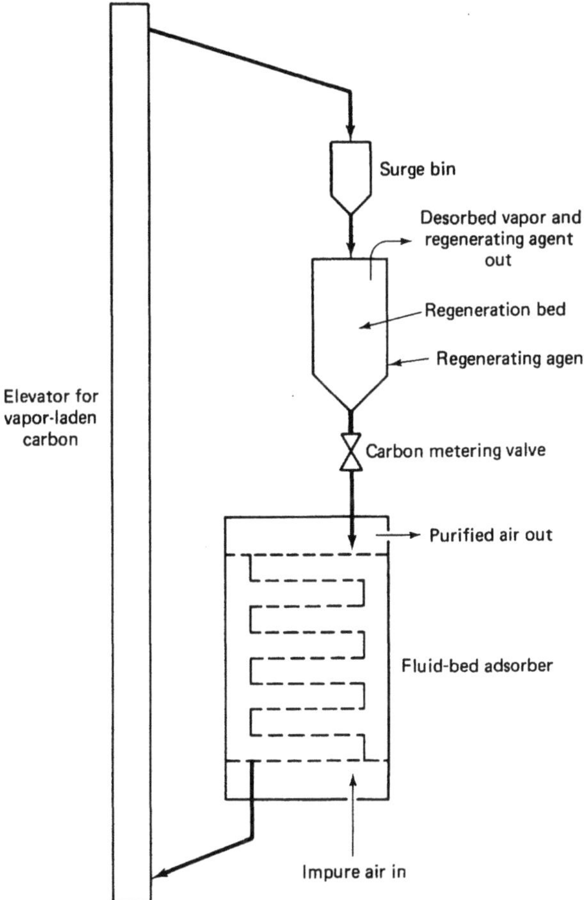

FIGURE 4-7 Fluidized-bed solvent recovery system.

into the regeneration phase has reached the lowest state of saturation in the three systems.

When large air volumes are treated and available space for the installation is at a premium, the smaller size and lower initial cost are definite advantages over the stationary-bed system. A serious disadvantage is that of high attrition losses of the carbon caused by the fluidization of the beds. Because of the attrition, filtration of the effluent airstream may be required. Influent airstream need not be highly filtered because plugging of the fluidized beds with particulate matter is minimal.

Auxiliary Equipment

No adsorber system could operate alone without sufficient auxiliary equipment and components required to collect, transport, and filter vapor-laden airstreams being delivered to the adsorber. Proper design of these components is necessary to provide proper service to the adsorber. The ducts and piping need to be sized properly for required air velocities to optimize the efficiency of the adsorber. If the air velocity is too high, the stationary-bed adsorber may become a fluidized-bed adsorber; or low flows may create severe channeling through the beds. The fan is the catalyst for forcing the gas stream in and out of the unit, so it is important that careful attention to design and sizing be given to this equipment [8].

Because there are several adsorber configurations, a filter's location can be varied: before the inlet airstream (in a stationary bed) to reduce possible contamination to the adsorbent, or after the fluidized-bed adsorber to reduce particulate emissions. Monitoring of the filter efficiency can be accomplished by measuring the pressure drop across the filter using either a manometer or pressure gauge that will read from 1 to 20 in. of water.

Compressed-air systems are necessary in some adsorber systems for valve and damper operation. For best results, the air supply should be kept contaminant-free through the use of a filter installed close to the adsorber. The compressed air supply should be equipped with an in-line filter, pressure regulator, and lubricator.

Several devices are installed in series following adsorption for recovery of the contaminant after regeneration. Condensers and separators are examples of recovery devices. The condenser is installed just after the system for removal of the heat from the vapors. There are two basic types of condensers: surface condensers and contact condensers. A surface condenser is shown in Figure 4-8 [5]. In this type of condenser the coolant does not contact the vapors or condensate. Most surface condensers are of tube-and-shell configurations. Water flows through the tubes, and vapors condense on the shell side. In contact condensers the coolant, vapors, and condensate are intimately mixed. These condensers are more flexible, simpler, and considerably less expensive to install. Sizing of this condenser is also more straightforward than the design for surface condensers.

Separators (decanters) are installed following the condenser to separate the contaminant from the water. Figure 4-9 shows a typical separator used with acti-vated-carbon-bed adsorbers. Separators work on the principle of gravitational forces,

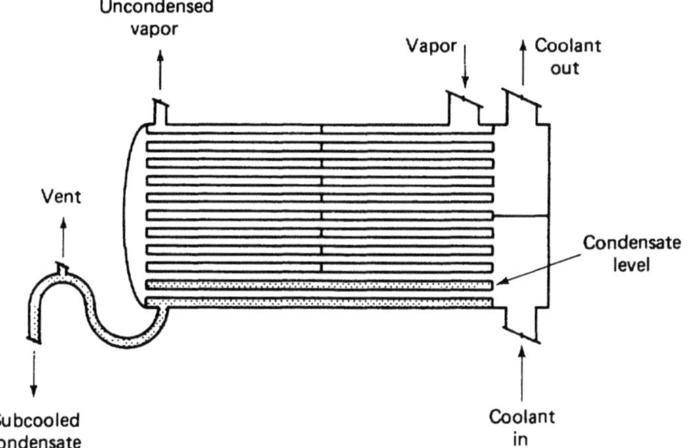

FIGURE 4-8 Maintaining a condensate level above the lower tubes to provide subcooling in a horizontal tube-and-shell condenser.

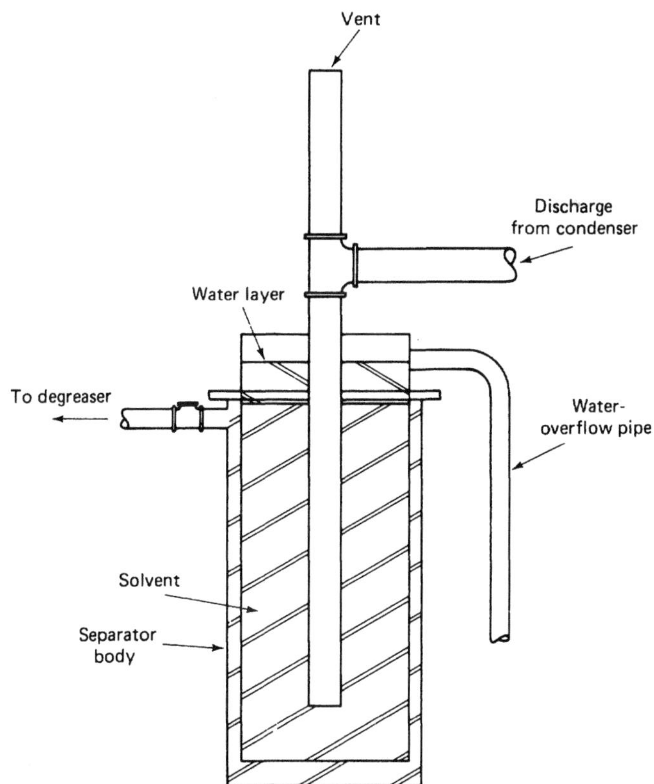

FIGURE 4-9 Water separator.

where the heavier material to be separated is removed from the bottom of the canister and the lighter material is removed through a line located at the top of the canister. Water separators are more effective with single solvent applications and only when the solvent is immiscible in water.

4.2 DESIGN PROCEDURES

The equipment necessary to constitute a complete adsorption system must be designed and sized adequately to achieve optimum emission reduction efficiencies and maximum solvent reclamation and recovery. The sizing and overall design of the adsorption system depends on the properties and characteristics of both the inlet gas stream and the adsorbent. Design of stationary-bed activated-carbon systems must include consideration of the following operating parameters.

1. Gas stream
   a. Adsorbate concentration
   b. Temperature
   c. Temperature rise during adsorption
   d. Pressure
   e. Flow rate
   f. Presence of adsorbent contaminant material
   g. Gas density at operating temperature and pressure
   h. Gas viscosity at operating temperature and pressure
2. Adsorbent
   a. Adsorption capacity as used on-stream
   b. Temperature rise during adsorption
   c. Isothermal or adiabatic operation
   d. Adsorbent life in presence of particulate matter
   e. Possibility of catalytic effects causing adverse chemical reaction in the gas stream and of formation of solid polymerizates on the adsorbent bed, with consequent deterioration

The adsorbent is generally used in a fixed bed through which the contaminated air is passed. Depending on the concentration and market value, the contaminant is either recovered or discarded when the loading of the adsorbent requires regeneration. The evaluation of the dynamic capacity of an adsorbent bed is essential when determining adsorption rates.

Adsorption of the vapor by the activated carbon occurs in two stages. Initially, as vapors enter the beds of charcoal, *rapid adsorption* occurs, but a stage is reached in which the carbon continues to remove the material but at a *decreasing rate of adsorption*. Eventually, the vapor concentration leaving the carbon equals that of the inlet. At this point, the carbon is saturated; that is, it has adsorbed the maximum amount of vapor that it can adsorb at the specific temperature and pressure. Figure 4-10 shows the various stages of adsorption through the adsorbent beds and the resulting outlet concentrations [4]. The adsorber bed rapidly becomes ineffective for removal of solvents after the breakpoint (C_3) is reached as the outlet concentration (C_4) climbs toward the value of the inlet concentration (C_0). At this breakpoint (C_3) the carbon bed has reached saturation and should be regenerated or stripped of solvent before passing additional vapors through the bed. This saturation value is different for each vapor and carbon. It is determined experimentally, at a given temperature and pressure, by passing a vapor-laden air stream through a known amount of carbon until the carbon ceases to increase in weight. Under these conditions, the carbon is then saturated with the adsorbate and must be stripped. The retentive capacity of an activated carbon is also a useful figure in the adsorption process. It represents the amount of adsorbate that a carbon, initially saturated, can retain when pure air is passed through the carbon at a constant temperature and pressure. This indicates the weight of the particular gas or vapor that the carbon can completely retain. This is called the retentivity of the carbon and is expressed as the ratio of the weight of the adsorbate retained to the weight of the carbon.

The service time is one of the main parameters to be considered in sizing adsorption equipment and can be calculated from the following equation [4]:

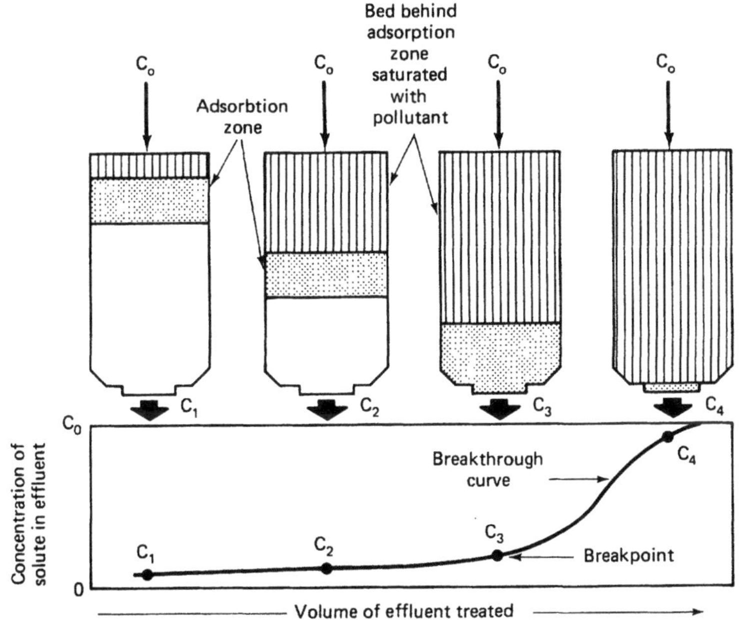

FIGURE 4-10 Stages of adsorption and resulting outlet concentrations.

$$t_B = \frac{A}{VC_I}(X_{st}L_{st} + X_ZL_Z)$$

where t_B = service time to breakthrough concentration, min

   A = area of adsorbent bed, ft^2 or cm^2

   V = total gas volumetric flow rate, ft^3/min or liters/min

   C_I = inlet concentration (gaseous pollutant)

   X_{st} = adsorptive capacity per unit adsorbent bed volume in the
   saturated zone, lb/ft^3 or g/cm^3

   X_Z = mean adsorptive capacity per unit bed volume in the adsorption
   zone, lb/ft^3 or g/cm^3

   L_{st} = length of saturated bed, ft

   L_Z = length of adsorption zone, ft

The service time t_B is dependent on the adsorption capacities of the saturated bed and the adsorption zone, as expressed in the equation. L_{st} is large relative to L_Z in deep or long beds (with activated-carbon adsorbents). With a maintained solvent air velocity of 100 ft/min, L_Z is approximately 3 in. and the adsorbent bed depth or length can range from 16 to over 36 in. L_{st} can be calculated accurately if the adsorption isotherm is determined previously.

While the air velocity through deep beds is maintained at 100 ft/min, thin beds of 2 in. and less have velocities of nearly 40 ft/min. For thin beds a different equation is used to determine service time.

$$t_B = \frac{A}{VC_I}X_ZL_Z$$

where $L_Z \simeq 2$ in.

After determining the service time necessary for a particular application, there are several other factors to be considered:

1.   The adsorbent particle size
2.   The physical adsorbent bed depth
3.   The gas velocity
4.   The temperature of the inlet gas stream and the adsorbent
5.   The contaminant concentration to be adsorbed
6.   The contaminants concentration not to be adsorbed, including moisture

7. The removal efficiency
8. Possible decomposition or polymerization of contaminants on the adsorbent
9. The frequency of operation
10. Regeneration conditions
11. The system pressure

These factors are considered next in more detail.

The Adsorbent Particle Size

The shape and size of the adsorbent particle affects several parameters, in particular the pressure drop and diffusion rate). The pressure drop is lowest when the particles are uniform in size and spherical. The pressure decrease will vary with Reynolds number. The mass transfer rate increases inversely with $d^{3/2}$ and the internal adsorption rate inversely with d^2. With all other parameters considered equal, the pressure drop increases with smaller particle size, yet provides a more efficient adsorptive environment.

The Physical Adsorbent Bed Depth

The actual adsorbent bed depth affects the adsorption mass transfer in two ways: (1) the bed depth is sized to be deeper than the transfer zone, which is unsaturated; and (2) any multiple of minimum bed depth yields a greater than proportional increase in capacity. The mass-transfer zone (MTZ) can be determined as follows [2]:

$$\text{MTZ} = \frac{\text{total bed depth}}{t_{st}/(t_{st} - t_B) - X}$$

where t_B = time required until breakpoint

t_{st} = time required to saturation

X = degree of saturation in the MTZ

The Gas Velocity

The velocity of gas stream through the adsorption bed is determined by the "crushing velocity" of the adsorbent. Crushing-velocity data are available from manufacturers of adsorbents. The length of the MTZ is directly proportional to the velocity; usually, the higher the velocity, the longer the unsaturated zone.

An example is shown in Figure 4-11, where the effects of velocity on the length of MTZ are demonstrated for ethanol on activated carbon.

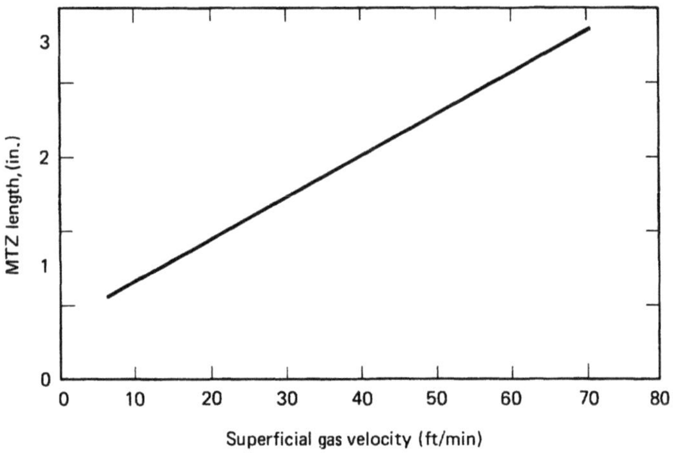

FIGURE 4-11 Effect of gas velocity on length of MTZ.

The Temperature of the Inlet Gas Stream and the Adsorbent

Temperature yields an inversely proportional effect on adsorption capacity, such as when temperature increases, the adsorption capacity decreases. The adsorption process is exothermic. As the adsorption activity moves through the bed, a temperature front follows and heat is transferred to the gas stream. When the gas leaves the area of adsorption activity, the heat exchange reverses (gas will transfer heat to the bed). The temperature differential during the adiabatic operation of the adsorber can be determined as follows, beginning with two assumptions:

$$\Delta T = \frac{6.1}{(C_p/C_I) \times 10^5 + 0.51(C_p/A/C_{st})}$$

where

ΔT = temperature rise, °F

C_{st} = saturation capacity of bed at $T + \Delta T$, °F

C_I = inlet concentration, ppm

C_p = heat capacity of adsorbent, Btu/ft³/°F (see Table 4-3)

TABLE 4-3 Specific adsorbent heat capacity values (ambient conditions, Btu/ft³/°F).

Activated carbon	0.25
Alumina	0.21
Molecular sieve	0.25

1. Thermal equilibrium exists between the gas and the bed.
2. The effluent-gas-stream temperature is similar to the bed temperature.

The Adsorbate Concentration

There is direct correlation between adsorbate concentration and adsorption capacity. The MTZ length is inversely proportional to the adsorbate concentration. Considering these two observations only, a deeper or longer bed will remove a lower-concentration contaminant and a higher concentration of the same contaminant with equal efficiency. In considering recovery, the value of the adsorbate and the concentration are primary factors. In handling combustible gas, it is mandatory that the concentration of the adsorbate be well below the lower explosive limit (LEL).

All gases present will be adsorbed on the adsorbent surface to some degree. The competition between the gases for available surface area lowers the adsorption capacity for the preferred adsorbate to be removed. Under ambient conditions, air is adsorbed only slightly (10 to 20 ml STP/g) on commercial adsorbents, but moisture and CO_2 (carbon dioxide) adsorb more significantly. Even though activated carbon is less sensitive than other adsorbents to moisture, its adsorption capacity is limited considerably when compared to adsorption from a dry airstream.

The Removal Efficiency

Varying percentages of removal are required with different applications. Of course, deeper beds are required to achieve complete removal than for a partial removal efficiency of 60 to 80%.

Adsorbate Decomposition and Polymerization

Decomposition and polymerization of solvents can occur when in contact with adsorbents. The decomposed product takes on properties different from those of the original substance. These different properties include corrosivity and the ability to be adsorbed at a lower capacity. Polymerization can significantly lower the adsorption capacity and render it nonregenerable by conventional methods. When adsorption of acetylene on activated carbon is accomplished at higher temperatures, polymerization occurs.

Most factors influencing adsorption, which have been discussed individually above, have a combined or interrelated effect on the adsorption system. The effects of some of these parameters are illustrated and shown in Figure 4-12 [4]. Curve A shows the theoretical saturation under equilibrium conditions as if it were obtained from isotherm data for gasoline adsorbed on activated carbon. This saturation canoccur only if the gas velocity is slow enough to let the bed become uniformly charged.

In practice, curve A does not occur, because the gas velocity is at a high rate

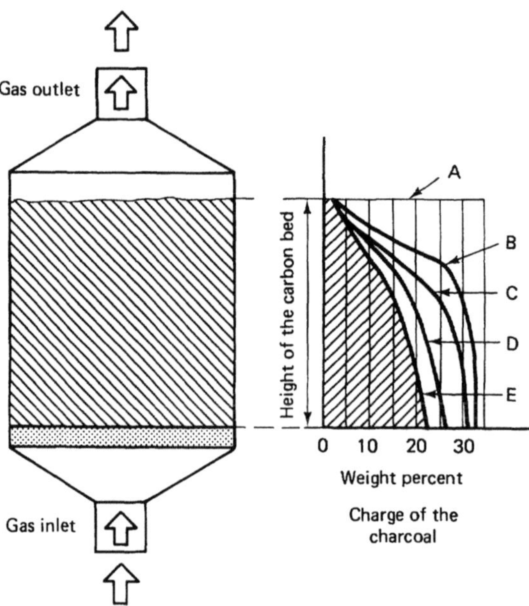

FIGURE 4-12 Progress of adsorption through a carbon bed.

where, theoretically, the carbon charge decreases from 100% to 0% (breakthrough) at the outlet. Curve B depicts the length of the bed zone, which is considerably reduced during saturated adsorption. Curve C illustrates that as temperature increases due to the heat of adsorption, the adsorption capacity decreases. Curve D represents the presence of another contaminant, such as moisture, and its effect also reduces adsorption capacity. Curve E is the result of these effects, and the area under the curve divided by the bed height yields the organic capacity of the carbon in the adsorber for gasoline.

Operation Frequency

Even with the most precise of design calculations and selection of equipment, the frequency of operation greatly affects an adsorption system's performance. If operation is continuous, the best conditions normally result; if operation is intermittent, the performance of the system will be impaired. In Figure 4-13 (an MTZ diagram), a system under normal operation is shown by curve A. Continued circulation of the carrier gas in the absence of the adsorbate causes diffusion of the adsorbate residue through the bed by desorption into the carrier and readsorption until the low concentration has lengthened the MTZ (curve B, dashed line).

Short periods of intermittent operation do not affect the overall performance significantly, but continued over long periods, particularly in overloaded systems, can cause a serious reduction of capacity.

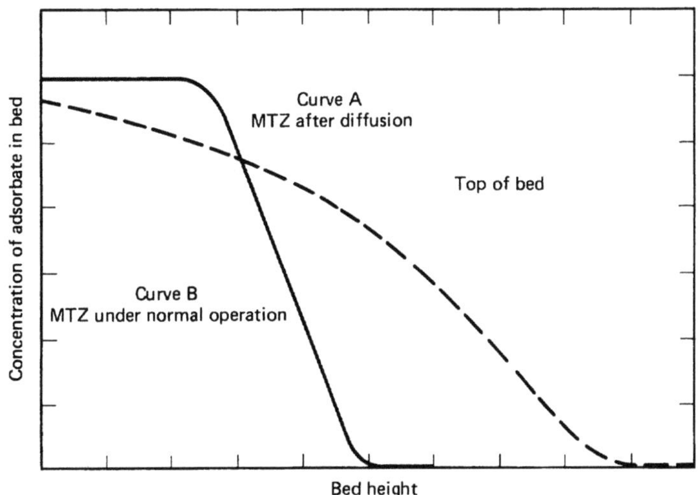

FIGURE 4-13 Effect of diffusion on MTZ.

Regeneration Conditions

When deciding whether to employ a regenerative system, several factors should be considered. The principal one is that of economics. It is important to establish if recovery of the adsorbate will be cost-effective or if regeneration of the adsorbent is the prime consideration.

If solvent recovery is the main objective, the design should be based on past experimental data to establish the ratio of sorbent fluid to recoverable adsorbent at different working capacities. A well-designed system will have steam consumption in the range of 1 to 4 lb of steam per pound of recovered solvent.

The steam entering the bed not only introduces heat but creates adsorption and capillary action of the moisture, which supplies additional heat for the desorption process. Certain parameters should be considered in the design of a stripping process:

1. Minimize the time required for the regeneration. If continuous adsorption and recovery are required, multiple systems have to be installed.

2. The short regeneration time requires a higher steaming rate, thus increasing the heat duty of the condenser system.

3. The steaming direction should be in a direction opposite to that of adsorption to prevent the possible accumulation of polymerizable substances.

4. To enable fast stripping and efficient heat transfer, it is necessary to sweep out the carrier gas from the adsorber and condenser system as fast as possible.

5. A larger fraction of the heat content of the steam is used to heat the adsorber vessel and the adsorbent; thus, it is essential that the steam condense quickly

in the bed. The steam should contain only a slight super heat to allow condensation.

6. It is advantageous to use a low-retentivity carbon to enable the adsorbate to be stripped out easily. When empirical data is not available, the following heat requirements have to be taken into consideration: heat to the adsorbent and vessel; heat of adsorption and specific heat of adsorbate leaving the adsorbent; latent and specific heat of water vapor accompanying the adsorbate; heat in condensed, indirect steam; and radiation and convection heat loss.

During the desorb cycle, condensation and adsorption will take place in the adsorbent bed, increasing the moisture content of the adsorbent. Also, a certain portion of the contaminant will remain; this is referred to as the "heel." To achieve a minimum efficiency drop from successive adsorption cycles, a drying and cooling cycle should occur before returning to the adsorb mode. When using high adsorbate concentrations, it may be desirable to leave some moisture; with other contaminants, a moisture-free bed is desired.

Pressure Effects

Generally, the adsorption capacity of an adsorbent increases with increasing pressure if the partial pressure of the contaminant increases. However, at high pressures (over 500 psig) a decrease in capacity will be observed.

The design of fixed-bed adsorption systems also requires the capability of estimating pressure drop through the bed. Ergun derived a correlation to estimate the pressure drop for the flow of a fluid through a bed of packed solids when it alone fills the voids in the bed. This correlation is given by the relationship

$$\frac{\Delta P \, g_c d_p \varepsilon^3}{z \, 2\rho u^2 (1 - \varepsilon)} = \frac{75(1 - \varepsilon) + 0.875}{Re}$$

where ΔP = pressure drop of gas, lb_f/ft^2

    z = depth of packing, ft

    g_c = conversion constant, $4.18(10^8)$ ft lb/lb_f hr^2

    d_p = effective particle diameter (the diameter of a sphere of the same surface/volume ratio as the packing in place), ft

    $= 6(1 - \varepsilon)/a_p$

    ε = fractional void volume in dry-packed bed, ft^3 voids/ft^3 packed volume

$$\rho = \text{gas density, lb/ft}^3$$

$$u = \text{superficial velocity of gas through bed, ft/hr}$$

$$\text{Re} = \text{Reynolds number, dimensionless}$$

$$= d_p \rho u / \mu$$

$$\mu = \text{gas viscosity, lb/ft hr}$$

There is a simpler form of the Ergun equation:

$$\frac{\Delta P}{z} = \frac{f C_T G^2}{\rho d_p}$$

where

$$C_T = \text{pressure-drop coefficient, ft hr}^2/\text{in}^2$$

$$f = \text{friction factor}$$

$$G = \text{superficial mass velocity, lb/hr ft}^2$$

$$= \rho u$$

$$\Delta P = \text{pressure drop, psi}$$

The friction factor, f, is determined from Figure 4-14 as a function of the modified Reynolds number. The pressure drop coefficient, C_T, is also determined from the same figure, which has C_T plotted as a function of ε. For molecular-sieve pellets, the effective particle diameter can be obtained from

$$d_p = \frac{d_c}{\tfrac{2}{3} + \tfrac{1}{3}(d_c / l_c)}$$

where

$$d_c = \text{particle diameter, ft}$$

$$l_c = \text{particle length, ft}$$

The suggested values for ε and d_p for various sizes of molecular sieve are:

	ε	d_p
$\frac{1}{8}$-in. pellets	0.37	0.0122
$\frac{1}{16}$-in. pellets	0.37	0.0061
14 × 30 mesh granules	0.37	0.0033

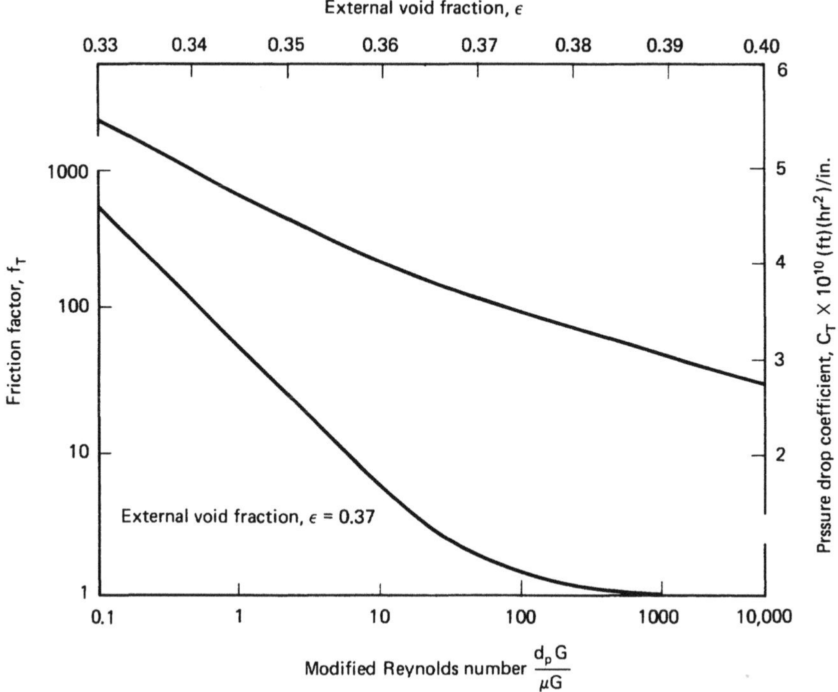

FIGURE 4-14 Friction factor as a function of Reynolds number.

Example Problem [4]

A 10,000-scfm (60°F, 1 atm) degreaser ventilation stream at 70°F and 20 psia contains 2000 ppm by volume trichloroethylene (TCE). The contaminated airstream enters an adsorber which is to recover 99.5% by weight TCE. The adsorbent is activated carbon with a bulk density of 36 lb/ft³. The adsorption column cycle is set at 4 hr in the adsorption mode, 2 hr heating and desorbing, 1 hr cooling, and 1 hr standby. If the activated carbon is capable of adsorbing 28 lb TCE vapor/100 lb carbon before breakthrough, size the vessel.

Solution: First calculate the actual volumetric flow rate.

$$10,000 \text{ scfm } (60 \text{ min/hr})(530/520)(14.7/20) = 4.5 \times 10^5 \text{ ft}^3/\text{hr}$$

The TCE volume flow rate is

$$(2000 \times 10^{-6})(4.5 \times 10^5) \text{ ft}^3/\text{hr} = 900 \text{ ft}^3/\text{hr}$$

The TCE mass flow rate is

$$900 \text{ ft}^3/\text{hr}(131.4 \text{ lb TCE/lb}_m)/(379)(520/530)(20/14.7) = 416.5 \text{ lb/hr}$$

The TCE to be adsorbed during the 4-hr cycle is then

$$(416.5)(0.995)(4.0) = 1658 \text{ lb}$$

The activated carbon required is

$$1658(100 \text{ lb carbon}/28 \text{ lb TCE adsorbed})/36 \text{ lb/ft}^3 = 164.4 \text{ ft}^3$$

Using a 6-ft-ID horizontal vessel 20 ft long, the height of adsorbent would be

$$164.4/(6)(20) = 1.37 \text{ ft}$$

The throughput velocity is approximately 60 ft/min. Using a vertical unit with a superficial velocity of 100 ft/min, the cross-sectional area available for flow is

$$4.5 \times 10^5/(100)(60) = 75 \text{ ft}^2$$

The diameter in this case becomes

$$(0.785)D_T^2 = 75$$

$$D_T = 9.5 \text{ ft}$$

with the height of column given by

$$164.4/75 = 2.2 \text{ ft}$$

A horizontal unit is suggested.

4.3 INSTALLATION PROCEDURES

After control equipment design has been completed, the installation becomes an important factor in the future convenience of operating and maintenance procedures on the air pollution control system. Improper installation may contribute to excessive energy consumption and unnecessary losses of volatile organic materials to the environment and to the workplace. This may result in deterioration of the air resource and the water quality and may present excessive employee exposure to toxic and hazardous chemicals. Because of these consequences, everything possible must be done to ensure correct operation and ease of maintenance of all air pollution control equipment.

Location

Whether the adsorber is a nonregenerative odor-control device or a multiple-bed carbon adsorption system, it is desirable to locate the control equipment as close as possible to the emission source. For optimum results, most carbon adsorber manufacturers recommend duct runs less than 100 ft. It is also important to select a location that will provide accessibility for observation, operation, and maintenance of the equipment.

Regenerative adsorption systems designed for reclamation and recovery of solvents have precise operation requirements and should be located in a restricted area. Security of the operation must be provided. The security may be as elaborate as total enclosure of the equipment or as simple as a handrail with restricted-entry signs.

If the equipment, especially steam stripping adsorbers, are to be located in an unheated area (outdoors or on the roof) where temperatures may be below freezing, adequate insulation must be provided on all water, steam condensate, and compressed-air piping and reservoirs.

Another important aspect in carbon adsorption system installation, and one that is often overlooked, is the preparation of the floor. To improve cleanup operations, bare concrete floors should be sealed and coated with a material such as epoxy paint that is impervious to the particular solvent being used in the process. The use of floor tile beneath the equipment is usually not suitable because of the ability of many organic solvents to penetrate the tile and dissolve the adhesives.

Noise Abatement

Most of the objectionable noise from a carbon adsorber originates with the air-handling portion of the system. The fan seldom generates enough noise to damage hearing; however, the location of the fan might prove to be objectionable to adjacent employees and/or neighbors. Exhaust ducts will carry sound from the system into adjacent workplaces and may produce uncomfortable working conditions at a remote area. Nuisance levels of noise may be projected toward neighboring residences by roof-top fans.

A loud, high-pressure fan may be located in an area of already high noise levels within a factory and may not be a significant contributor to the resultant sound pressure levels. However, if a fan is located in a relatively quiet area or is connected by ductwork into quiet areas, then even low-sound-level fan installations may be objectionable.

In fan installations where resultant sound levels are excessive or bothersome, acoustical treatment of the facility may be appropriate. To minimize noise levels, the fan and motor assembly should be mounted on isolation materials such as springs or rubber-in-shear pads. This will produce the added benefit of increased bearing life since the vibrational energy is absorbed in the springs and rubber rather than in the bearings. Isolation of the blower housing from the inlet and outlet ducts with

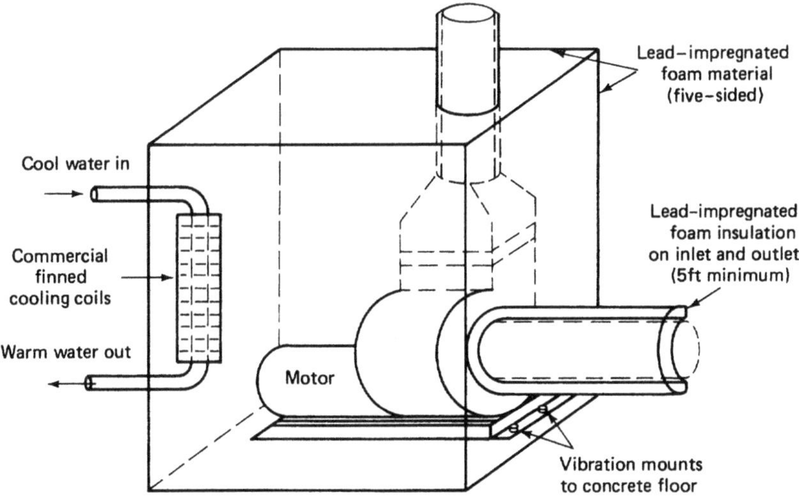

FIGURE 4-15 Sound abatement.

flexible connections will further reduce noise transmission through the ductwork into adjacent areas. The application of adsorption materials along the length of adjacent ductwork, especially around transitions and elbows, will significantly reduce noise losses from the system. Some installations may require complete acoustical enclosure around the blower assemblies in order to reduce radiated sounds from the fan (Figure 4-15). It is important to remember when dealing with sound enclosures that sound travels in a straight line from the source and will travel through even the smallest opening. Therefore, the enclosure must be sealed to eliminate noise leakage. Auxiliary cooling of the motor may be required in these cases and can be accomplished by running chilled water, available on the adsorber condenser, through finned coils within the enclosure to carry away the heat. The amount of cooling water required depends upon the size and efficiency of the blower.

Exhaust Duct Installation

Sound exhaust system design must be followed by good installation practices. Recommended exhaust duct installation procedures are included in Figure 4-16 to point out some basic principles of effective exhaust system installations [9]. The purpose of the duct, of course, is to convey the materials from the emission source to the air pollution control process. The duct velocity must be adequate to convey particulate matter without settling out in an unwanted location within the ductwork, while minimizing airflow to reduce unnecessary evaporation of volatile materials at the source. If evaporated volatile solvents enter the duct airstream at a warm temperature and subsequently cool while traveling to the adsorber, condensation of the solvents may occur. Care must be taken to eliminate traps in the exhaust system where moisture and solvents may condense and collect. Moisture conden-

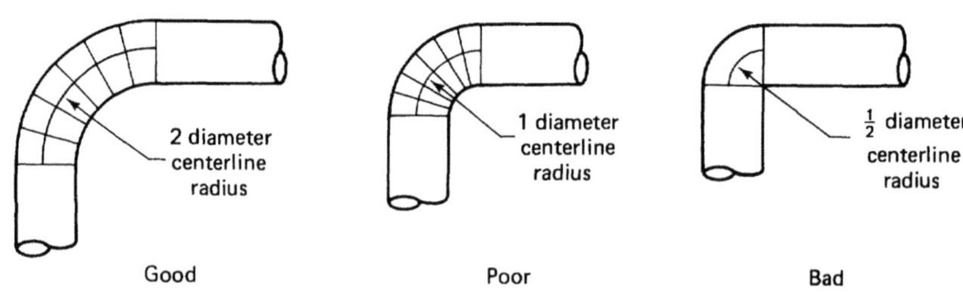

Elbow radius

Elbows should be 2 or $2\frac{1}{2}$ diameters centerline radius except
where space does not permit

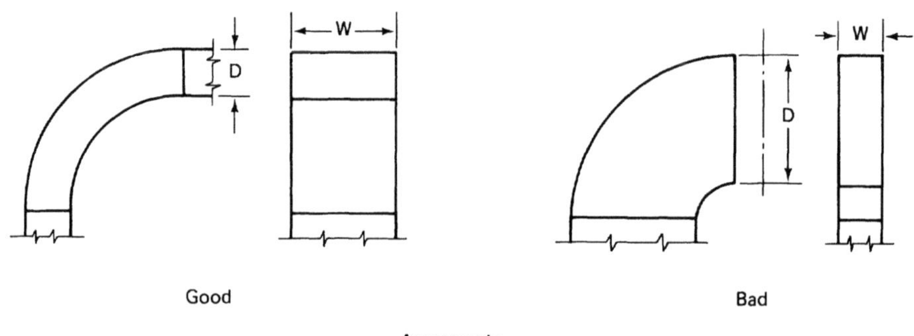

Aspect ratio

Keep AR $\dfrac{W}{D}$ high in using rectangular duct.

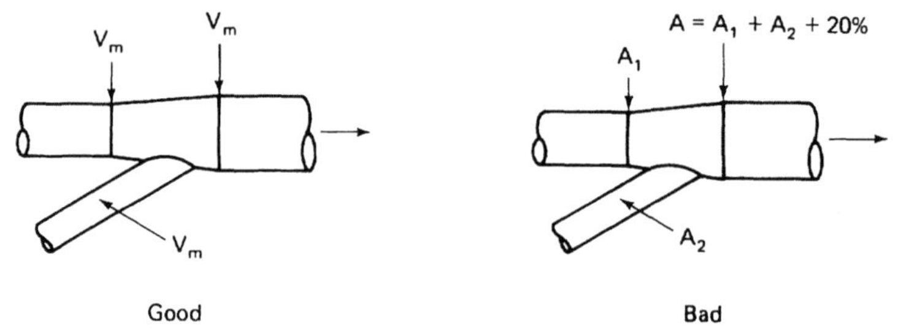

Proper duct size

V_m, = minimum transport velocity

A, = cross-sectional area

Size the duct to hold the selected transport
velocity or higher

FIGURE 4-16 Principles of duct design.

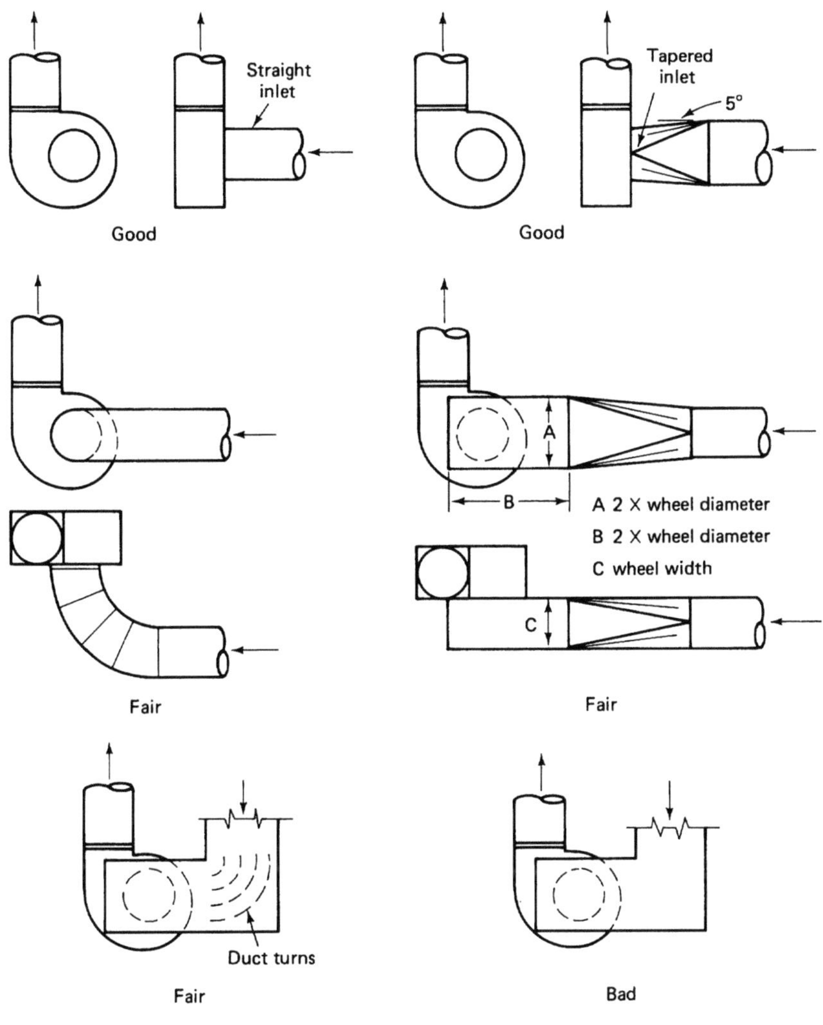

Fan inlet

A straight inlet is best: if an elbow inlet is necessary, provide an inlet box and duct turn vanes to eliminate air spin or uneven loading of the fan wheel. Inlet boxes should not be used for dust-laden air.

Figure 4-16 (continued)

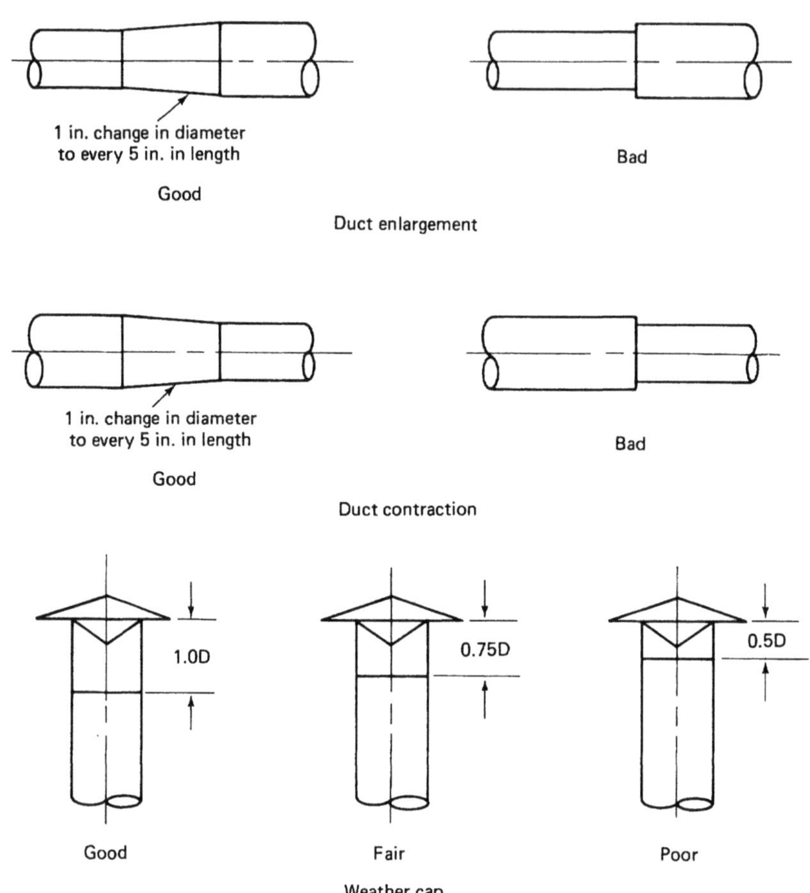

Figure 4-16 (continued)

sation is especially troublesome on the discharge side of the adsorber in processes using the steam strip cycle for regeneration. While the carbon beds are drying, moisture is driven out of the beds and condensation will occur in the outlet stack. All trap areas in the stack must be drained periodically to remove water from the duct; otherwise, premature rusting of ductwork will occur.

Particulate matter that is entrained in the airstream must be removed prior to entry into the adsorption beds. A suitable air filter must be installed in the airstream to remove this matter or severe plugging of carbon will occur.

Clean-out doors should be provided throughout the duct system for ease of maintenance on internal exhaust duct cleaning. On multiple-emissions-source exhaust systems, blast gates should be installed for balancing airflows in ducts as operating conditions change. It is extremely important to eliminate leaks, especially on the pressure side of the fan. The exhaust duct connections and seams should be either welded or soldered.

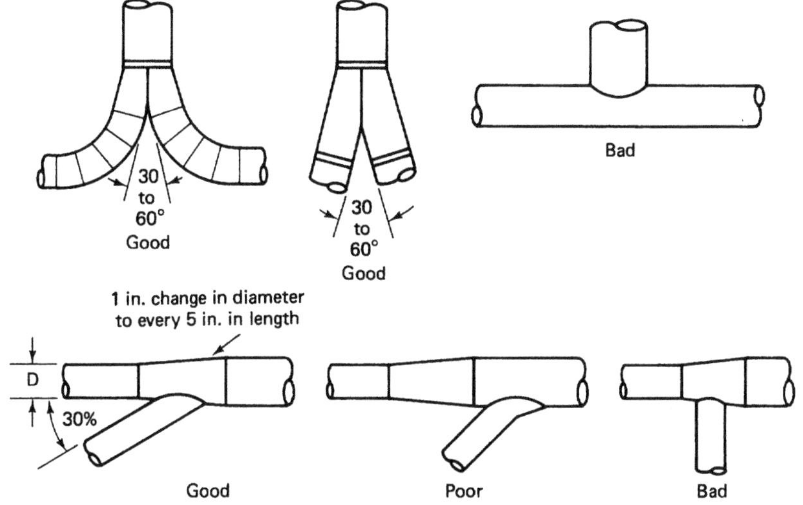

Branches should enter at gradual expansions and at an angle of 30° or less (preferred) to 45° if necessary.

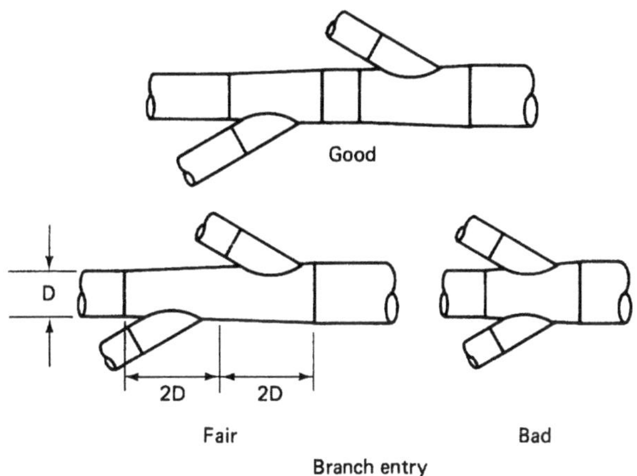

Branches should not enter directly opposite each other.

Figure 4-16 (continued)

When the outlet from the adsorber is discharged through the roof, care must be taken to eliminate the possibility of blowing exhaust gases into an adjacent air intake downwind from the adsorber. An adequate rain hood should be installed on the stack to eliminate rainfall from entering the system. Some carbon adsorber manufacturers suggest that the outlet from the adsorber be discharged directly into the workplace. However, extensive monitoring of the effluent stream must accompany this practice.

In all inlet and outlet ducts to the adsorption equipment, provide necessary openings in ductwork to accomplish cleanout of the duct, and monitoring and sampling of the airstream. Suitable pressure gauges should be provided across filters and carbon beds to indicate free flow of the airstream. Airflow switches may be installed in the airstream to sound a warning when the flow is lost. A thermometer is usually installed in the inlet duct to the adsorber beds so that the incoming temperature may be periodically recorded on the outlet duct. It is a good idea to consult a handbook on ventilation design and practice prior to installing the exhaust system and components.

Cooling-Water Supply

An adequate supply of cooling water is necessary to condense the vapors of water and solvent while steam-stripping the carbon bed. The demand for this cooling water exists only while the steam strip cycle is operational, and the installation should call for an automatic shutoff valve to interrupt flow at all other times. The water inlet and outlet temperatures should be provided at the appropriate location. If water is being delivered at a temperature below 60°F, the incoming supply line should be insulated to eliminate condensate dripping. The outlet water temperature may reach 120°F on some heat exchanges where a high-boiling-point solvent is being condensed.

Where flammables are being adsorbed or where a cool carbon adsorber bed is required for efficient adsorption, cooling-water coils may also be installed within the carbon beds. Installation of controls such as a flow control valve or a globe valve for regulation of flow on outlet piping with thermometer wells on inlet and outlet is appropriate on all cooling-water lines. If a connection is made to a potable water supply, adequate backflow-prevention devices must be installed in the supply to eliminate the possibility of contamination of drinking water.

Steam Supply

On steam-stripping regenerative carbon adsorption systems, adequate steam must be available to strip beds of adsorbed solvents at about 3 to 15 psi. Where a high-pressure steam supply is available and reduction of pressure is required, more precise control seems to be obtained when a double pressure-reducing valve arrangement is used. For example, first-stage reduction will lower steam pressure from 100 psi to 30 psi, then second-stage reduction will lower steam pressure from 30 psi to the desired tank pressure of about 5 psi. Steam traps should be installed where required, as the adsorption process stripping cycle requires a dry steam. Do not discharge steam traps into the same closed container as the water from the decanter, because steam pressure during blowout will retard decanter flow. All steam piping must be properly insulated, not only to conserve steam, but to reduce the burn hazard. The carbon tanks themselves may also be insulated, for the same reasons. Provide a steam-supply cutoff valve in a readily accessible location in case of emergency. Do not use the shutoff valve to regulate steam flow because of severe eroding of the valve components by restricted steam flow.

Compressed Air Supply

Dry, clean compressed air is required to operate controls and dampers on carbon adsorption units. A shutoff valve should be installed in adjacent locations and provide a filter, regulator, lubricator, and pressure gauge to ensure correct delivery pressure. Prior to assembly, the shafts on the air cylinders should be lightly oiled to eliminate damage to the seals.

Electrical Supply

Provide necessary electrical service to the system according to the manufacturer's recommendations and in accordance with the National Electrical Code. Special considerations should be given to installations using flammable solvents.

Internal Solvent and Water Piping

The principle of operation of the decanter is based on the differing densities of the liquids being separated. The greater the differences in density, the easier the separation of the liquids involved. In a process containing one solvent such as PCE, the heavier solvent will flow from the bottom of the decanter (see Figure 4.9) and the lighter water will flow from the outlet near the top of the cylinder. It is important that the water discharge and the solvent discharge be totally unrestricted in their flow from the decanter. Any back pressure acting on either line from restricted flow or improper venting may cause solvent to flow into the water discharge pipe or water to flow into the solvent discharge pipe. The same back-pressure condition within the decanter will produce erratic flow conditions and contamination of both wastewater and reclaimed solvent. The various water drain lines should be kept separated so that all flows are independent and will not interfere with other flow conditions. It is strongly recommended that a suitable recording flowmeter be installed on the solvent recovery piping. A sight glass on the reclaimed solvent tank is also desired with float-level controls on the solvent or water storage tanks to reduce overfills and spills. In all piping of water, solvent, compressed air, etc., unions and valves should be installed for ease of shutoff, disassembly, and repair.

Provide appropriate fire extinguishers for controlling local electrical and chemical fires, as well as flame arresters inside ductwork transporting flammable solvents. Follow all established safety precautions, and use common sense in all installation procedures. A job well done during installation will return many times the effort in both operation and maintenance time savings in the future.

4.4 OPERATION

Utilizing adsorption as an air pollution control measure on sources emitting volatile organic hydrocarbons has proven to be extremely effective if proper design is applied and rigid operating procedures are established and followed. It is always desirable to check over the process being controlled to determine that normal operation is being experienced. Of course, the blower must be running with the fan turning in the correct direction. Three-phase motors that are running backward may be corrected by changing two of the phase wires to the motor. Initial startups will probably require a system balance of the exhaust ducts. If multiple processes are being exhausted into the same system, adjustments of individual slide dampers may be required to obtain correct airflow in each branch duct. Airflow switches are inexpensive and should be placed in each branch to sense airflow and allow the process to operate when the airflow is adequate. Airflows below 100 ft/min through the carbon beds will provide adequate retention time for solvents in the airstream to be adsorbed on the charcoal. Excessive flows will reduce carbon efficiencies and allow volatiles to escape into the atmosphere. Excessive airflows are also detrimental to process operations where unnecessary solvents are evaporated and lost from process tanks and delivered to the carbon beds, posing additional loads on the system. Figure 4-17 shows the effect of excessive airflow rates on fluorocarbon losses from a typical vapor degreaser. Additional expenses to replace conditioned factory air that is wasted is significant, and every effort should be made to minimize airflow from the manufacturing plant.

Prior to startup, the prefilter, ahead of the carbon tanks, should be checked to verify that it is properly secured to the housing. This filter should be closely monitored during the first few days of operation. It may tend to accumulate dirt or lint from the construction and installation activities, causing the filter to become ineffective and restricting solvent entry into the carbon beds. After about 3 days of operation, replace or clean the dirty filter bag and observe the manometer gauge that registers air pressure between the blower and the carbon bed. There is no exact pressure differential that is proper for all machines, as the duct restrictions of specific installations will yield different gauge readings. Each system should stabilize and provide consistent manometer readings. However, during the first 2 or 3 days of operation, the carbon beds will be settling, so the readings will not reflect final operating conditions. With both carbon tanks set on the adsorption cycle, place a mark on the manometer dial face to indicate a normal reading. Any changes in the reading can be used to help maintain the correct air flow to the adsorber. A change of 1 in. of water pressure on this gauge is significant, and the problem should be corrected. If the reading drops by 1 in. or more, check the following points ahead of the blower: (1) dirty filter bag, (2) improper slide damper setting, and (3) obstruction in the duct. If the reading increases by 1 in. or more, check the following: (1) appropriate beds should be in the adsorption cycle when the reading is noted; (2) the adsorber tank dampers must be open; (3) there must be no restriction in the exhaust duct; and, (4) the filter bag may be loose or missing.

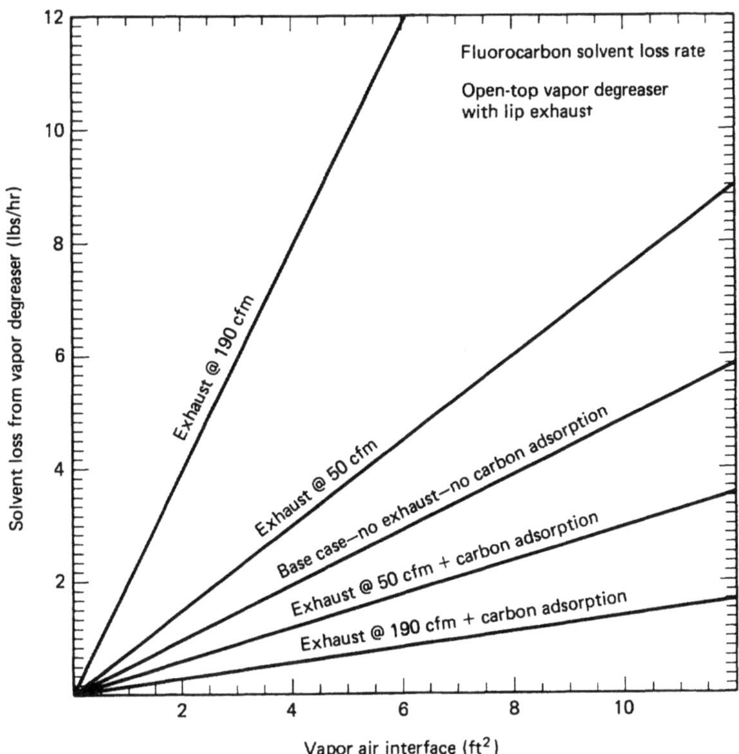

FIGURE 4-17 Fluorocarbon solvent-loss rate: open-top vapor degreaser with LIP exhaust.

A black discharge may be noticed coming from the adsorber exhaust upon startup. This is due to the carbon dust contained in all new charcoal and should subside after a couple of cycles. The duration of the adsorption cycle may be preset by manually set timers within the control panel of the unit. The precise time from startup to saturation depends on a number of factors. No simple, universal formula can be given to cover every situation. The type of solvent, airstream temperature, airflow rates, variations in concentration, and total amount of solvent evaporated from the process per unit of time must be considered. The initial adsorb cycle should be sufficiently long to completely saturate the carbon beds with solvent. All subsequent cycles, then, should be shortened because total regeneration cannot be effectively achieved.

On startup, initial cycle time and timer settings can be estimated by using the experimental data for carbon efficiencies contained in Table 4-4.

During actual operation the inlet and the outlet solvent concentrations may be continuously monitored with an organic vapor analyzer (OVA) to determine the breakpoint on the adsorption cycle. Timers can then be adjusted to discontinue adsorption and begin regeneration. A desirable feature of activated carbon in the control of solvent emissions is its ability to recover the adsorbed solvents on steam

TABLE 4-4   Physical properties of common VOCs.

	Boiling Point (°F)	Molecular Weight	Soluble in Water	Flammable	Lower Explosive Limit[1] (vol %)	Carbon Adsorption Efficiency[2]
Acetone	133	58.1	Yes	Yes	2.15	8
Benzene	176	78.1	No	Yes	1.4	6
Butyl acetate	259	116.2	No	Yes	1.7	8
Butyl alcohol	241	74.1	Yes	Yes	1.7	8
Carbon tetrachloride	170	153.8	No	No	—	10
Ethyl acetate	171	88.1	Yes	Yes	2.2	8
Ethyl alcohol	165	46.1	Yes	Yes	3.3	8
Heptane	209	100.2	No	Yes	1	6
Hexane	156	86.2	No	Yes	1.36	6
Isobutyl alcohol	241	74.1	Yes	Yes	1.68	8
Isopropyl alcohol	205	60.1	Yes	Yes	2.5	8
Methyl alcohol	153	32	Yes	Yes	6	7
Methylene chloride	104	84.9	Yes	No	—	10
Methylethyl ketone	174	72.1	Yes	Yes	1.81	8
Methylisobutyl ketone	237	100.2	Yes	Yes	1.4	7
Perchloroethylene	250	165.8	No	No	—	20
Toluene	231	92.1	No	Yes	1.27	7
Trichloroethane	189	131.4	No	No	—	15
Trichlorotrifluoroethane (113)	117.6	186.3	No	No	—	8
Naphtha	208	—	No	No	0.81	7
Xylene	292	106.2	No	Yes	1	10

[1] Lower explosive limit: The lowest concentration value of a vapor that will support propagation of a flame upward through a cylindrical tube.

[2] Carbon adsorption efficiency: Efficiencies are based on 200 cfm of 100°F solvent-laden air per 100 lb of carbon per hour at concentrations above 15 ppm.

regeneration. Regeneration may be accomplished by passing hot gas through the carbon bed. Low-pressure steam at 3 to 15 psig is the usual source of heat and is sufficient to remove most solvents. Normally, the flow of steam passes in a direction opposite to the flow of gases during adsorption. With this arrangement, the steam passes upward through the carbon. The steam flow through the bed is only one-fifth to one-tenth of the air velocity and is too low to initiate boiling or cratering of the bed. This countercurrent flow is an advantage in regeneration because a solvent gradient exists across the adsorbent bed and, depending on the concentration of adsorbate and bed depth, the inlet side of the bed may be saturated before the outlet reaches the breakpoint. Thus, with countercurrent regeneration, the solvent, driven out of the adsorbent from the outlet side by the incoming steam, will in turn start to remove vapor at the inlet before it becomes heated because it is already saturated. This results in lower steam consumption and more efficient operation.

Steam consumption per pound of solvent recovered varies with strip time and with the particular solvent adsorbed. Figure 4-18 provides a typical relationship

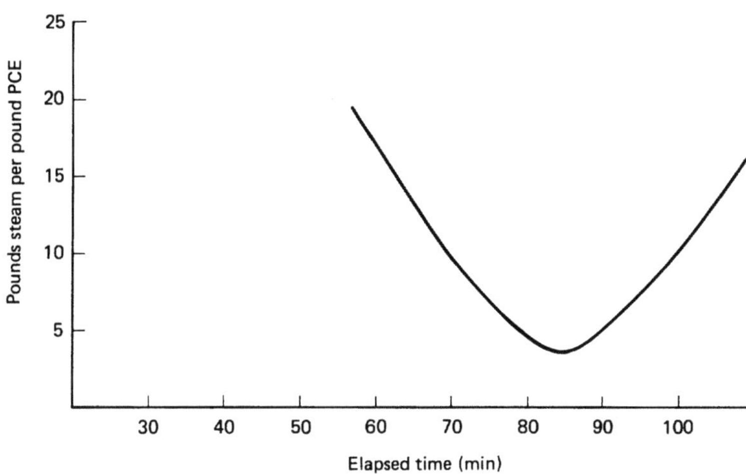

FIGURE 4-18 Steam usage per pound PCE.

between elapsed time of steam stripping vs. the pounds of steam used per pound of solvent recovered in a typical carbon adsorption system. The low point on the curve (maximum yield per energy expended) for a typical 1500-lb carbon bed has been found to occur between 80 and 90 minutes for a 4-psi steam regeneration pressure. As the steam strip time is extended, more steam per pound of solvent recovered is required and a point is reached at which expended steam cost exceeds recovery benefits. Therefore, it is more economical to operate the strip cycle to retrieve only part of the adsorbed solvent: that part recovered in less than 90 minutes, leaving a "heel" of solvent within the tank. All carbon systems should therefore be operated with the heel left in the beds in order to maximize solvent recovery vs. steam usage and/or energy consumption. In general, the adsorption/desorption curve follows the typical "hysteresis" pattern, as shown in Figure 4-19.

After the solvent is steam-stripped, the carbon beds are not only hot but are saturated with water. Cooling and drying are accomplished by opening the bed to the incoming airstream, allowing the entrapped water to evaporate and adsorb the heat of the carbon and to be swept away to the atmosphere by the airstream.

The solvent vapors and steam that emerge from the tank during stripping are then condensed in a water-cooled condenser. The cooling required for condensation may typically be 5 gal/min at an inlet temperature of 60°F. Temperature rise in the condenser will be approximately 40°F, making the outlet water temperature about 100°F. High cooling-water temperatures or low water-flow rates will slow the condensation process, thereby increasing the steam-strip cycle time and adding unnecessary cost to the process. The final step in the regeneration cycle is the separation of the steam and solvent condensate by using a gravity-type decanter that is (1) adequately vented to eliminate "air locks," (2) properly leveled and drained into an open container or tank to reduce back pressure, and (3) precharged

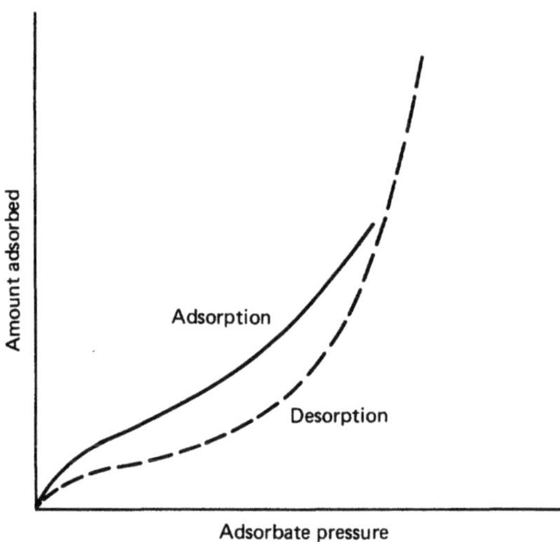

FIGURE 4-19 Adsorption isotherms showing hysteresis.

with a sufficient amount of solvent to prevent initial steam condensate from entering the solvent discharge pipe and contaminating the solvent storage tank.

It is important to sample both the wastewater discharge and the recovered solvent periodically to maintain the necessary effluent quality. Solvent solubility in the wastewater stream for the chlorinated solvents is quite low (0.015 g/100 g for PCE). Any additional solvent discharges into municipal sewer systems or to local streams should be eliminated. The integrity of the recovered solvent can be determined through tests for moisture content and acidity. Moisture content should be less than the solubility of water in solvent. Acid acceptance tests will reflect losses of solvent inhibitors that have been added to reduce breakdown of the solvent.

Multiple Compound Solvents

The adsorption phenomenon becomes somewhat more complex if the gases or vapors to be adsorbed consist of several compounds. Carbon absorption of the various components in a mixture is not uniform, and generally, these components are adsorbed in an approximately inverse relationship to their relative volatilities. Hence, when air containing a mixture of organic vapors is passed through an activated-carbon bed, the vapors are equally adsorbed at the beginning; but as the amount of the higher-boiling constituent retained in the bed increases, the more volatile vapor revaporizes. After the breakpoint is reached, the exit vapor consists largely of the more-volatile material. At this stage, the higher-boiling component has displaced the lower-boiling component, and this is repeated for each additional component (see the graph in Figure 4-20). Two or more VOCs in the airstream, as a general rule, will have the following effects: (1) The adsorption of organic compounds having higher molecular weights will tend to displace those having lower molecular weights. Lighter compounds will tend to be separated or partitioned from the heavier compounds and will pass through the bed at a faster rate. This will increase the mass-transfer zone and may require additional carbon bed depth, or shorter operating cycles. (2) Carbon retentivity may be reduced. (3) Efficiencies of any given system will tend to be lower on a multiple organic application. (4) The lower explosive limit (LEL) of the mixture will vary directly with the LEL of the individual components. Safety considerations may dictate more or less dilution air to reduce the flammability potential.

Decanter operation will also be affected by multiple organics in the system. Where multiple solvents are involved, the organic layer will consist of a mixture of the compounds roughly paralleling their feedstream concentrations. Some compound mixtures of immiscible organics may possess a variety of densities resulting in multi-layer decanter separations requiring multiple decantation for recovery. The recovered organics may also be a compound mixture and their value as a reclaimed solvent will depend upon the following questions: (1) Can the mixture be reused, as decanted? (2) Is it possible to reconstitute the mixture for reuse by solvent additions? (3) If mixtures cannot be reused, is it possible to sell it to a central refiner who would fractionate it and resell it as used solvent? (4) If the mixture cannot be made suitable for reuse, can it be used as a fuel for generation of thermal energy? (5) Is the recovered mixture of high-enough value to consider on-site distillation equipment?

Water-soluble organics present a special problem to steam regeneration systems. Because of gross water contact in the decanter, the water-soluble materials will be carried away in the wastewater stream. If losses of these organics are detrimental to the process or to water quality, alternative stripping mechanisms should be considered. Complex distillation columns may also be employed to recover the water solubles and return them to the source.

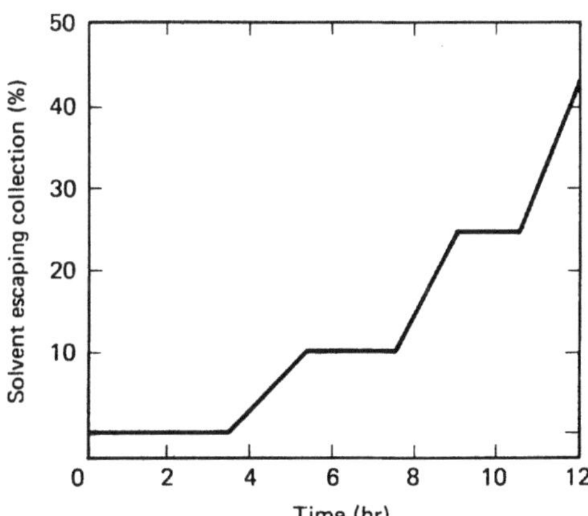

FIGURE 4-20 Adsorption efficiency, three-component lacquer solvent. (From *Experimental Program for the Control of Organic Emissions From Protective Coating Operations*, Report No. 8, Los Angeles County Air Pollution Control District, Los Angeles, Calif., 1961.)

The solvent corrosion factor should be considered in selection of the materials of construction for the adsorption equipment. Corrosion should be prevented or retarded if long-term operation of the equipment is expected. Many volatile organic compounds are not particularly corrosive and may be handled in conventional carbon steel tanks. Some solvents are subject to hydrolysis and/or other chemical reactions and to the formation of corrosive by-products. These must be dealt with by the use of coatings or base metals such as stainless steel or other materials that have a high resistance to oxidation. Most carbon steel carbon adsorption systems will employ some type of protective coating to isolate the activated carbon from direct contact with the interior of the adsorption vessel. This preventive measure is necessary to reduce galvanic action between the carbon and the tank wall that will result in excessive pitting and corrosion in the tank. The cycle time and/or manner in which the system is to be utilized will also be a factor. Systems that are to be regenerated only once or twice a day will tend to be dryer than those that cycle hourly and may be satisfactorily designed using internal coatings rather than expensive base materials. More expensive materials of construction should be provided only in those inaccessible areas such as vapor lines, small valves, drain connections, and damper housings, where the satisfactory applications of coatings would be difficult. Where corrosive compounds are utilized and where systems are to be cycled on a frequent basis, the more resistant types of construction materials should be considered.

All hydrocarbon materials that may be introduced into the adsorption system are characterized as either aliphatic or aromatic. Aromatic compounds exhibit a special type of unsaturation having to do with resonance and the stable nucleus provided by the six-carbon benzene-ring configuration. Aromatics of the simpler type will resist the formation of by-products and may be used with mild steel. Many of the aliphatic compounds that we deal with are the saturated alkanes or paraffins, and do not usually react with aqueous solutions of acids, bases, or oxidizing agents and therefore are compatible with mild steel constructions.

The major families of corrosive hydrocarbons are the halogenated ketones or aldehydes and the esters. Esters are the product of reaction between an acid and an alcohol. Such reactions are reversible, and the hydrolysis of an ester will yield the acid and alcohol from which they were produced. In use, most esters involve the formation of acetic acid and will require processing in stainless steel equipment.

The halogenated compounds represent a saturated hydrocarbon in which a hydrogen atom is replaced by a halide. The halogen atom is easily displaced from its associated carbon atom, and the formation of other compounds—often acidic—can be expected. Many halogenated materials are stabilized by the addition of corrosion retardants and inhibitors and can generally be processed in carbon steel equipment with an impervious surface coating. Monel, hastelloy, or titanium are sometimes used in connection with difficult halogenated compounds.

Ketones and aldehydes are both characterized by the presence of a carbonyl group. This carbon/oxygen group is subject to chemical reactions that may produce a number of corrosive by-products requiring handling in equipment of stainless steel construction. The ketones form the major category of "reactive organics," and it is the presence of the carbonyl group, with its ability to undergo chemical reaction, that causes the problem in carbon adsorption systems. Often these reactions are exothermic, and these reactive organic operations will require additional hardware and modified control sequences to control maximum temperature within the beds. Live steam injection or permanently installed cooling coils are effective in controlling temperatures that may approach the lower explosive limit of the solvent.

Undesirable contaminants in the airstream may also produce detrimental effects if not removed prior to their introduction into carbon beds. Contaminants may be broken down into three categories: (1) particulates, (2) entrained liquids, and (3) high-boilers. Almost all industrial applications will require solid-media-type filtration systems for the removal of airborne dust, lint, and general dirt in a particle size down to 3 to 5 μm. Solid-media filters of the type generally available are made of cloth or fiberglass and are usually satisfactory for this application. Automatically operated filters of the moving-media type, controlled by pressure drop across the filter material, usually offer a satisfactory solution for excessive particulate contamination.

In other applications, where fine particles of resin or other solids used in the coating process have diameters of 1 μm or less, special filtration systems will be required. Electrostatic precipitators, for instance, may find application in this area.

Entrained liquid is also a form of contaminant. There are a number of mist eliminators available, most of which are designed to be mounted in the solvent airstream.

Problems with high-boilers have been mentioned previously and are listed in Table 4-5. These are volatile organics that have boiling points in excess of 500°F, such as resins, placticizers, and/or compounds that react chemically on the carbon to form solid or polymerization products that will not be removed during steam desorption.

Some flammable materials will require special considerations in carbon adsorption applications. When solvents are being adsorbed that have low flash points, as indicated in Table 4-4, extreme care must be taken to ensure against carbon bed

TABLE 4-5 OCs not suitable for carbon adsorption.

Reactive compounds
   Organic acids
   Aldehydes
   Ketones (some)
   Monomers (some)
High-boilers
   Plasticizers
   Resins
   Long-chain hydrocarbons: C_{14} and up
   Phenols
   Glycols
   Amines

fires. It was previously mentioned that the phenomenon of adsorption is exothermic or "heat-liberating." If the carbon beds are not allowed to stay cool during adsorption and if sufficient oxygen reaches the flammable solvent in the bed, the potential exists for ignition, and a fire in the carbon bed may occur. Purging with an inert gas such as nitrogen rather than air will eliminate sufficient oxygen from entering the unit and reduce potential fire hazards. Carbon beds may also be cooled by direct water injection. The water will serve as a heat sink until sufficient flammable solvents displace the water molecules on the carbon beds during normal adsorption operation.

4.5 MAINTENANCE

Effective air pollution control utilizing carbon adsorption must be accompanied by a routine maintenance program. The program should provide for scheduled inspections of all equipment components as well as necessary monitoring of operating parameters to ensure correct operation and optimum performance of the control equipment. Monitoring of the inlet and outlet gas streams, quality of reclaimed solvent, wastewater quality, etc., will provide valuable information about the performance of the carbon adsorber and is covered in detail in Section 4.6.

Routine maintenance of air pollution control equipment can also be important because equipment failure can be expensive in terms of lost production, lost solvents, degradation of our air resources, and potential health effects on employees. Each component must operate properly to ensure steady, efficient output and desired results from the system.

Establishing an equipment maintenance program need not be elaborate or complicated. The actual work in routine inspection and servicing may, in large part, be performed by shop personnel operating the control equipment. Of course, extensive repairs or rebuilding should be accomplished by skilled and trained maintenance personnel. System components of the carbon adsorber that require routine maintenance fall into four major catagories: (1) air handling, (2) adsorbing, (3) stripping, and (4) reclaiming.

The function of the air-handling apparatus is to collect, transport, and deliver particulate-free, solvent-laden air to the adsorber. Any leaks in the duct works on the suction side of the fan will introduce excessive ambient air into the system, resulting in a reduction of solvent concentration and poor adsorber efficiency. Air-duct leakage on the pressure side of the fan will discharge unwanted solvent vapors into the workplace. Leakage checks should be performed periodically, especially at flexible connections, at joints in the ductwork, on the fan and filter housing, and around the adsorber bed dampers. Accurate collection velocity data should be established by routinely checking the capture velocity at the source using a vane-type velocity meter or a thermal anemometer. Correct operation of flow-indicating devices within the ducts can be verified by mechanically stopping flow to the duct or by turning off the fan motor. Ventilation system imbalance may also occur from time to time and may require periodic adjustments to dampers to rebalance the system. The particulate filter bag installed in-line ahead of the adsorber beds should be equipped with a differential pressure gauge to indicate dirty or stopped-up filter media. The bag should be changed or cleaned when the differential pressure increases by 1 in. (water) or more. Observance of the operation of the disk dampers within the adsorber tanks should occur periodically to determine correct seating and opening of the valves.

Maintenance inspections on the adsorption cycle equipment are somewhat more complex than for the air-handling apparatus. The integrity of the activated carbon must be maintained to ensure efficient removal of solvent vapors from the airstream. As the carbon particles erode with time and the capillaries become plugged with contaminants and polymers, the granules gradually lose their ability to adsorb and retain solvent molecules. Carbon adsorptability and retentivity should be tested regularly by opening up the bed and extracting carbon samples from the top, center, and bottom layers of the bed. Laboratory analysis will reveal the effectiveness of the carbon bed. Most manufacturers of activated charcoal will perform adsorptability and retentivity tests for their customers. If the carbon fails the tests, all the adsorbent should be removed from the system and be regenerated or replaced.

Maintenance requirements on the desorption-cycle equipment involves primarily the steam supply, valving, and timer controls. Steam pressure on the carbon tanks should be regulated to minimize steam stripping pressure (3–10 psi on PCE). Lower steam pressure will require too much run time to adequately strip the beds, while pressure that is too high will tend to fluidize the bed and create excessive erosion of the carbon granules. Steam traps must be operative or water will be carried into the carbon beds and retard proper stripping action. Periodically, the steam pressure relief valves located in the supply line and in the main carbon tanks should be checked for correct pressure settings by increasing steam pressure until the valves "pop off." Steam leaks around gaskets and operating dampers should be corrected by replacing the gaskets and seals. Gasket materials that are in contact with the solvent vapors must withstand that particular solvent's chemical properties. Often, the gasket material supplied with the system may not be suitable for the solvent presented to the adsorber and a substitute material may be required. Leaks around the carbon tanks may create additional problems of corrosion around the leak. The appearance of corrosion throughout the system may indicate a loss of inhibitors in the solvent that have been lost through numerous steam stripping cycles. Makeup inhibitors or makeup solvent may alleviate this condition. Boiler feedwater treatment may require some modification if "carryover" chemicals are introduced into the carbon beds and create corrosion problems.

The apparatus utilized to reclaim the solvent requires little maintenance. The automatic cooling-water valve should be checked for proper opening and closing operation. Automatic mechanical valve shafts and other mechanisms should be lightly oiled. The condenser, in time, may become inefficient because of excessive buildup of solubles from the cooling water. Acidizing of the water jacket or tubes may be required to renew condenser efficiency. Inadequate separation of water and solvent in the decanter may indicate a plugged vent line. All vent lines and drain lines must be unrestricted for correct operation of the system.

In general, normal maintenance procedures should be followed in routine cleaning of electrical contacts, lubrication of all bearings, compressed-air components, air cylinder shafts, replacement of obviously broken or worn parts, and housekeeping practices around the adsorber. In the final analysis, common sense is the best maintenance tool available in view of the fact that a large percentage of carbon adsorber equipment failures can be traced to neglect, improper operation, or just plain abuse.

4.6 IMPROVING OPERATION AND PERFORMANCE

Optimizing performance of air pollution control equipment such as carbon adsorbers involves consideration and monitoring of the following aspects: (1) operation of manufacturing process controls to minimize solvent emissions; (2) quality of the solvent/air inlet stream; (3) characteristics of the inlet stream, such as concentration, temperature, and flow; (4) duration of the adsorption cycle; saturation, and working bed capacities; (5) quality and quantity of available steam for regeneration; (6) duration of steam-strip cycle; (7) saturation and retentivity of the carbon; (8) quantity and quality of cooling water; (9) effectiveness of the water/solvent separator; (10) quality of reclaimed solvent; (11) quality of wastewater; and, (12) quality of exhaust stream from the adsorber bed. Each of the above is now considered.

1. *Operation of the manufacturing process*. Prior to applying operating controls to end-of-the-line air pollution control equipment such as the carbon adsorber, consideration must be given to process controls that will minimize solvent losses to the airstream and/or workplace. Listed below are some important considerations.

a.   Consider substitution of the solvent by one of lower volatility or environmental impact.

b.   Minimize local exhaust ventilation from the process.

c.   Utilize adequate freeboard height to reduce losses.

d.   Provide canopies, hoods, and enclosures over tanks if possible, and minimize all openings to process.

e.   Provide a parts-drying chamber within the process if possible.

f.   Cover tanks when not in use.

g.   Minimize the travel rate of parts into and out of the cleaning tank vapor zone.

h.   Solvent spraying should be performed in the vapor zone, preferably with a gentle flush rather than an atomized spray.

i.   Allow parts to drain properly prior to exiting the cleaning tank.

j.   Maintain a cold-air blanket above tanks with either chilled-water coils or direct-expansion refrigeration coils.

k.   Do not clean porous or adsorbent materials.

l.   Do not vigorously boil solvent sump.

m.   Do not agitate the solvent bath with compressed air.

n.   Use compressed air blow-off as a last resort.

o.   Do not direct ventilating fans on process.

p.   Establish and maintain filling, draining, startup, shutdown, operation, and maintenance procedures.

After application of appropriate design parameters to the cleaning process, consideration can be given to end-of-the-line controls.

2. *Quality of solvent/air inlet stream.* The large surface areas initially offered by the activated carbon granules will gradually be lost in the adsorption/desorption process. The effective life of the carbon depends upon the quality of the incoming airstream presented to the beds. Airborne contaminants must be removed from the vapor-laden stream prior to introduction to the adsorbent to eliminate "plugging" of capillaries within the granules. Periodic and routine maintenance on the bag filter will ensure longer life to the carbon beds. Some solvents offered to the carbon beds may decompose or polymerize when contacting the adsorbent. Polymerization will significantly lower the adsorption capacity of the adsorbent and also reduce its regenerating capabilities by conventional means. These polymerizing solvents must be avoided.

3. *Characteristics of inlet stream.* In many instances exhaust ventilation design is concerned primarily with contaminant control at the point of emission to protect worker health and reduce employee exposures to hazardous chemicals. As a result, many exhaust systems have been designed to satisfy only these requirements, with little regard for system efficiencies. Earlier discussions of process design characteristics indicated a desire to minimize exhausted air losses and excessive carryout of solvent vapors. Lower airflow will not only improve the cleaning process efficiency but will reduce fan cost, ventilation system cost, and power consumption, and will reduce previously conditioned room air losses to the atmosphere. Cost comparisons of annual operating and maintenance cost and cost to replace exhausted air from a building are shown in Figure 4-21. The costs shown are general in nature and will vary considerably from industry to industry and as the materials of construction vary to suit the process design criteria. Annual conditioned-air loss through exhaust ventilation reflects cost associated with year-around comfort-air-conditioned factories. The operating costs are based on a 2-shift, 5-day operation. In addition to minimizing airflow, attempts should be made to present a cool gas to the adsorber beds. An increase in temperature results in a corresponding decrease in the quantity of gas adsorbed. The effect of temperature on gas adsorption is shown in Figure 4-22 [5]. It would therefore be an advantage in promoting adsorption to reduce mechanically the inlet gas temperature to the carbon beds if adequate cooling is not

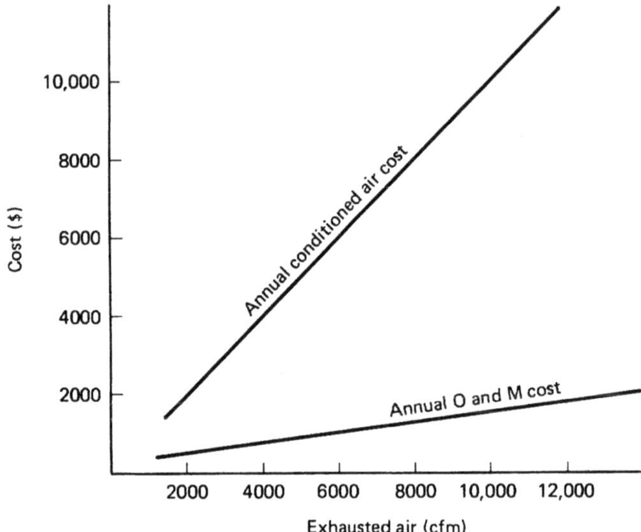

FIGURE 4-21 Annual fan operating cost.

achieved in the transporting ductwork from the process to the adsorber. By minimizing the airflow, reducing the temperature, and thereby raising the solvent concentration, maximum utilization of the adsorptive process can be realized.

4. *Duration of the adsorption cycle.* Working bed capacities vary considerably, depending upon the particular solvent being reclaimed and its regeneration characteristics. To maximize performance of the carbon adsorber, the duration of the adsorption cycle should be extended to just below the breakpoint of the beds. Breakthrough can be determined using organic vapor analyzers simultaneously on the inlet and outlet streams of the adsorber bed. Breakthrough history can then be determined on the particular process being controlled, and regeneration can be initiated only when absolutely necessary.

5. *Steam quality.* Low steam pressure and wet steam will reduce the effectiveness of the stripping operation. Steam traps must be continually checked for correct operation. Because of the intermittent steam demand, condensate in the supply lines

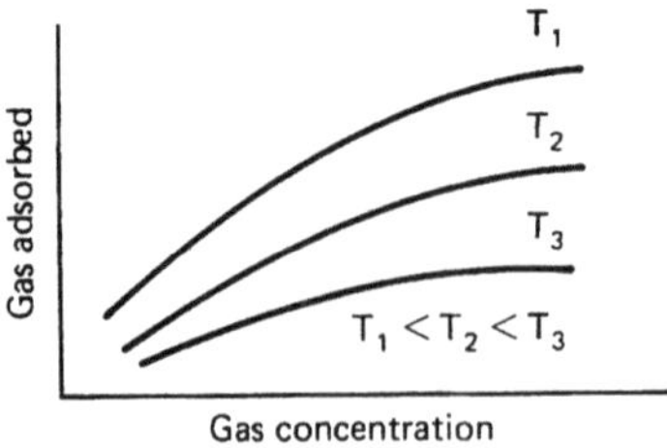

FIGURE 4-22 Effect of temperature on gas absorption.

must be removed prior to stripping. Automatic steam traps should ensure the availability of dry steam for the desorption of the beds.

6. *Duration of steam-strip cycle*. To provide efficient recovery of solvents, steam stripping must not be excessively long, to try to recover too much of the "heel," or too short, requiring more frequent regeneration. Desirable strip-cycle times can be assured by continuous monitoring of reclaimed solvent quantities that flow from the decanter.

7. *Saturation and retentivity of carbon*. Ensure good operation and performance of the carbon beds by periodically testing the charcoal for its ability to adsorb solvent and to retain the solvent after the manufacturing process flow is discontinued. Adsorbent manufacturers will assist in this test. The adsorptability and retentivity of the carbon will be affected by airborne contaminants that enter the beds and plug the capillaries in the granule. Figures 4-23 and 4-24 show an individual carbon granule after approximately 8 years of adsorbing perchloroethylene from a wave solder cleaner [2]. Figure 4-23 shows a granule eroded from thousands of steam-stripping cycles. Small impurities can be seen on the surface of the granule in Figure 4-24. These two conditions have reduced the efficiency of this carbon bed and retentivity tests indicate a necessity to perform extensive regeneration (a service provided by activated-carbon suppliers) or total replacement of the adsorbent.

8. *Cooling-water quantity*. Monitor cooling water inlet and outlet temperatures. Usually, the steam-strip cycle is operated on a timer mode and cooling water is

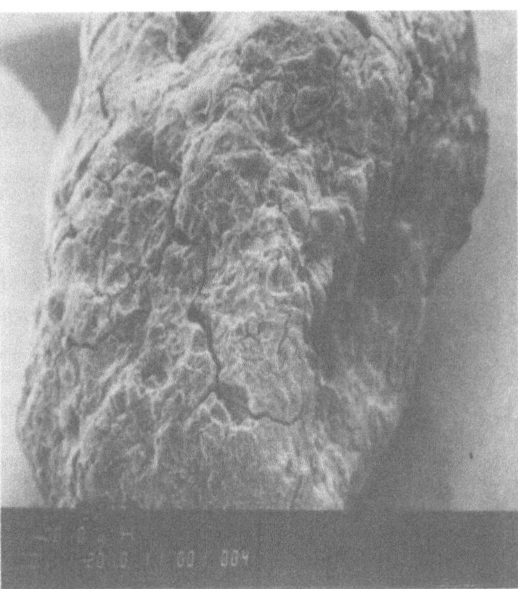

FIGURE 4-23 Individual carbon granule after long adsorption of perchloroethylene.

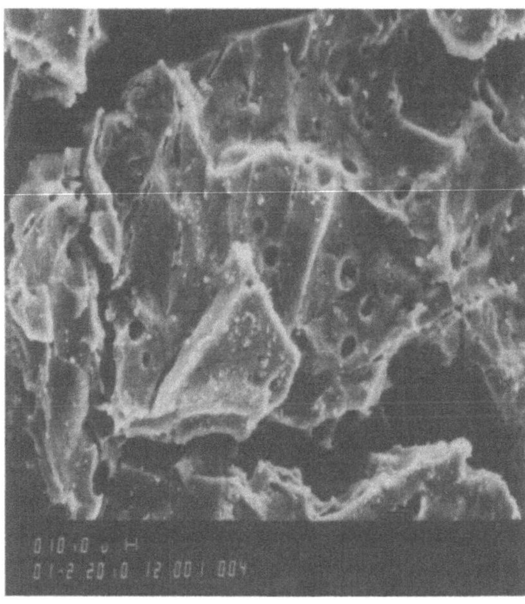

FIGURE 4-24 Surface impurities on carbon granule.

required to condense all the vapors presented to the heat exchanger in the same given amount of time. Inadequate cooling water will result in solvent losses to the airstream. The inlet and outlet temperatures will vary depending upon the particular solvent being reclaimed. The cooling-water supply must be sufficient to condense vapors of steam as well as solvent vapors that are presented to the condenser for the duration of the steam-strip cycle.

9. *Effectiveness of the water/solvent separator.* Vent properly, provide unrestricted flow of solvent and water discharging from the decanter, and check the quality of both solvent and wastewater.

10. *Quality of reclaimed solvent.* Laboratory analysis is required to determine losses of inhibitors, decomposition of solvent, acidity, or contamination.

11. *Quality of wastewater.* Wastewater analysis can also detect the presence of pollutants in the stream. Because of the rapid separation of water and solvent, gross amounts of nonsoluble solvents can be detected by observation. Dissolved contaminants must be detected through laboratory analysis.

12. *Quality of exhaust stream.* The final analysis required to determine efficient operation of the adsorber is in the exhaust stream. By using an organic vapor analyzer the concentration of organic solvent in the airstream can be determined. The total quantity of solvent losses in pounds per unit time should be recorded to establish a historical file on the system. Deviations from the file data could then be noticed, rectified, and carbon adsorber efficiencies maintained.

Monitoring

Improvements in the operation and performance of carbon adsorption systems can be obtained through a well-defined program of operation and maintenance, a program that encompasses the total manufacturing process, beginning with the solvent used and the design and operation of the process and its exhaust system thus should then be extended to the operation of end-of-the-line equipment. To ensure proper operation and maintenance, a formal monitoring program should be established. Implementation of this program will provide an accurate data file of such parameters as inlet concentrations, outlet concentrations, solvent removal efficiency, cycle times, and solvent quantities reclaimed.

Two methods recognized for monitoring of organic emissions are the source test screening method, which specifies the organic vapor analyzer instrument, and the source testing verification method, which specifies an integrated bag as the sampling device with a gas chromatograph or infrared spectrophotometer analysis.

The organic vapor analyzer (OVA) is one of the most popular instruments employed for organic gas stream concentration measurement. Both portable and bench-top OVAs are available and both have proven to be reliable for carbon adsorption monitoring. Figures 4-25 and 4-26 show both types of units [10]. The

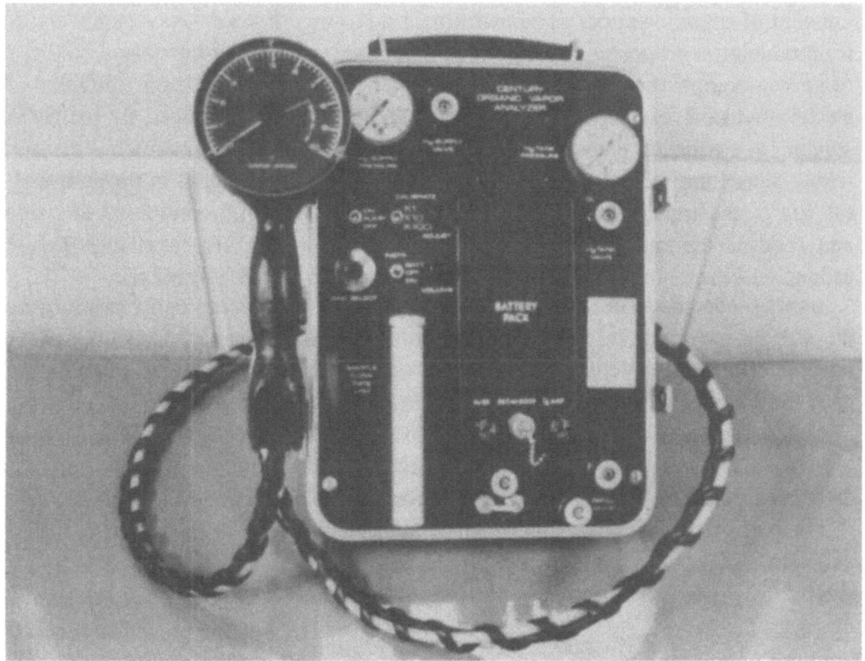

FIGURE 4-25 Portable OVA.

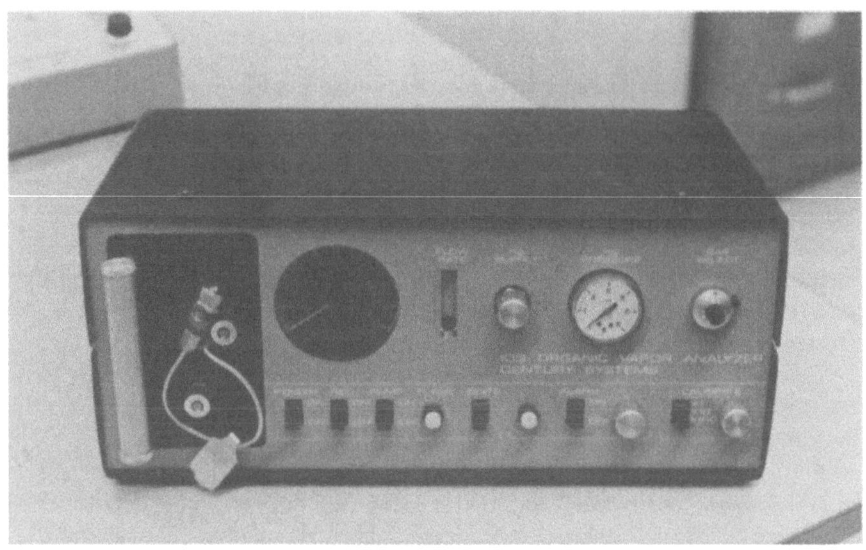

FIGURE 4-26 Bench-top OVA.

OVA works on the principle of hydrogen flame ionization for detection and measurement of organic vapors. The instrument measures organic vapor concentrations by producing a response to an unknown sample, which can be related to the gas of known composition to which the instrument has previously been calibrated. An electric field drives the charged carbon ions, which are produced by incinerated solvent, to a collector electrode, which in turn generates a current which is measured with a logarithmic electrometer preamplifier. The output signal is proportional to the log of the input or the ionization current. An amplifier conditions this signal and sends a readout to the external monitor. Units are also available that will produce a linear output signal over a smaller range of concentrations.

A strip-chart recorder can be attached to the OVA to record concentrations over the sample-time interval. The chart recorder can make the OVA semipermanent, because once the strip chart and OVA are calibrated, they can be left to continually monitor the organic vapor concentration in parts per million (ppm).

Monitoring of the organic vapors in the airstream is done for several reasons: to demonstrate compliance, to establish organic vapor removal efficiencies, and to determine optimum time intervals for adsorption and regeneration cycles.

The organic vapor removal efficiency at any given time is calculated as follows:

where C_I = inlet organic vapor concentration

          C_O = outlet organic vapor concentration

          E_R = organic vapor removal efficiency

1. Simultaneously monitor the concentration of the inlet airstream and outlet airstream.

2. Insert the values into the equation for instantaneous removal efficiency:

$$\frac{C_I - C_O}{C_I} = E_R; \qquad E_R \times 100 = \% \text{ removal}$$

For an average removal efficiency for a defined time interval t, use the following procedure:

1. Monitor the concentration of the inlet and outlet airstreams for time interval t using the strip-chart recorder (Figure 4-27 shows an example of the output of the recorder).

2. Determine the average inlet and outlet concentration, whether through time integration or extrapolation.

3. Insert the determined values into the equation for a time-averaged removal efficiency:

$$\frac{C_{I(t)} - C_{O(t)}}{C_{I(t)}} = E_{R(t)}; \qquad E_{R(t)} \times 100 = \% \text{ removal}$$

The OVA measures total hydrocarbons only; therefore, in situations where more than one solvent exists, the infrared spectrophotometer should be employed to preferentially monitor each respective solvent. A more elaborate system is the "in-

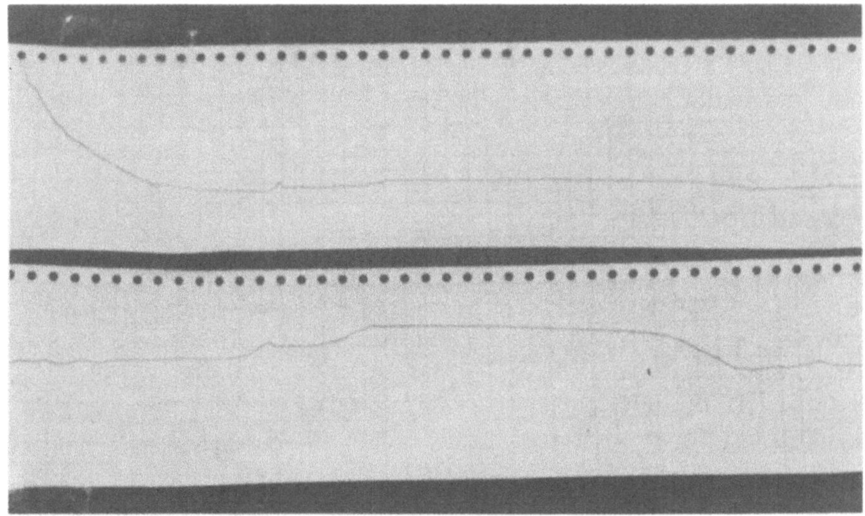

FIGURE 4-27 Output of strip-chart recorder.

line" gas chromatograph. These instruments are permanently installed and monitor the hydrocarbon concentration of gas streams on a regular cycle (minimum cycle time is 15 minutes for extracting a sample and analysis through the gas chromatograph column).

Solvent concentration monitoring must be accompanied by adequate airflow determination to calculate solvent losses and to perform the required material balance on the system. Airflow may be determined using the standard pitot tube and inclined water-filled manometer; in cool, dry gas streams a thermal anemometer may be used.

Wastewater discharge and recovered solvent monitoring are also important to maintain the necessary effluent quality. Discharges to municipal sewers or to local streams should be minimized to reduce solvent losses. The integrity of the solvent to be reused is dependent upon individual cleanliness requirements; however, samples of the solvent should be tested periodically for moisture content and acidity as well as for cleanliness. Often, visual inspection will provide needed information pertaining to either solvent or wastewater quality. Discolorations in solvents that have been returned to the process may signal a need for further investigation. When water and immiscible solvents are present together, a clear boundary line may be visible.

Sight glasses on holding tanks can be misleading if two immiscible solvents are present in the tank, such as water and perchloroethylene. Because of the density difference, the PCE will assume the bottom position in the tank and, in conventional sight-glass installations, will fill from the bottom of the tank. The presence of water may not be reflected at all in sight-glass readings. Because of this density difference, the sight-glass levels may also be misleading, not reflecting the true level of liquid within the tank.

Certainly, a viable monitoring program is mandatory in "tracking" the performance of any air pollution control equipment. The costs to implement such a program will be returned many times over through improvements in source operations, recovery of organic solvents, and in protection of the air resource through reduced emissions.

4.7 CONCLUSIONS

A well-designed and maintained carbon adsorption system will normally capture in excess of 95% of the organic input to the bed. EPA studies indicate that carbon adsorption applied to solvent cleaning processes will actually achieve a reduction in total solvent emissions of 40 to 60%. One reason for the difference between the theoretical and actual losses is found in the inadequacy of most ventilation systems and their inability to captivate all of the solvent vapors and deliver them to the carbon bed. Many solvent losses occurring in manufacturing processes are in areas of dragout on parts, leaks, spills, and disposal of waste solvents, none of which are greatly affected by the ventilation system. Improved ventilation design can increase an adsorber's overall emission-control efficiency. A higher ventilation rate alone, however, will not necessarily be advantageous because increased turbulence could disrupt the air–vapor interface, causing an increase in emissions, all of which would not be captured by the collection systems. The effectiveness of the ventilation system can also be improved through use of drying tunnels and other devices which decrease losses due to dragout.

Poor operation has been found to decrease the control efficiency of carbon adsorption systems. Examples are dampers that do not open and close properly, use of carbon that does not meet specifications, poor timing of the desorption cycles, and excessive inlet flow rates. Desorption cycles must be frequent enough to prevent breakthrough of the carbon beds, but not so frequent as to cause excessive energy consumption. The positive aspects of carbon adsorption are well known. There are, however, a few negative aspects that must be considered.

Air impact   No significant adverse air impact should result through the correct use of the carbon adsorber, although negligence with maintenance and operation of control devices could increase emissions in individual cases. Carbon adsorption systems operating with spent or saturated adsorbent, coupled with excessive ventilation rates, will lead to excessive solvent losses to the atmosphere. Excessive steam-strip cycles in high condensing water temperatures will also allow solvent losses to the atmosphere to increase. Consideration must also be given to the emissions from steam-generating boilers as the demand for steam increases during the desorption cycle.

Water impact   The largest impact on water quality is a result of the steam condensate discharged from the carbon adsorber. Steam used for desorption is in direct and intimate contact with the solvent vapors, and as the steam condenses into water, it will take on a small percentage of the solvent as well as the water-soluble stabilizers. When the contaminated water is discharged into the waste stream, the stabilizers and dissolved solvent are carried along as pollutants.

Solid waste impact   There appears to be no significant solid waste impact resulting from routine carbon adsorption operation; however, periodic replacement of the activated charcoal is required and may present a disposal problem. Most major activated-carbon manufacturers are equipped to reactivate spent carbon and thus eliminate any solid waste impact from normal applications.

Energy impact   A carbon adsorber consumes a large amount of energy because of the steam required for desorption; however, this energy expenditure is far less than the energy required to manufacture replacement solvent.

Noise impact   Consideration must be given to increased noise levels due to the carbon adsorber when choosing an in-plant location for the system. Additional noise-suppression technology may be required for some installations.

Generally speaking, good operation practices and a proper maintenance program will eliminate many of the problems mentioned as environmental impacts, and allow the carbon adsorption equipment to operate effectively as intended: to reduce the discharge of pollutants into the atmosphere and to return recovered solvents to the manufacturing process.

4.8 NOMENCLATURE

$^\circ C$	degrees Celsius
C_I	inlet organic vapor concentration, ppm
C_O	outlet organic vapor concentration, ppm
C_p	specific heat of adsorbent, Btu/ft^3/$^\circ$F
C_{st}	saturation capacity of bed at $T + \Delta T$, $^\circ$F
C_t	pressure-drop coefficient, ft hr^2/in^2
C_T	pressure-drop coefficient, ft hr^2/in^2
C_3	breakpoint concentration, ppm
C_4	outlet concentration, ppm
d_c	particle diameter, ft
d_p	effective particle diameter, ft
E_R	organic vapor removal efficiency
f	friction factor
$^\circ$F	degrees Fahrenheit
G	superficial mass velocity, lb/hr ft^2
g	conversion constant, $4.18(10^8)$ ft lb/lb$_f$ hr^2
l_c	particle length, ft
L_{st}	saturated bed length, ft
L_z	length of adsorption zone, ft
MTZ	mass transfer zone
Re	Reynolds number
t_b	service time to breakthrough concentration
T_i	inlet-water temperature, $^\circ$F
T_m	mean temperature difference, $^\circ$F
T_o	outlet-water temperature, $^\circ$F
U	overall coefficient, Btu/hr ft^2 $^\circ$F
v	gas viscosity, lb/ft hr
VP	velocity pressure, in. of water
X	degree of saturation in MTZ
X_{st}	adsorptive capacity per unit of bed volume in saturated zone, lb/ft^3

X_z mean value of X_{st}, lb/ft^3

Z depth of packing, ft

ΔP pressure drop, psi

ΔT temperature rise, °F

ε frictional void volume in dry packed bed, ft^3 voids/ft^3 packed volume

ρ gas density, lb/ft^3

u   superficial velocity of gas through bed

4.9 REFERENCES

[1] CHEREMISINOFF, P. N., AND ELLERBUSCH, *Carbon Adsorption Handbook*. Ann Arbor Science Publishers, Ann Arbor, Mich., 1978.

[2] BROTHERS, E. W., *Electron Scanning Microscope Photographs*. Western Electric Co., June 1980.

[3] OAQPS, *Control of Volatile Organic Emissions From Solvent Metal Cleaning*. EPA-450/2-77-022, November 1977.

[4] Buonicore, A.J., and Theodore, L., *Industrial Control Equipment* for *Gaseous Pollutants*, Volume I, CRC Press, Inc., Boca Raton, Florida, 1975.

[5] OAQPS, *Air Pollution Engineering Manual*, 2nd ed. EPA, May 1973.

[6] MSA RESEARCH CORP., *Package Sorption Device System Study*, EPA, April 1973.

[7] VIC MANUFACTURING CO., *Installation, Operation and Maintenance for Vic Air Pollution Control System*.

[8] CROSS, F., AND HESKETH, H. E., *Handbook for the Operation and Maintenance of Air Pollution Control Equipment*. Technomic Publishing Co., Inc., Westport, Conn., 1975.

[9] AMERICAN CONFERENCE OF GOVERNMENTAL INDUSTRIAL HYGIENISTS *Industrial Ventilation*, 14th ed. Edward Brothers, Inc., Ann Arbor, Mich., 1976.

[10] Century Systems Corp., Arkansas City, Kans.

5

Incinerators

Joseph P. Bilotti, P.E.

York Research Corp.
Stamford, Connecticut

and

Timothy K. Sutherland

Peabody Process Systems
Stamford, Connecticut

5.1 DESCRIPTION OF CONTROL DEVICE

Fume incinerators can be used to control the emission of gaseous or small-particulate air pollutants which are either combustible or which thermally decompose at high temperatures. Three rapid oxidation methods are used to destroy combustible contaminants:

1. Thermal incineration
2. Catalytic incineration
3. Flares (direct flame combustion)

The thermal and flare methods are characterized by the presence of a flame during combustion. Catalytic incinerators serve the same purpose but utilize a metallic catalyst to promote the required rapid reduction.This unit operates at lower generating temperatures and with higher reaction rates.

The main uses of fume incinerators are for the control of odor and smoke with some applications in hydrocarbon and carbon monoxide (CO) emission control. Generally, the gas stream that the incinerator is applied to contains a low concentration of combustible emissions. This comes about because of the safety requirement to keep the concentration of combustible gases within the fume generating equipment at some percentage below the lower flammability limit. If the gas stream, for whatever reason, is of a high enough concentration of combustibles, it can be used directly as a fuel.

The design of all types of fume incinerators must take into consideration the "three T's" of combustion: time, temperature, and turbulence. These three variables will govern the speed and completeness of the combustion reaction. For complete combustion to occur, the oxygen must come into intimate contact with the combustible molecule at a sufficient temperature, and for a sufficient length of time, in order that the reaction be completed. Incomplete reactions may result in the generation of aldehydes, organic acids, carbon, and carbon monoxide.

Thermal Incinerators

Description Thermal incinerators are the most widely used method to control the release of hydrocarbon fumes, especially smokes and solvents from coating processes. Typically, thermal incinerators are constructed of a steel outer shell lined with refractory material. The purpose of the refractory lining, which is typically 4 to 8 in. thick, is to protect the steel shell from direct exposure to the effects of high temperatures and corrosive materials, and also to improve the thermal efficiency of the unit by limiting heat losses. The refractory, which will be discussed in detail later in this section, serves as a thermal insulator. It will lower the temperature from greater than 2000°F on the inside of the combustion chamber to a temperature of 180 to 400°F at the outer surface of the steel shell. In this temperature range, the steel shell will continue to retain its strength while also remaining hot enough to prevent the condensation of water vapor on the surface, thereby reducing corrosive attack.

These incinerators are equipped with a burner at one end that fires a fuel, typically natural gas. There is also a fume inlet located near the burner where the gas stream to be oxidized enters the incinerator. The burner may utilize the air in the process waste stream as the combustion air for the fuel, or it may use a separate source of outside air for this purpose. Figure 5-1 depicts a typical incinerator schematic.

Refractory Refractory material consists of various types of fireclays, usually containing alumina or silica, and is available in several forms, including bricks and castable or plastic material. Bricks are manufactured in standard sizes (4½ in. wide × 4 in. high × 9 in. long is typical) but may be precut at the factory to form any desired shape. They are best used to form regular surfaces. Castable refractory is usually supplied in a dry powder form which is mixed with water and cast in place

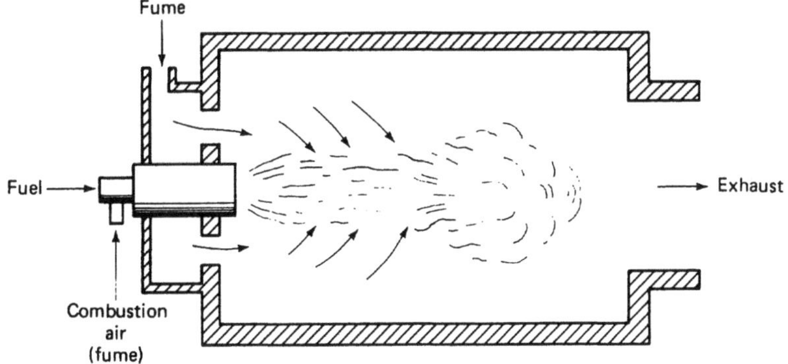

FIGURE 5-1 Thermal incinerator schematic.

in the field. Hence, initial cost and shipping charges are less than those for brick, and irregular surfaces are more easily formed. However, castable, as well as plastic, refractory requires the use of a suitable anchoring system. Anchors are normally recommended by the refractory manufacturer for a particular material and service condition. Normally, they are made of alloy steel and are welded to the steel surface before lining. This adds considerably to the cost of installing a castable or plastic type of refractory.

Plastic refractory material is essentially a premixed castable with binders added to improve its plastic-handling qualities. It is normally used only to cover small areas, for patching work, or to form intricate shapes. Castable or plastic refractory linings are usually homogeneous (i.e., consist of the same-density material throughout). Brick installations, however, are often composed of layers of different-quality bricks. On the hot side, plain firebrick is normally used. Because it is very dense (approximately 150 lb/ft^3), it has a high degree of "refractoriness" and is able to withstand the furnace environment for long periods. Between the firebrick and steel shell, one or more layers of "insulating firebrick" may be placed. Insulating firebrick is much less dense than firebrick (50 to 80 lb/ft^3) and so has much better insulating properties. Finally, a layer of "insulating block," which is a ceramic material, may also be used between the insulating firebrick and the steel shell. This material has excellent insulating properties but cannot be exposed for long periods to temperatures above 1000°F.

A refractory lining is almost always installed in the field. Refractory material is heavy and rather brittle and so must be completely supported during shipment if installed at the factory. The cost of such support is usually prohibitive.

A brick lining is installed much as any other brick structure. Skilled masons are required. Bricks are trimmed if necessary and cemented in place with special mortar. Castable refractory may require the use of forms, much as in concrete work, and the material may be either poured or "gunned" into place. Care must be taken in placing castable refractory to assure that no voids are formed.

Ceramic blanket insulation Blanket-type insulation is made from ceramic fiber materials. They are very lightweight and have excellent thermal insulating properties. In some cases a 6-in. refractory lining, weighing 100 lb/ft^2 could be replaced by a 2-in.-thick blanket weighing less than 10 lb/ft^2. This will result in a significant weight reduction and a smaller incinerator. Because the blanket material is so lightweight and flexible, it can be shipped installed in the incinerator. This saves in field-installation and shipment costs.

The installation of insulating blankets requires first that a stud system be added to the incinerator shell. The insulation blanket is then impaled on the studs and secured on the stud with a locking washer. The stud-and-washer system is made of high-alloy steel and represents the most expensive part of the installation.

Insulation blanket is not usually suitable for use when fuel oil is used as an auxiliary fuel. This is because during upset conditions fuel oil may saturate the blanket. When the burner is relit, the fuel oil will ignite and destroy the blanket. Blankets are also very pervious to fumes, which may cause a corrosion problem in particular applications.

The major advantage of insulation blankets is that no warm-up period is required for the incinerator, resulting in a large fuel savings over refractory linings. Replacement of the insulation blanket does not usually require replacement of the stud-and-washer system and is therefore relatively inexpensive.

Catalytic Incinerators

Description A catalytic incinerator is an alternative to a thermal incinerator as a means for oxidizing gaseous or oxygenated hydrocarbons to carbon dioxide and water. Contact of a waste stream with a catalyst bed allows oxidation reactions to occur rapidly in the temperature range 700 to 900°F, in contrast to the 1300 to 1600°F typically required of a thermal incinerator. The oxidation reaction that occurs at the surface of the catalyst produces the same products, carbon dioxide and water, and liberates the same heat of combustion as a thermal incinerator would.

The basic elements of a catalytic incinerator are shown schematically in Figure 5-2. The heat required to bring the waste stream up to the required oxidation temperature is usually supplied by a natural gas fuel burner. This type of incinerator is again constructed of a steel outer shell lined with a refractory material. The wall thickness of the refractory for a catalytic incinerator can be less than that of the thermal incinerator simply because the operating temperatures are not as great.

A catalyst bed is located at a distance downstream from the burner sufficient to allow for a uniformly preheated and distributed mixture of combustion products and waste gas stream. The catalyst bed in commercial units is typically a metal mesh mat, ceramic honeycomb, or other ceramic matrix structure with a surface coating of finely divided platinum or other platinum-family metals, such as chromium, vanadium, nickel, or cobalt. This metal coating is the actual catalyst for the

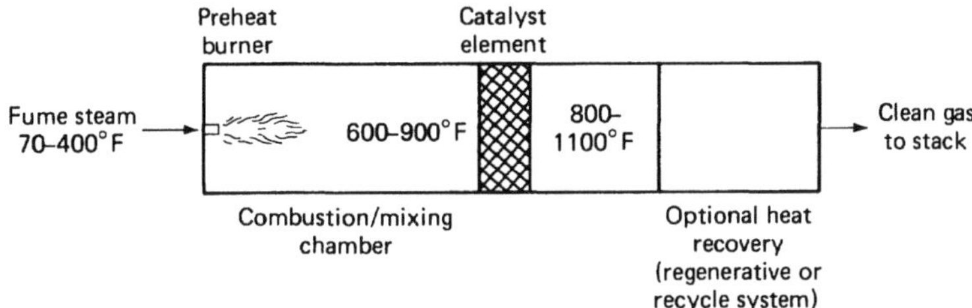

FIGURE 5-2 Schematic of catalytic incinerator [2].

oxidation reactions. The matrix serves to support the catalyst on a high geometric surface area and promotes good contact between waste streams and catalyst. Typically, a volume of only 0.5 to 2 ft^3 catalyst per 1000 scfm of waste stream is required for 85 to 95% conversion of hydrocarbons [1].

Flares

Description   The flare system is used primarily as a safe method for disposing of excess waste gases. All process plants that handle hydrocarbons, hydrogen, ammonia, hydrogen cyanide, or other toxic or dangerous gases are subject to emergency conditions that require the immediate release of large volumes of such gases for protection of plant and personnel. Flares are used for this purpose. In operation, the gas containing the organics is continually fed to and discharged from a stack, with the combustion occurring near the top of the stack and characterized by a flame at the end of the stack.

In general, three types of flares are in use today: elevated, ground, and forced draft. The elevated flare is made up of several components, as shown schematically in Figure 5-3. The ground or "bilevel" flare, depicted in Figure 5-4, is employed where noise, luminosity, or smoke are objectionable. They are, for all practical purposes, enclosed in combustion chambers which produce smokeless burning and conceal the flare flame. However, this type of flare has been outlawed in most states. The forced-draft flare, shown in Figure 5-5, uses air provided by a blower to supply primary air and the turbulence necessary to provide smokeless burning of exhaust gases without the use of a stream injection system as used in the elevated flare.

Typical maximum flare capacity are as follows:

Type	*Capacity (1000 lb/hr) [3]*
Ground flare	80–100
Elevated flare	1000–2000
Elevated, forced draft	100

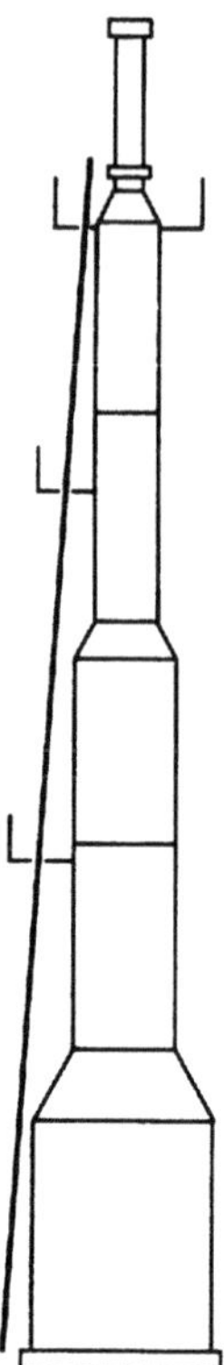

FIGURE 5-3 Elevated flare schematic.

5.2 DESIGN PROCEDURES

Thermal Incinerators

The greatest variation among different afterburner designs is in how well they achieve the goal of raising *all* the fume to the required temperature for the required combustion residence time. Most cases of poor performance are due to nonuniform temperatures and flows, which allow some of the pollutants to escape without adequate treatment.

Time and temperature requirements will be discussed together since they are, to some degree, interchangeable—a higher temperature allows use of a shorter residence time, and vice versa. Temperature is the expensive variable, with time a direct function of the combustion chamber volume. For example, the residence times for a particular application could be doubled for only a 20 to 30% increase in total capital costs. However, variation of the temperature is directly related to fuel usage, which is the major component of daily operating costs. Therefore, it

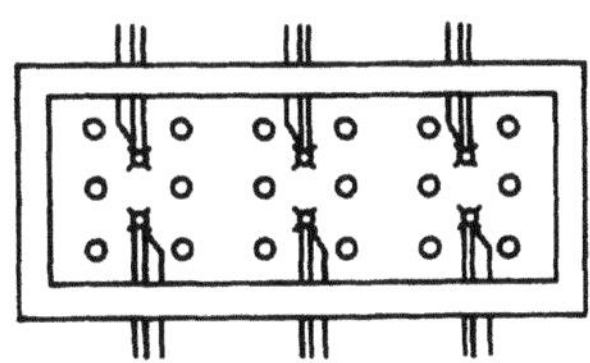

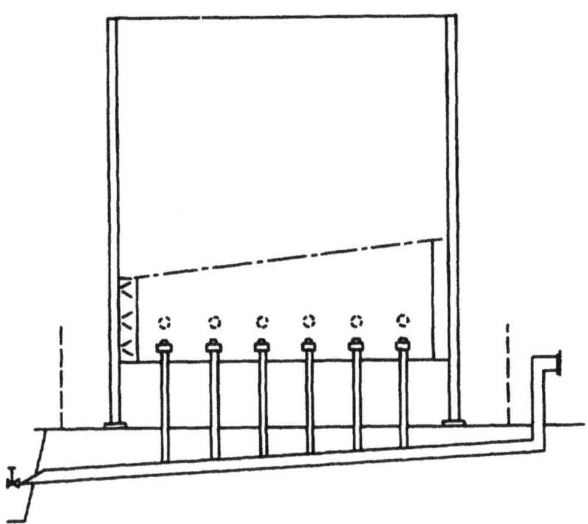

FIGURE 5-4 Ground flare (From National Air Oil Burner Company; [3].)

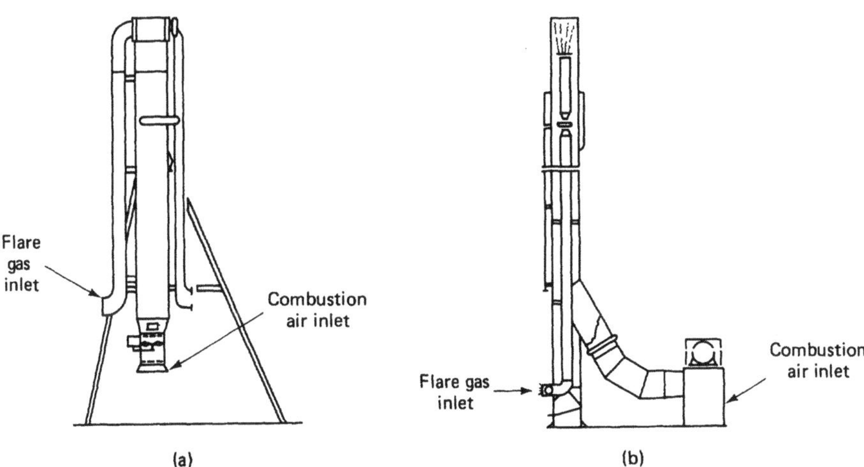

FIGURE 5-5 Two designs for forced-draft flare systems: (a) biaxial forced-draft unit; (b) coaxial forced-draft unit [3].

can be seen that depending on specific considerations (i.e., fuel cost, hours of operation a year, available space, etc.), it would be more economical to increase residence time and reduce operating temperature whenever possible.

However, there is a practical lower limit to the temperature reduction because oxidation rates are very strongly temperature-dependent. Figure 5-6 indicates schematically the general effects of temperature and residence time on oxidation rates in an incinerator. Over a narrow temperature range, the rate increases from essentially zero to rates measured in milliseconds or less. At high temperatures, complete conversion is controlled more by concentrations of pollutant and oxidant than by the temperature-dependent rate.

Table 5-1 shows typical ranges of residence times and operating temperatures for each of the major pollution-abatement categories for which thermal incineration is applicable. The range of conditions shown for each category is relatively wide. To some extent this range results from differences in oxidation rates of specific pollutants due to their physical or chemical characteristics. However, the major factor contributing to the width of the range of conditions is the variability of thermal incinerator designs in the effectiveness with which they mix fume with combustion products, and avoid bypassing or short-circuiting of any significant part of the fume stream around the hot reaction zone.

In general, the following steps are used to develop the design of a thermal incinerator:

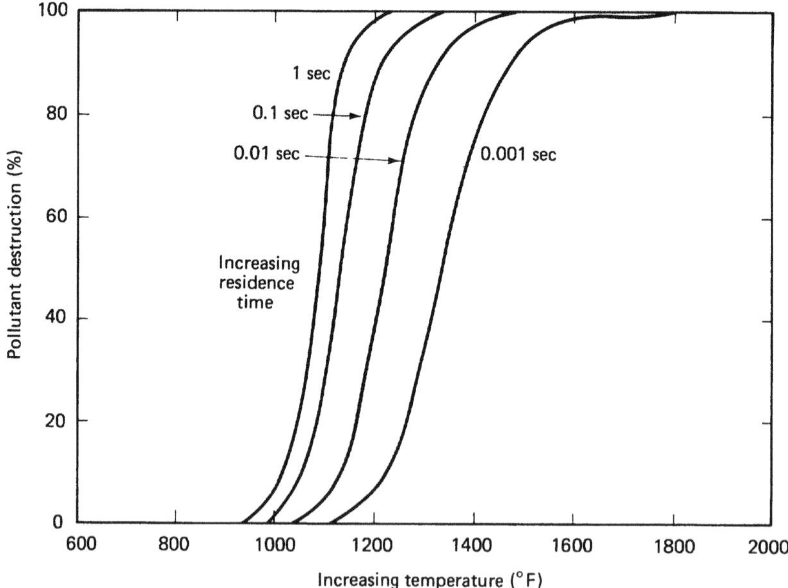

FIGURE 5-6 Coupled effects of temperature and time on rate of pollutant oxidation [2].

TABLE 5-1 Conditions required for satisfactory performance in various abatement applications [2].

Abatement Category	Afterburner Residence Time (sec)	Temperature (°F)
Hydrocarbon emissions (90% + destruction of HC)	0.3–0.5	1100–1250[1]
Hydrocarbons + CO (90% + destruction of HC + CO, as in LAAPCD rule 66)	0.3–0.5	1250–1500
Odor		
(50–90% destruction)	0.3–0.5	1000–1200
(90–99% destruction)	0.3–0.5	1100–1300
(99% + destruction)	0.3–0.5	1200–1500
Smokes and plumes		
White smoke (liquid mist)		
(Plume abatement)	0.3–0.5	800–1000[2]
(90% + destruction of HC + CO)	0.3–0.5	1250–1500
Black smoke (soot and combustible particulates)	0.7–1.0	1400–2000

[1] Temperatures of 1400–1500 °F may be required if the hydrocarbon has a significant content of any of the following: methane, cellosolve, substituted aromatics (e.g., toluene, xylenes).

[2] Operation for plume abatement only is not recommended, since this merely converts a visible hydrocarbon emission to an invisible one, and frequently creates a new odor problem due to partial oxidation in the afterburner.

Step 1: Calculate the combustion chamber volume:

$$V = \left(\frac{\text{acfm}}{60}\right)t$$

where V = combustion chamber volume (following burner and prior to heat exchangers, dilution, etc.), ft^3

   acfm = exhaust gas flow rate to incinerator (at average incinerator operating temperature), actual ft^3/min

   t = total average residence time, sec

Step 2: Calculate the fuel required to sustain combustion. Most industrial combustion calculations for fuel requirements include a two-step calculation.

1. The heat load (Btu/min) required to raise the temperature of the waste gas stream to the desired combustion temperature.

$$Q = mC_p\,\Delta T$$

where Q = heat load, Btu/min

m = mass flow rate of waste gas stream into incinerator, lb/min

ΔT = temperature rise required to bring waste gas stream from temperature upstream of incinerator to required combustion temperature, °F

C_p = average heat capacity of composite air and waste gas stream, Btu/lb-°F

2. The energy released (Btu/scf of fuel) and available for heating the gas stream to combustion temperature. This form is often referred to as the available heat for the fuel.

Dividing (1) by (2) yields the fuel ratio in scfm. This rough calculation provides a result that is often satisfactory for engineering (design) purposes.

Catalytic Incinerators

Variations and unknowns in many of the parameters in the describing equation(s) make it nearly impossible to provide a generalized procedure for the design of a catalytic incinerator.

Flares

The following information is recommended for the proper design of a flare:

1. Type of material to be flared
2. Average molecular weight, M
3. Percent unsaturation
4. Lower (net) heating values, (Btu/scf)
5. Specific heat ratio, $k = C_p/C_v$
6. Mass flow rate at maximum discharge, lb/hr
7. Average vapor temperature, °F
8. Flowing pressure, psig
9. Percent toxic, odorous, or noxious gases

Determine flare burner diameter   The parameters applicable to the calculation of a proper flare burner diameter include mass flow rate and discharge conditions, the type of flame holder used by the manufacturer, and density and heat-capacity ratios, which serve to define the sonic velocity in the fluid stream.

The pressure drop, rather than flow velocity, is usually the controlling factor in burner design. The flare burner is usually limited to a pressure drop of 3 psig

(60 in. water). When sizing the burner, the maximum discharge rate is converted to an equivalent volume of air in scfh. The following orifice equation is used to calculate the burner diameter [3]:

$$V_e = 1656KA \sqrt{\Delta P}$$

where V_e = volume equivalent flow rate, scfh

   K = orifice factor, dimensionless (normally 0.9)

   A = area of flare burner tip, ft^2

   ΔP = allowable pressure drop at tip, in. water

Maximum discharge velocity The sonic velocity is determined by assuming that the gas behaves as a perfect gas, and therefore the following equation applies [3]:

$$V_a = \frac{g_c KRT}{M}$$

where V_a = acoustic velocity, or velocity of sound in the fluid, ft/sec

   g_c = dimensional constant, 32.17 lb-ft/lb$_f$hr -sec^2

   K = ratio of specific heats, C_p/C_v

   R = gas constant, 1546 ft-lb$_f$/°R lb-mole

   M = average molecular weight

5.3 INSTALLATION PROCEDURES

Fume incinerators are generally shipped full assembled. They may, however, be shipped in two sections if too large to be shipped as one unit. Installation involves mounting the unit on a concrete slab, bolting the sections together if required, and connecting the incinerator's utility lines (electric, oil, gas, and air) to plant supply lines. Small or specially designed lightweight units may be roof-mounted. If necessary, additional structural support may be provided. As previously stated, normally the incinerator's fume inlet is connected to the discharge side of the fan, which exhausts the fume-generating equipment. The incinerator's combustion-chamber outlet is connected to a stack or to heat-recovery equipment preceding the stack.

   The concrete support pad should extend several feet beyond the incinerator on all sides, should be at least 8 in. thick, and should be properly bedded on a gravel drainage bed laid on firm, compact earth.

The incinerator discharge duct must be 100% tight. Discharge gases may be in the neighborhood of 1400°F and duct leakage could lead to fires. The fan and ducting to the incinerator should also be insulated to prevent loss of heat by the fumes. The fan must be firmly secured.

In areas where below-freezing temperatures are encountered, steam or electric tracer lines are required for air and fuel lines to prevent freezing of entrained water. Control valves must also be protected against freezing.

Installation is generally by a contractor engaged by the incinerator owner, with supervision being provided by a representative of the manufacturer.

The following are excerpts from safety startup instructions by one firm to its field personnel:

Before energizing the electrical control cabinet, make sure the lockout tags have been removed by the person who attached them. After the control cabinet is energized, but before pushing any start buttons, check to see if all persons are standing clear of the fan. Push the start button but do not allow the motor to gain full speed. Push the stop button and check for direction of travel. If the direction is correct, push the start button and allow the motor to reach full speed. Make a visual check of the fan for rubbing or other malfunctions. Do not attempt to run the motor with the belt guard removed.

The local gas company should be notified before opening the hand valve allowing gas into the gas valve train. Most gas companies reserve the option to visit the plant at the time of startup and supervise the opening of the valve and inspection of the gas valve train. Prior to attaching the gas to the gas train, it is possible to check the train for leaks with compressed air or nitrogen. Since the operating gas pressure will be relatively low, caution must be exercised to prevent rupture of diaphragm valves. High and low gas pressure switches and pressure reducing stations have diaphragms that are easily damaged by high pressures. These devices should be removed before air pressure is applied for test purposes. All leaks must be repaired, regardless of size, before attempting startup. Generally, one and one half times the operating pressure is the accepted pressure for testing; however, never test below 3 pounds (per square inch).

Installation of catalytic incinerators differs from installation of thermal incinerators only in that the catalytic bed must be inserted in its holder and temperature and/or pressure instruments relative to operating conditions preceding and following the catalyst be connected. The installation procedures presented for thermal incinerators and stacks (see Chapter 12) apply, with little or no modification, to flares.

5.4 OPERATION

There are three distinct categories of incinerator operation:

1. Automatic
2. Semiautomatic
3. Manual

The categories refer specifically to the method of startup and then flame supervision, and have been adopted by the various insurance agencies (such as Factory Mutual) and by NFPA (the National Fire Protection Association). Most incinerator burners and fuel systems are designed in accordance with one of these insurance agency's guidelines, whether or not insurance is actually obtained from them. Therefore, these publications should be consulted before attempting to operate the equipment.

Typically, the following procedure must be followed to start up any thermal incinerator. Each function, or group of functions, is interlocked electrically so that each must be accomplished before the next step can occur:

1. Prepurge
   a. Power on to control system.
   b. Fan motor on.
   c. Airflow verified.
   d. Fuel valves closed.
2. Purge
   a. Airflow to maximum rate.
   b. Purge timer starts (length of time set to allow furnace to be purged of any combustible fumes. Eght complete air changes are considered sufficient).
3. Ignitor sequence
   a. Airflow to light-off rate.
   b. Auxiliary fuel pressure verified as sufficient; temperature verified if required.
   c. Ignitor electric spark mechanism starts.
   d. Ignitor timer starts.
   e. Ignitor fuel valve opens, fuel is ignited in furnace by sparking mechanism.
   f. Ignitor flame sensed by flame scanner.
   g. Ignitor timer ends time cycle, shuts off sparking mechanism (if no flame sensed by flame scanner, ignitor fuel valve immediately closes, and postpurge cycle starts).

4. Main burner sequence
   a. Main burner fuel-control valve verified in light-off position.
   b. Atomizing fluid (air or steam) valve opens (needed only on certain oil-fired units).
   c. Burner timer starts cycle.
   d. Main burner fuel shut off valves open—allowing fuel to main burner.
   e. Main flame ignites from ignitor flame.
   f. Burner timer ends time cycle, ignitor fuel valve closes (if main flame not sensed by flame scanner, fuel shut off valves immediately close, and postpurge cycle starts).
   g. Interlocks verifying light-off positions on fan and control valves are now bypassed, allowing fuel and air rates to be adjusted.
   h. Main burner is now in service subject to the following conditions:
      (1) Fan on.
      (2) Airflow proven.
      (3) Flame detected.
      (4) Fuel pressure adequate.
      (5) Furnace temperature not high.

5. Normal operation
   a. Fuel and airflow are now adjusted to provide optimum furnace temperature for incinerator. This may be done manually or automatically. On refractory-lined units, warm-up must be done gradually to prevent thermal shock to the refractory.
   b. Once the incinerator reaches its operating temperature, fumes can be introduced into the furnace.
   c. Proper incineration operation is maintained by controlling incinerator outlet temperature and fume, auxiliary fuel, and airflow rates.

6. Normal shutdown
   a. Normal shutdown is accomplished by closing the fume flow valves, then the auxiliary fuel valves. The systems then proceed to the postpurge cycle.

7. Postpurge
   a. Fan adjusted to maximum flow rate.
   b. Postpurge timer starts time cycle.
   c. Keeps air flowing until five complete air changes have occurred in the furnace.
   d. Postpurge timer ends cycle, fan turned off, incinerator is shut down.

An automatic system requires only that an operator throw a switch to start operation. The control mechanism then starts the incinerator, including the fans and other related equipment, in a presequenced order; brings it up to temperature; and, maintains proper incinerator operation for as long as required. Shutdown is usually initiated by manually throwing another switch; the control system will then carry out all the operations necessary to shut down the incinerator.

A distinguishing feature of the "automatic" system is its ability to "recycle." This includes recycling to reattempt light-off if the previous attempt failed, and shutting down or starting up again as fume flow dictates. This is not a feature of the semiautomatic system.

Semiautomatic operation is the most common method. It involves considerably more human input than does the completely automatic operation. In this type of operation, the operator must perform certain operations when called for by the control system. This input is especially required during the startup sequence. Semiautomatic control is a broadly applied term and may refer to a system that requires the operator to manually operate valves or merely to push certain buttons. However, in all cases, proper sequencing is enforced by the control systems. Emergency shutdown, however, is completely automatic.

In the "manual" type of control system, all the actions required by the startup sequence are performed manually. Proper sequencing must be followed, but the control logic allows the operator sufficient time to accomplish his tasks. Again, emergency shutdown is automatic.

The length of time required to bring the incinerator from "cold" to its normal operating temperature varies with the type and thickness of the lining and with the actual operating temperature. Typically, the warm-up period may require several hours. For a thick lining (6 to 12 in. of refractory plus any insulating block), the manufacturer's recommendation is usually to start initially with an outlet temperature of 200°F and hold for 1 hour; thereafter, increase the outlet temperature at the rate of 100°F per hour until the outlet temperature is 600°F. Then the outlet temperature may be increased at the rate of 200 to 400°F per hour until the final operating temperature is reached. This process may take an entire shift and wastes considerable amounts of fuel; it is one of the principal disadvantages of using a refractory-lined furnace.

In actual practice, however, the manufacturer's recommended warm-up rate schedule is seldom followed, especially for relatively thin linings and for incinerators that operate only on an 8- to 16-hour shift. Warm-up periods of an hour or less are certainly not uncommon, regardless of lining thickness. The penalty for this is increased inspection and maintenance (patching with plastic) of the lining; usually twice a week is sufficient. Certainly, reduced refractory life can also be expected from such a procedure, but this must be balanced against the wasted fuel usage required for longer warm-up periods. Using short warm-up periods, one can expect to replace the lining approximately once a year compared with a 10- to 15-year interval using the recommended procedure. A shortened warm-up period should *never* be used for the initial curing period or if the refractory has been wetted or soaked with water. A quick warm-up period under these circumstances will result in the immediate loss of the lining.

No special cool-down procedure is usually followed. Usually, the incinerator and fans are simply turned off and allowed to cool naturally. Under these conditions, it will take from 8 to 24 hours for the refractory to reach ambient temperature; however, the lining will usually be cool enough for inspection after a 4- to 12-hour period. Leaving the fan on during the cool-down period usually does not cause a serious problem but should be avoided, as it does reduce refractory life. Water should *never* be sprayed on the refractory to cool it down. Although this may seem obvious, it nevertheless happens quite frequently if the maintenance people have not been properly trained.

Normal operation of a fume incinerator should be quite simple. A controller should be incorporated into the design to maintain the outlet temperature at a fixed value by varying the auxiliary fuel input. Combustion air (assuming limited air in the fume stream) is usually controlled by the average amount of fumes to be incinerated. This adjustment is normally set manually if the fume flow rate and fume heat content is fairly constant. If the fume heat content or flow rate is likely to be highly variable, a more sophisticated control system may be appropriate— one that analyzes combustion efficiency at the outlet as well as fume flow and outlet temperature, and varies both auxiliary fuel input and combustion air accordingly. But normal operation of even a complicated control system usually requires nothing more than an occasional monitoring of the instruments.

5.5 MAINTENANCE

Most fume incinerators are custom-designed within certain basic parameters. There-fore, they are likely to be accompanied by a very complete instruction manual which should include the manufacturer's basic maintenance instructions from all the subsuppliers. That is, the incinerator manufacturer will have bought equipment, such as the refractory, valves, and controls, from other suppliers. The operating and maintenance instructions for this equipment will be quite extensive and complete because it has been written by the original manufacturer, who is concerned only with particular items. The instruction manual is therefore a very useful document from a maintenance viewpoint and should be followed explicitly. (The instruction manual may not be so useful from a *system* operations viewpoint because fume incinerator systems are usually custom-designed and so *system* problems cannot always be anticipated, which often results in some field modifications to the op-erating instructions.)

There are, however, some general maintenance guidelines that can be discussed. The refractory should be inspected on a regular basis. Cracks that may develop, especially in brick joints, and thermal shock damage (spalling) should be repaired with a suitable plastic of the same thermal properties. The burner should be inspected at regular intervals for signs of warpage or corrosion. Moving parts should be lubricated with graphite or a similar high-temperature lubricant. Lubricants that carbonize under any circumstances should not be used. Also clear any dirt, mortar, carbon, or other foreign matter from the burner area. Inspect the pressure seals around any parts projecting through the burner or incinerator shell. Usually, these are asbestos rope packing glands and should be fairly tight after the adjusting/ retaining screws have been loosened. Lubricate these seals only with flake or powder graphite. This should never be mixed with oil, as the oil will carbonize. On burners using gas as an auxiliary fuel, the gas jets should be free of corrosion and should be cleared of any deposits.

The outer shell of the incinerator should be inspected, especially when new or when a new lining has been installed for signs of thermal shock. That is, welds, especially at the outlet, should be checked for hairline cracks, which are the first signs of poor thermal design.

The auxiliary fuel piping train should be inspected in accordance with the manufacturer's instructions. Electrically operated valves and interlock switches should be inspected frequently for conditions that might cause "shorting" (e.g., dirty contacts, moisture leaks, deteriorating insulation). Air-supply lines and filters (to air-operated valves) should be inspected for dirt or blockages. The valves themselves are usually provided with air-signal and air-supply pressure gauges, and these should be checked occasionally for accuracy.

If there are shut-off dampers in the ductwork to or from the incinerator, their seals should be checked frequently.

Maintenance procedures for catalytic incinerators should include catalyst cleaning every 3 months to a year. Cleaning is usually accomplished by blowing clean compressed air through the catalyst element, by vacuuming, or by washing with water or a mild detergent not containing phosphates. Iron oxide deposits can be removed by soaking with a mild organic acid followed by a water rinse.

5.6 IMPROVING OPERATION AND PERFORMANCE

Improving the performance of a thermal incinerator basically involves the optimization of fume combustion. Ideally, no more combustion air should be used than is required for complete combustion of the fumes and the auxiliary fuel. The auxiliary fuel should be used only in amounts required to maintain the design furnace temperature.

An incinerator operating efficiently should have only 1 to 2% O_2 and 0 to 1% combustibles in the outlet gases. Monitors are available which can indicate these parameters to the operator as well as provide automatic control of the incinerator when required.

5.7 CONCLUSIONS

Thermal incinerators, catalytic incinerators, and flares may each be used in appropriate situations to control smoke, odor, and combustible fumes. They are not particularly complicated devices, but their efficiency depends, to a large degree, upon proper operation and maintenance. Improper operation and maintenance can lead to unsafe combustion conditions and premature failure of the insulation or catalyst, which can significantly increase the system's lifetime costs. It is absolutely essential, therefore, that operation and maintenance personnel be skilled in their tasks and thoroughly familiar with a detailed instruction manual. The procedures given in the manual should be followed closely.

5.8 NOMENCLATURE

A	area, ft^2
acfm	flow rate, actual cubic feet per minute
C_p	heat capacity at constant pressure, Btu/lb-°F
C_v	heat capacity at constant volume, Btu/lb-°F
g_c	dimensional constant, 32.17 lb-ft/lb$_f$-sec^2
K	orifice factor, dimensionless
k	ratio of heat capacities, C_p/C_v
M	molecular weight
m	mass flow rate, lb/min
ΔP	pressure differential, in. water gauge
Q	heat flow rate, Btu/min
R	universal gas constant, 1546 ft-lb$_f$/°R lb-mole
ΔT	temperature differential, °F
t	residence time, sec
V	volume, ft^3
V_a	acoustic velocity, ft/sec
V_e	volume equivalent flow rate, scfm

5.9 REFERENCES

[1]   BUONICORE, A. J., AND THEODORE, L., *Industrial Control Equipment for Gaseous Pollutants*, Vol. 2. CRC Press, Inc., Boca Raton, Florida, 1975.

[2] ROLKE, R. W., ET AL., *Afterburner Systems Study.* EPA-R2-72-062, 1972.

[3] KLETT, M. G., AND GALESKI, J. B., *Flare Systems Study.* EPA-600/2-76-079, 1976.

[4] CROSS, F. L., AND HESKETH, H. E., *Handbook for the Operation and Maintenance of Air Pollution Control Equipment,* Chap. 5. Technomic Publishing Co., Inc., Westport, Conn., 1975.

6

Condensers

William F. Connery

Doyle & Roth Mfg. Co., Inc.
New York, New York

6.1 DESCRIPTION OF CONTROL DEVICE

Condensation can be accomplished by increasing pressure or decreasing temperature (removing heat). In practice, air pollution control condensers operate through extraction of heat. Condensers differ in the means of removing heat and the type of device used. The two different means of condensing are direct contact (or contact), where the cooling medium with vapors and condensate are intimately mixed and combined, and indirect (or surface), where the cooling medium and vapor/condensate are separated by a surface area of some type.

Contact condensers are simpler, less expensive to install, and require less auxiliary equipment and maintenance. The condensate/coolant from a contact condenser has a volume 10 to 20 times that of a surface condenser. This condensate cannot be reused and may pose a waste disposal problem unless the dilution of the pollutant is sufficient to meet regulatory requirements. Some typical contact condensers are shown in Figure 6-1.

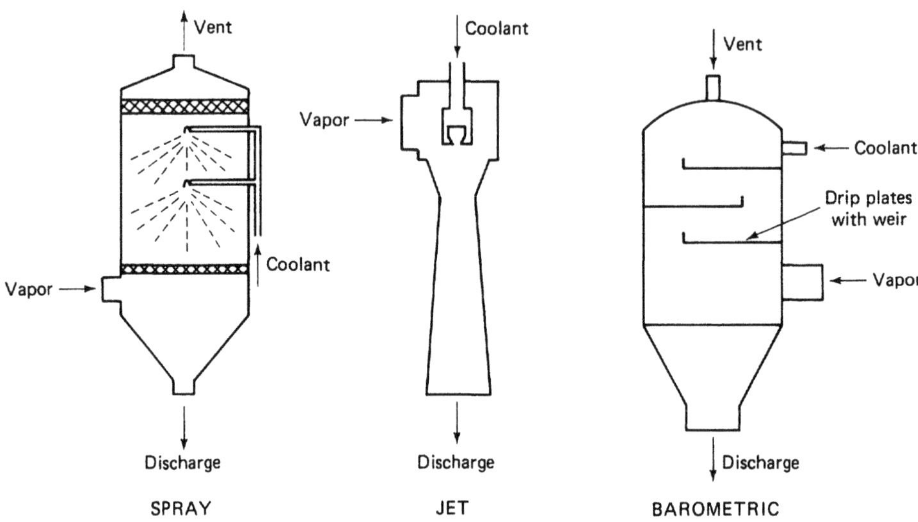

FIGURE 6-1 Contact condensers: (a) spray; (b) jet; (c) barometric.

Surface condensers form the bulk of the condensers used for air pollution control. Some of the applicable types of surface condensers are: shell and tube, double pipe, spiral plate, flat plate, air-cooled, and various extended surface tubular units. This chapter focuses on shell-and-tube condensers, because they are so widely used in industry and have been standardized by the Tubular Exchanger Manufacturers Association (TEMA).

Condensing can be accomplished in either the shell or tubes. The economics, maintenance, and operational ramifications of the allocation of fluids are extremely important, especially if extended surface tubing is being considered. The designer should be given as much latitude as possible in specifying the condenser. Details of TEMA designations for shell-and-tube heat exchangers are shown in Figure 6-2. Figure 6-3 shows some typical complete units with the component parts identified.

Air-cooled condensers consist of a rectangular bank of high finned tubes, a fan, a plenum for even distribution of air to the rectangular face of the tube bank, a header for vapor inlet and condensate outlet, and a steel supporting structure. Fins are usually aluminum 0.5 to 0.625 in. high applied to a bare tube by tension winding, or soldering, or cold-extruded from the tube itself. Bimetallic tubes can be used to provide process corrosion resistance on the inside and aluminum extruded fins on the outside. Headers for the tube side frequently contain removable gasketed plugs corresponding to each tube end and are used for access to the individual tube-to-tube sheet joint for maintenance.

Nearly all tubular exchangers employ roller expanding of the tube ends into tube holes drilled into tube sheets as a means of providing a leakproof seal. TEMA requires that tube holes be of close tolerance and contain two concentric grooves.

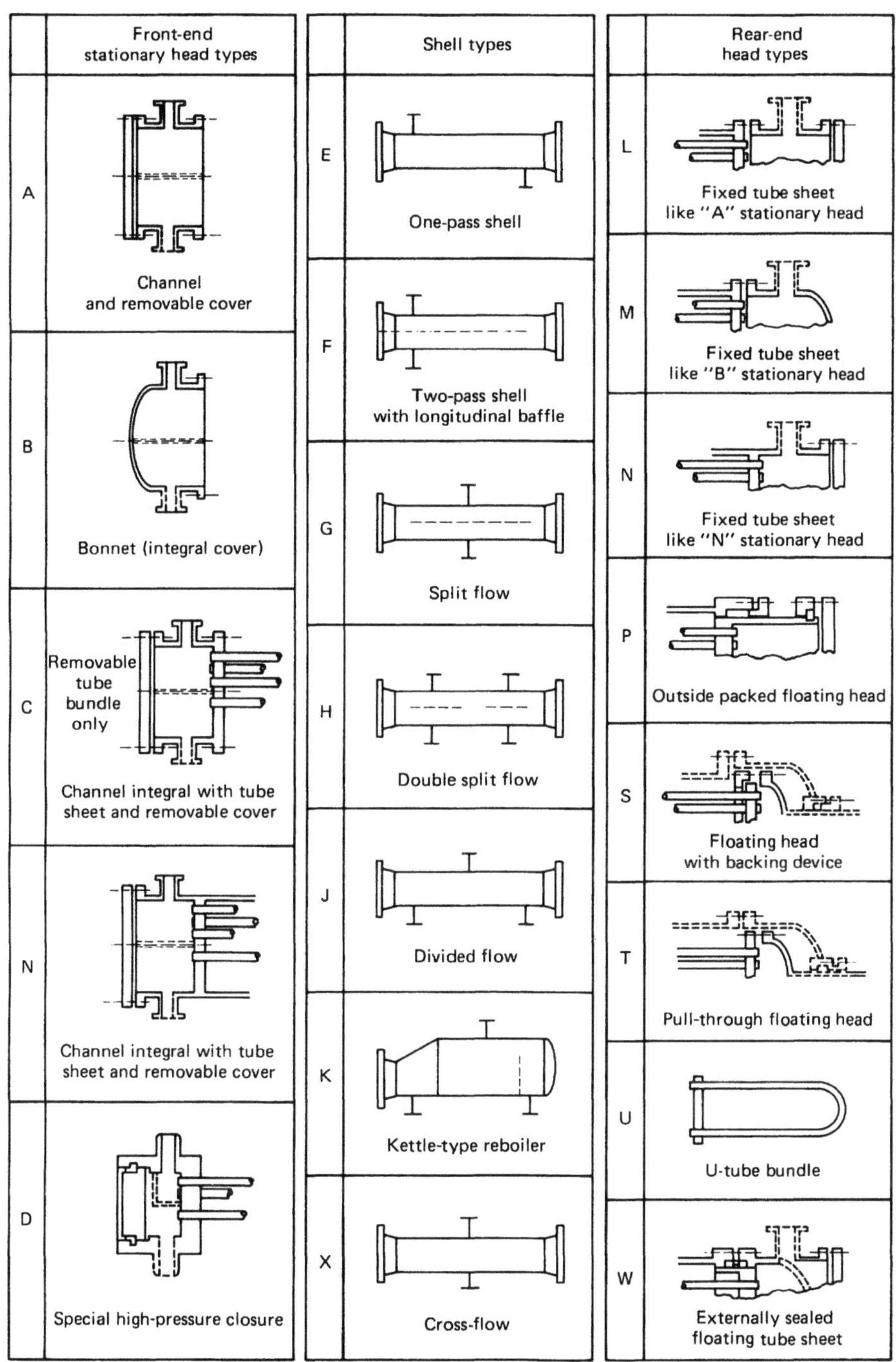

FIGURE 6-2 Shell-and-tube condensers. (© 1978 Tubular Exchange Manufacturers Association.)

These grooves give the tube joint greater strength but do not increase sealing ability and may actually decrease it. Research is currently being done to clarify what is the best procedure to follow for a given set of materials, pressures, temperatures, and loadings. Welding or soldering of tubes to tube sheets may be performed for additional leak-tightness or in lieu of roller expanding.

6.2 DESIGN PROCEDURES

Contact Condensers

Design of contact condensers involves calculating the quantity of coolant required to condense and subcool the vapor and proper sizing of discharge piping and hotwell. Calculation of coolant flow rate is as follows:

1. Stationary head—channel	20. Slip-on backing flange
2. Stationary head—bonnet	21. Floating head cover—external
3. Stationary head flange—channel or bonnet	22. Floating tubesheet skirt
4. Channel cover	23. Packing box
5. Stationary head nozzle	24. Packing
6. Stationary tubesheet	25. Packing gland
7. Tubes	26. Lantern ring
8. Shell	27. Tierods and spacers
9. Shell cover	28. Transverse baffles or support plates
10. Shell flange—stationary head end	29. Impingement plate
11. Shell flange—rear head end	30. Longitudinal baffle
12. Shell nozzle	31. Pass partition
13. Shell cover flange	32. Vent connection
14. Expansion joint	33. Drain connection
15. Floating tubesheet	34. Instrument connection
16. Floating head cover	35. Support saddle
17. Floating head flange	36. Lifting lug
18. Floating head backing device	37. Support bracket
19. Split shear ring	38. Weir
	39. Liquid level connection

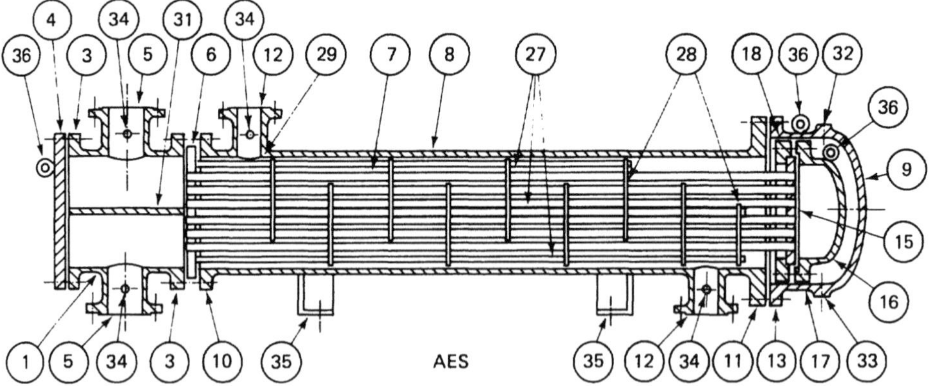

FIGURE 6-3 Nomenclature of heat-exchanger components: typical shell and tube. (© 1978 Tubular Exchange Manufacturers Association.)

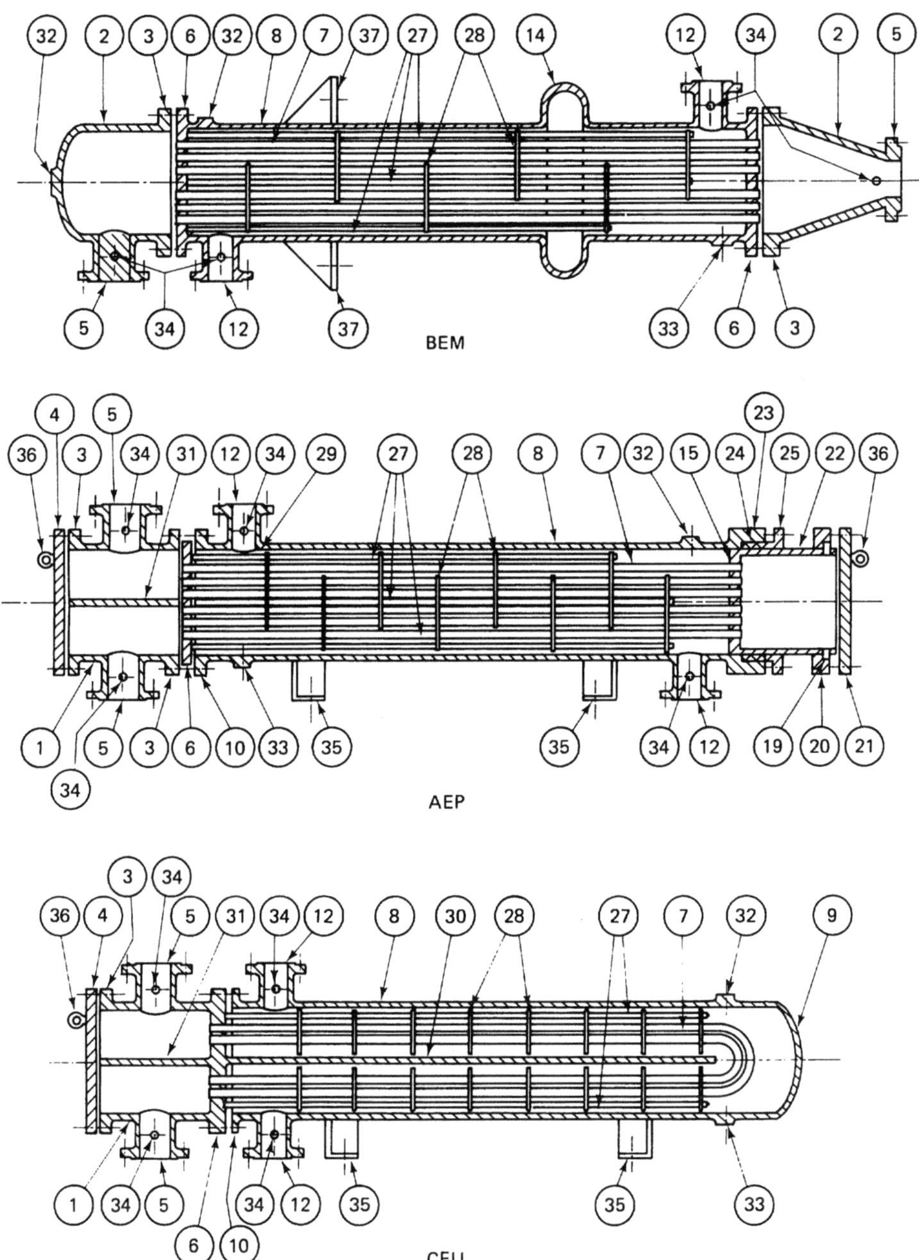

FIGURE 6-3 (con't.)

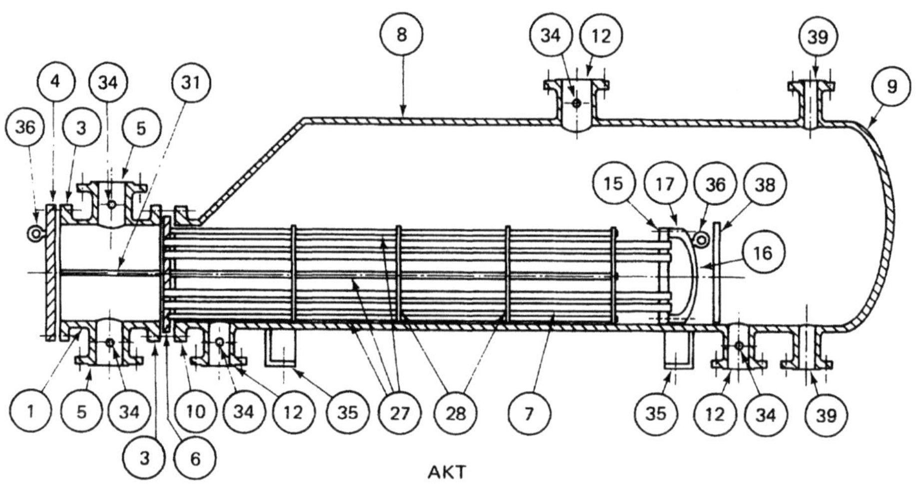

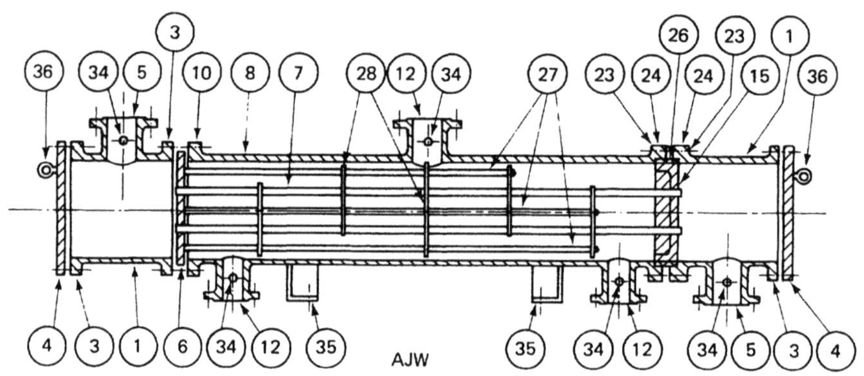

FIGURE 6-3 (con't.)

$$M_c = \frac{M_v H_v + M_1 C_{pl}(T_i - T_d)}{C_{pc}(T_d - t_i)}$$

where M_v, M_1, M_c = flow rate of vapor, liquid condensate, and coolant, respectively, lb/hr

C_{pl}, C_{pc} = heat capacity of condensate and coolant, Btu/lb °F

T_i, T_d = inlet temperature of vapor and discharge temperature, °F

t_i = inlet temperature of coolant, °F

H_i = latent enthalpy of vaporization of condensate, BTU/lb

Use of a contact condenser requires consideration of coolant availability, liquid waste disposal, or treatment facilities. Contact condensers are relatively efficient scrubbers as well as condensers and have the lowest equipment cost. Contaminants that are not ordinarily condensable or even particulates can be removed from the vapor stream. Reputable manufacturers of the various types of contact condensers should be consulted for sizing and layout recommendations.

Surface Condensers

The heat transfer for a surface condenser is governed by the following relationship:

$$Q = U_o A T_m \qquad \text{or} \qquad A = \frac{Q}{U_o T_m}$$

where Q =   total heat load, Btu/hr

U_o =   overall heat-transfer coefficient, Btu/hr °F ft^2

T_m =   mean temperature difference, °F

A =   surface area, ft^2

The first determination is the heat load. In the simplest case it is the latent heat to condense plus the subcooling load. If a condensing range exceeding 10 to 20°F exists, gas cooling or desuperheating and intermediate liquid subcooling must be accounted for. The next step is specifying the coolant flow rate and temperatures to balance the vapor heat load. Temperature crosses where the outlet coolant temperature is higher than the vent and condensate outlet temperature should be avoided. Temperature crosses can be performed in condensers, but they exclude various flow arrangements that could improve efficiency and cost.

After the terminal temperatures of the vapor and coolant have been determined, a mean temperature difference can be established. For isothermal or linear condensing and liquid subcooling the logarithmic mean temperature difference (LMTD) is applicable. The LMTD is calculated as follows:

$$\text{LMTD} = \frac{t_g - t_l}{\ell_n(t_g/t_l)}$$

where t_g =   greater temperature difference between the hot and cold streams

t_l =   lesser temperature difference between the hot and cold streams

ℓ_n =   natural log

Fluids are usually in counterflow, but in some cases, as with a fluid near its freezing point, parallel flow is used to afford a greater degree of safety from freezing the condenser. Frequently, there is a combination of parallel and counter flow as in a two-pass tube side, single-pass shell tubular condenser. The procedure is to apply a correction to the LMTD. The correction factors have been reduced to a series of graphs available in most heat-transfer texts [1]. Strictly, the correction factor applies only to the subcooling or desuperheating zones, but it is sometimes applied to condensers.

If condensation is nonlinear and has a range exceeding 20°F, the LMTD is not applicable and a condensing curve or calculations at several points over the temperature range is required to determine a correct T_m. A mixture of miscible vapors or a vapor and noncondensables are examples where the LMTD is not applicable.

The remaining unknown to calculate the required surface is the overall heat-transfer coefficient, U_o. The overall heat-transfer coefficient is the inverse of the sum of the resistances to heat transfer and is given by the following relationship:

$$U_o = \frac{1}{1/h_o + 1/h_{io} + r_w + f_o + f_{io}}$$

where h_o = outside or shell-side coefficient, Btu/hr °F ft^2

           h_{io} = inside or tube-side coefficient corrected to the outside area

           r_w = resistance of the surface area or tube wall, hr °F ft^2/Btu

           f_o = fouling resistance on outside or shell side of surface area

           f_{io} = fouling resistance on inside corrected to outside of surface area

Individual tube-side coefficients or resistances are corrected to the outside area. This results is an effective reduction in the individual heat-transfer coefficient. Finned tubing can have an outside-to-inside surface area ratio from 3 to as much as 25. Correcting the inside heat-transfer coefficient to the outside area is done as follows:

$$h_{io} = \frac{h_i A_i}{A_o}$$

where A_i and A_o are the inside and outside surface areas

It becomes apparent, therefore, that the use of fin tubes will be most beneficial when there is a major difference in the inside and shell or outside heat-transfer coefficients and the higher coefficient is inside the tubes.

Finned tubes present a resistance to heat transfer not applicable to bare tubes. It results from a temperature gradient in the fin itself. This resistance, due to conduction through the fin, is a function of the fin geometry and material, and is known as fin efficiency.

Determination of heat-transfer coefficients requires an evaluation of the vapor–liquid equilibrium; the simplest case is isothermal or moderate condensing range. The primary resistance to heat transfer is the liquid film, and the Nusselt theory and appropriate equations as described by Kern [1] are applicable. The general equation is as follows:

$$h \left(\frac{\mu_f^2}{k_f^3 \rho_f^2 g} \right)^{1/3} = 1.5 \left(\frac{4G'}{\mu_f} \right)^{-1/3}$$

where μ_f = viscosity of condensate film, lb/ft hr

   k_f = thermal conductivity of film, Btu/hr ft °F

   ρ_f = density of condensate film, lb/ft^3

   g = acceleration of gravity, ft/hr^2

   G' = condensate loading, lb/hr linear ft

There are various semiempirical modifications to this theoretical value to differentiate between horizontal, vertical, inside tubes, and outside tubes. The literature should be consulted for the actual loading to be used [1,2].

If there is noncondensable gas present or a vapor mixture of miscible and immiscible components with a large condensing range, diffusion of the condensing vapor through another vapor or a gas must be considered. The presence of a noncondensable gas in a vapor stream reduces the film coefficient; the amount of reduction is related to the size of the gas cooling load relative to the total load. The greater the ratio of sensible heat to total, the closer the heat-transfer coefficient approaches that of cooling the gas only. The calculation procedure is essentially that established by Colburn and Hougen [3]. It is a trial-and-error method performed for at least five points along the condensing curve. Following are some typical condensing coefficients:

Fluid	*h*
Steam	1500
Steam/10% gas	600
Steam/20% gas	400
Steam/40% gas	220
Pure light hydrocarbons	250
Mixed light hydrocarbons	175
Medium hydrocarbons	100
Medium hydrocarbons with steam	125
Pure organic solvents	250

The choice of a coolant will depend upon the particular plant and the efficiency required of the condenser. The most common coolant is the primary plant coolant, usually cooling tower or river water. It has been shown [4] that the vapor outlet temperature is critical to the efficiency of a condenser. In some instances, use of a chilled brine or a boiling refrigerant can achieve collection efficiency that will be sufficient without additional control devices. It is not unusual to specify multiple-stage condensing, where condensers, usually two, connected in series use cooling mediums with successively lower temperatures. For example, a condenser using cooling-tower water can be used prior to a unit using chilled water or brine—thereby achieving maximum efficiency while minimizing use of chilled water.

Careful consideration should be given to the fouling resistances f_o and f_{io} applied to the condenser design. TEMA lists some typical values to be used for various streams. These values are guidelines only and attention should be given to the economics of lower and higher values. The purpose of adding a fouling resistance is to provide consistent performance without excessive cleaning. Theoretically, increased fouling factors will provide greater safety allowance. In practice, however, they can actually decrease the safety allowance. Consider, for example, a typical application where a maximum pressure drop and overall length is specified. The most efficient design will use the length and pressure drop to the limit specified. If the fouling resistance specified is so large that this is not possible, velocities of the respective streams will be lower than desirable and fouling will be greater than in a design using a lower value of fouling resistance. Fouling resistances used indiscriminately can actually be "self-fulfilling." The most effective design feature to minimize fouling is high velocity and efficient use of pressure drop.

6.3 INSTALLATION PROCEDURES

Preparation of a condenser or heat exchanger for installation begins upon receipt of the unit from manufacturer. Condensers are shipped domestically using skids for complete units and boxes or crates for bare tube bundles. Units are normally removed from trucks using a crane or forklift. Lifting devices should be attached to lugs provided for that purpose, (i.e., for lifting of the complete unit as opposed to individual parts) or used with slings wrapped around the main shell. Shell supports are acceptable lugs for lifting, provided that the complete set of supports are used together; never use nozzles for attachment of lifting cables. Upon receipt of the unit, the general condition should be noted to determine any damage sustained during transit. Any dents, cracks, or connections out-of-square should be reported to the manufacturer prior to attempting to install the unit. Flanged connections, are blanked with plywood, Masonite, or equivalent covers and threaded connections are blanked with suitable pipe plugs. These closures are to avoid entry of debris into the unit during shipping and handling and should remain in place until actual piping connections are made.

Clearance Provision

Sufficient clearance is required for at least inspection of unit or in-place maintenance. Inspection of condensers requires minimal clearances for the following: access to inspection parts if provided, removal of channel or bonnet covers, and inspection of tube sheets and tube-to-tube sheet joints. If removal of tubes or tube bundles in place is anticipated, provision should be provided in equipment layout. Actual clearance requirements can be determined from the condenser setting plan.

Foundations

Condensers must be supported on structures of sufficient rigidity to avoid imposing excessive strains on nozzles due to settling. Horizontal units with saddle-type shell supports are normally supplied with slotted holes in one support to allow for expansion. Foundation bolts in these supports should be loose enough to allow movement.

Leveling

Condensers should be carefully leveled and squared to ensure proper drainage, venting, and alignment with piping. On occasion, condensers are purposely angled to facilitate venting and drainage, and alignment with piping becomes the prime concern.

Piping Considerations

The following guidelines for piping are necessary to avoid excessive strains, mechanical vibration, and access for regular inspection.

1. Sufficient support devices are required to prevent the weight of piping and fittings from being imposed on the condenser.
2. Piping should have sufficient expansion joints or bends to minimize expansion stresses.
3. Avoid forcing alignment of piping so that residual strains will not be imposed on nozzles.
4. If external forces and moments are unavoidable, their magnitude should be determined and made known to the manufacturer so that necessary stress analysis can be performed.
5. Provide surge drums or sufficient length of piping to condenser to minimize pulsations and mechanical vibrations.
6. Valves and bypasses should be provided to permit inspection or maintenance in order to isolate the condenser during periods other than complete system shutdown.
7. Plugged drains and vents are provided at low and high points of shell-tube sides not otherwise drained or vented by nozzles. These connections are functional during startup, operation, and shutdown, and should be piped up for either continuous or periodic use and never left plugged.
8. Instrument connections are provided either on the condenser nozzles or in the piping close to the condenser. Pressure and temperature indicators

should be installed to validate the initial performance of the condenser as well as to demonstrate the need for inspection or maintenance.

6.4 OPERATION

Safe Working Conditions

The maximum allowable working pressures and temperatures are indicated on the condenser's nameplate. These values must not be exceeded. Special precautions should be taken if any individual part of the condenser is designed for a maximum temperature lower than the condenser as a whole. The most common example is some copper-alloy tubing with a maximum allowable temperature lower than the actual inlet gas temperature. This is done to compensate for the low strength levels of some brasses or other copper alloys at elevated temperatures. An adequate flow of the cooling medium must be maintained at all times.

Condensers are designed for a particular fluid throughput. Generally, a reasonable overload can be tolerated without causing damage. If operated at excessive flow rates, erosion or destructive vibration could result. Erosion could occur at normally acceptable flow rates if other conditions, such as entrained liquids or particulates in a gas stream or abrasive solids in a liquid stream, are present. Evidence of erosion should be investigated to determine the cause. Vibration can be propagated by other than flow overloads (e.g., improper design, fluid maldistribution, or corrosion/erosion of internal flow-directing devices such as baffles). Considerable study and research has been applied in recent years to develop a reliable vibration analysis procedure to predict or correct damaging vibration. At this point, the developed correlations are considered "state of the art," yet most manufacturers have the capability of applying some type of vibration check when designing a condenser. Vibration can produce severe mechanical damage, and operation should not be continued when an audible vibration disturbance is evident.

Startup

Condensers should be warmed up slowly and uniformly; the higher the temperature ranges, the slower the warm-up should be. This is generally accomplished by introducing the coolant and bringing the flow rate to design level and gradually adding the vapor. For fixed-tube-sheet units with different shell-and-tube material, consideration should be given to differential expansion of shell and tubes. As fluids are added, the respective areas should be vented to ensure complete distribution. A procedure other than this could cause large differences in temperature between adjacent parts of the condenser and result in leaks or other damage. It is recommended that gasketed joints be inspected after continuous full-flow operation has been established. Handling, temperature fluctuations, and yielding of gaskets or bolting may necessitate retightening of the bolting.

Shut Down

Cooling down is generally accomplished by shutting off the vapor stream first and then the cooling stream. Again, fixed-tube-sheet condensers require consideration of differential expansion of the shell and tubes. Condensers containing flammable, corrosive, or high-freezing-point fluids should be thoroughly drained for prolonged outages.

6.5 MAINTENANCE

Inspection

Recommended maintenance of condensers require regular inspection to ensure mechanical soundness of the unit and a level of performance consistent with the original design criteria. A brief general inspection should be performed on a regular basis while the unit is operating. Vibratory disturbance, leaking gasketed joint, excessive pressure drop, decreased efficiency indicated by higher gas outlet temperature or lower condensate rates, and intermixing of fluids are all signs that a thorough inspection and maintenance procedures are required.

Complete inspection requires shutdown of the condenser for access to internals and pressure testing and cleaning. Scheduling can only be determined from experience and general inspections. Tube internals and exteriors, where accessible, should be visually inspected for fouling, corrosion, or damage. The nature of any metal deterioration should be investigated to determine properly the anticipated life of equipment or possible corrective action. Possible causes of deterioration include general corrosion, intergranular corrosion, stress cracking, galvanic corrosion, impingement, or erosion attack.

Cleaning

Fouling of condensers is the deposition of foreign material on the interior or exterior of tubes. Evidence of fouling in operation is increased pressure drop and general decreasing performance. Fouling can be so severe that tubes are completely plugged, resulting in thermal stresses and subsequent mechanical damage of equipment.

The nature of the deposited foul determines the method of cleaning. Soft deposits can be removed by steam, hot water, various chemical solvents, or brushing. Plant experience can determine which method to use. Chemical cleaning should be performed by contractors specialized in the field who will consider the deposit to be removed and the materials of construction. If the cleaning method involves elevated temperature, consideration should be given to thermal stresses induced in the tubes; steaming-out individual tubes can loosen the tube-to-tube sheet joints.

Mechanical methods of cleaning are useful for soft and hard deposits. There are numerous tools for cleaning tube interiors: brushes, scrapers, and various rotating cutter-type devices. The condenser manufacturer or suppliers of tube tools can be consulted in the selection of the correct tool for the particular deposit. When cutting or scraping deposits, care should be exercised to avoid damaging tubes.

Cleaning of tube exteriors is generally performed using chemicals, steam, or other suitable fluids. Mechanical cleaning is performed but requires that the tubes be exposed, as in a typical air-cooled condenser, or capable of being exposed, as in a removable bundle shell-and-tube condenser. The layout pattern of the tubes must provide sufficient intersecting empty lanes between the tubes, as in square pitch. Mechanical cleaning of tube bundles, if necessary, requires the utmost care to avoid damaging tubes or fins.

Testing

Proper maintenance requires testing of a condenser to check the integrity of the following: tubes, tube-to-tube sheet joint, welds, and gasketed joints. The normal procedure consists of pressuring of the shell with water or air at the nameplate-specified test pressure and viewing the shell welds and the face of the tube sheet for leaks in the tube sheet joints or tubes. Water should be at ambient temperature to avoid false indications due to condensation. Pneumatic testing requires extra care because of the destructive nature of a rupture or explosion or fire hazards when residual flammable materials are present. Condensers of the straight-tube floating-head construction require a test gland to perform the test. Tube bundles without shells are tested by pressuring the tubes and viewing the length of the tubes and back face of the tube sheets.

Corrective action for leaking tube-to-tube sheet joints requires expanding the tube end with a suitable roller-type tube expander. Good practice calls for an approximate 8% reduction in wall thickness after metal-to-metal contact between tube and tube hole. Tube expanding should not extend beyond ⅛ in. of the inner tube-sheet face to avoid cutting the tube. Care should be exercised to avoid over-rolling the tube, which can cause work hardening of the material, an insecure seal, and/or stress-corrosion cracking of the tube.

Defective tubes can either be replaced or plugged. Replacing tubes requires special tools and equipment. The user should contact the manufacturer or a contractor qualified in repair. Plugging of tubes, although a temporary solution, is acceptable provided that the percentage of the total number of tubes per tube pass to be plugged is not excessive. The type of plug to be used is a tapered one-piece or two-piece metal plug suitable for the tube material and inside diameter. Care should be exercised in seating plugs to avoid damaging the tube sheets. If a significant number of tube or tube joint failures are clustered in a given area of the tube layout, their location should be noted and reported to the manufacturer. A concentration of failures is usually caused by other than corrosion (e.g., impingement, erosion, or vibration).

6.6 IMPROVING OPERATION AND PERFORMANCE

Within the constraints of the existing system, improving operation and performance refers to maintaining operation and original or consistent performance. There are several factors previously mentioned which are critical to the design and performance of a condenser: operating pressure, amount of noncondensable gases in the vapor stream, coolant temperature and flow rate, fouling resistance, and mechanical soundness. Any pressure drop in the vapor line upstream of the condenser should be minimized. Deaerators or similar devices should be operational where necessary to remove gases in solution with liquids. Proper and regular venting of equipment and leakproof gasketed joints in vacuum systems are all necessary to prevent gas binding and alteration of condensing equilibrium. Coolant flow rate and temperatures should be checked regularly to ensure that they are in accordance with the original design criteria. The importance of this can be illustrated merely by comparing the winter and summer performance of a condenser using cooling-tower or river water. Decreased performance due to fouling will generally be exhibited by a gradual decrease in efficiency and should be corrected as soon as possible. Mechanical malfunctions can also be gradual, but will eventually be evidenced by near total lack of performance.

Fouling and mechanical soundness can only be controlled by regular and complete maintenance. In some cases, fouling is much worse than predicted and requires frequent cleaning regardless of the precautions taken in the original design. These cases require special designs to alleviate the problems associated with fouling. A leading PVC manufacturer found that carryover of polymer reduced the efficiency of its monomer condenser and caused frequent downtime. The solution was providing polished internals and high condensate loading in a vertical downflow shell-and-tube condenser. A major pharmaceutical intermediate manufacturer had catalyst carryover to a vertical downflow shell-and-tube condenser which accumulated on the tube internals. The solution was to recirculate condensate to the top of the unit and spray it over the tube-sheet face to create a film descending down the tubes to rinse the tubes clean.

Most condenser manufacturers will provide designs for alternate conditions as a guide to estimating the cost of improving efficiency via alternate coolant flow rates and temperatures as well as alternate configurations (i.e., vertical, horizontal, shell side or tube side).

6.7 CONCLUSIONS

Condensation, which has long been a valuable operation for the chemical, petrochemical, and related industries, is now, with the advent of strict environmental concern, an important air pollution control device. Its contemporary application consists of increasing efficiency of replacement equipment via more rigorous computerized design and consideration of vent streams, operation of condensers for streams previously vented to the atmosphere, and use of condensers as a precleaner for other air pollution control equipment.

More often than not, a vapor condenser is a secondary control device since it cannot achieve the collection efficiency required by industry using ordinarily available and economical cooling mediums. One advantage of condensation is that it can reduce the load, improve the performance, and extend the operating life of more expensive devices. Surface condensers frequently return product or an intermediate to be recycled into the process. Contact condensers can return a condensate of sufficient dilution to be directly disposed of.

Virtually all heat exchangers which for decades have been used for process condensing—shell and tube, double pipe, air-cooled, flat plate, spiral plate, barometric, jet, and spray—are applicable for air pollution control.

6.8 NOMENCLATURE

A	surface area, ft^2
A_i	surface area on inside of tube, ft^2
A_o	surface area on outside of tube, ft^2
C_{pl}	heat capacity of condensate, Btu/lb °F
C_{pc}	heat capacity of coolant, Btu/lb °F
f_{io}	fouling resistance on inside of tube corrected to outside, hr °F ft^2/Btu
f_o	fouling resistance on outside of tube, hr °F ft^2/Btu
g	acceleration of gravity, ft/hr^2
G'	condensate loading, lb/hr linear ft
h	individual heat-transfer coefficient, Btu/hr °F ft^2
h_i	tube-side heat-transfer coefficient
h_{io}	tube-side heat-transfer coefficient corrected to outside
h_o	shell-side heat-transfer coefficient
H_v	latent enthalpy of vaporization of condensate, Btu/lb
k_f	thermal conductivity of condensate film, Btu/hr °F ft
$LMTD$	logarithmic mean temperature difference
M_c	flow rate of coolant, lb/hr
M_l	flow rate of condensate, lb/hr
M_v	flow rate of vapor, lb/hr
Q	total heat load, Btu/hr
r_w	resistance of the surface area or tube wall, hr °F ft^2/BTU
t_i	inlet temperature of coolant, °F

t_g greater temperature difference, °F

t_i lesser temperature difference, °F

T_i inlet temperature of vapor, °F

T_d discharge temperature, °F

T_m mean temperature difference, °F

U_o overall heat-transfer coefficient, Btu/hr °F ft^2

ρ_f density of condensate film, lb/ft^3

μ_g viscosity of condensate film, lb/ft hr

6.9 REFERENCES

[1] KERN, D. Q., *Process Heat Transfer*. McGraw-Hill Book Company, New York, 1950.

[2] PERRY, J. H., AND CHILTON, T. H., *Chemical Engineers' Handbook*, 5th ed. McGraw-Hill Book Company, New York, 1950.

[3] COLBURN, A. P., AND HOUGEN, O. A., *Ind. Eng. Chem.*, 26(11), 1178–1182, 1934.

[4] CONNERY, W. F., et al., "Energy and the Environment," Proceedings of the Third National Conference, AIChE, 1975, pp. 276–282.

7

Mechanical Collectors

Herbert G. Kaplan

Research-Cottrell, Inc.
Somerville, N.J.

7.1 DESCRIPTION OF CONTROL DEVICE

A mechanical collector is a device that separates suspended particles from a gas by causing the gas stream to change direction while the particles, because of their inertia, tend to continue in their original direction and be separated from the gas.

The collection efficiency of mechanical collectors depends upon the motion of the gas stream and the physical properties of the suspended particulate. Separation forces are a function of the gas velocity, its rate of change in direction, and of particle size and specific gravity. The forces acting on a particle can be as high as 2500 times gravity.

Operation of mechanical collectors depends on separating forces produced by the gas stream motion. The forces developed are primarily a function of the gas stream's speed and rate of curvature, and of particle size and specific gravity. These forces acting on a particle to carry it into a collecting zone can be made as high as 2500 times gravity. The collection efficiency is a function of the separating forces.

Although mechanical collectors have been used for many years to collect suspended materials, a satisfactory theory relating the many operating variables has never been universally accepted. The performance of collectors is in some cases almost entirely based on experience and empirical data, probably because of the complexity of the aerodynamics involved in collector flow.

The relatively low capital and operating costs for mechanical collectors make them an attractive choice for particulate removal where the collection efficiency requirements are not high. These collectors can be gravity settling chambers, dynamic precipitators, or dry inertia separators.

Gravity Settling Chamber

The gravity-type collection device (Figure 7-1) depends solely on the gravitational pull on the particles to collect the dust. The separator consists of a housing and hopper in which the velocity of the gas stream is made to drop rapidly below the transport velocity of the dust particle. The velocity reduction is accomplished by the sudden expansion into the enlarged housing.

The simple settling chamber consists of a single compartment in which the dust settles. One of the factors affecting the selection of gravity separators is that the velocity of the gas stream should be kept as low as possible since the settling rate of the dust decreases with increased velocity and turbulence. Since it is impractical to keep the velocity in the laminar flow region, the velocity should be maintained at approximately 60 ft/min. Higher values up to 600 ft/min. are used, but with corresponding drops in efficiency.

Provided that a good velocity distribution is attained, the efficiency of a settling chamber is based on the following equation:

$$E_{ff} = \frac{100 U_t L}{HV} \tag{7-1}$$

where E_{ff} = efficiency, weight percentage of particles of settling velocity U_t

U_t = settling velocity of dust, ft/sec

L = chamber length, ft

H = chamber height, ft

V = gas velocity, ft/sec

Combining equation (7-1) with Stokes' law, the minimum particle size that can be completely removed is calculated by

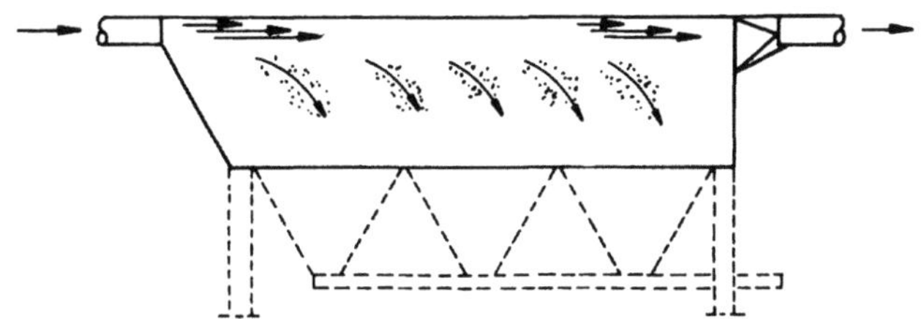

FIGURE 7-1 Gravity settling chamber.

$$D_p = \sqrt{\frac{18\mu HV}{gL(\rho_p - \rho)}} \tag{7-2}$$

where D_p = minimum particle size that can be completely removed, feet (particle diameter, actual or aerodynamic equivalent)

μ = gas viscosity, lb/ft/sec

H = chamber height, ft

V = gas velocity, ft/sec

g = acceleration due to gravity, 32.2 ft/(sec)^2

L = chamber length, ft

ρ_p = particle density, lb/ft^3

ρ = gas density, lb/ft^3

Equation (7-2) is only theoretical, and a sharp particle-size division cannot be obtained due to the turbulence, eddy currents, and agglomeration during settling and reentrainment.

The gravity settling chamber is usually of simple design and can be manufactured of almost any material, assuming that it is not affected by temperature. Maintenance is minimal, provided that corrosion control is maintained. The hoppers of a settling chamber should be kept empty. With a high inlet dust burden, the chamber's hoppers could fill up quite quickly and should be emptied frequently to eliminate the possibility of reentrainment.

Applications for gravity settling chambers are limited due to the large space requirements and relatively low efficiency. Gravity settling chambers have been applied on ore-handling systems, on power heating plants, and in various applications by the food industry.

Dynamic Precipitator

The dynamic precipitator in most cases is a combined fan and dust collector and is normally supplied with a hopper that also acts as a support base (Figure 7-2).

The precipitator is similar to a pressure blower, but contains two air passages: the primary air circuit (a), which passes through the scroll and emerges a clean air; and the secondary air circuit (b), containing the collected dust that passes into the dust chamber, which is connected to the dust hopper in the base of the precipitator. The rotation of the hyperbolically shaped blades creates the necessary force to draw in dust-laden air and causes it to turn in an arc of almost 180° before entering the primary air passage. The heavier dust particles are impinged upon the impeller disk (c), while the lighter particles are collected on the surfaces of the blades. Both light and heavy particles move outward by centrifugal force and converge on the outer edge of the impeller. At a point opposite the outer edge is a narrow angular opening through which the dust escapes. The blade tips extend into the dust chamber, creating a secondary air circuit that conveys the collected material through a port in the base and to the hopper below. The airtight hopper serves as a settling chamber in which the reduced air velocity permits the entrained dust to settle. The small amount of secondary air then returns to the dust chamber through the return air port. The dust concentration in the secondary circuit will automatically adjust itself so that the amount of dust settling in the hopper equals the amount of dust removed from the airstream.

The dynamic precipitator has the typical performance characteristics of a for-

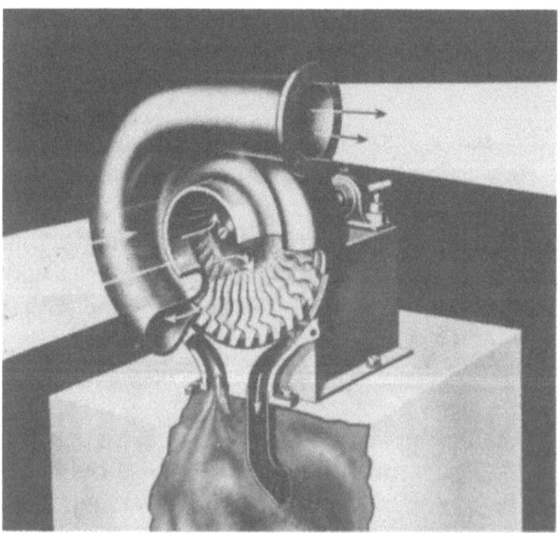
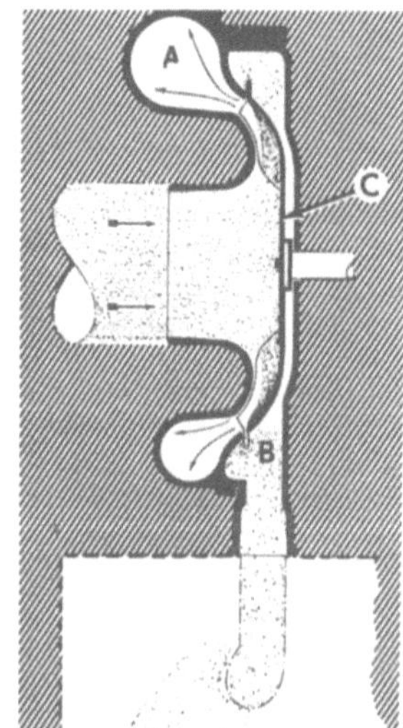

FIGURE 7-2 Typical dynamic precipitator.

ward-curve fan and thus has no pressure loss. The only loss experienced with this type of collector is mechanical inefficiency.

The dynamic precipitator collection efficiency is not appreciably affected by changes in air volume or operating speeds and remains essentially constant over the normal pressure–volume range. If applied in the recommended operating range, the efficiency will range from 40 to 50%.

The dynamic precipitator is well suited for dry dust particles 10 μm and above. It has been applied in connection with exhaust systems for grinding, cutting, broaching, and polishing in the metalworking and woodworking industries. The precipitator is usually not used where the dust is extremely fine, fibrous, or sticky. Extremely large particles should be avoided since they will bounce into the clean air chamber and bulky dense materials will float through to the clean air chamber.

Inertial Separators

There are numerous types of inertial separators, including baffle-type separators, chip trap separators, louvered-type separators, and cyclones.

The inertial separators depend on the deposition of particulates in a discontinuous fashion, as the gas passes around or through specially shaped baffles or apertures. Some inertial separators operate in the same range of efficiency as a cyclone. The simple baffle-type inertial separator collects only very large particles.

The only difference between a baffle-type separator (Figure 7-3) and the settling gravity type is the baffle, which imparts an abrupt change of direction for the dust-laden air. The particles are dropped from the airstream by their inertia.

A typical chip trap, Figure 7-4 shows the operation of the baffle-type collector. This type of collector relies on an abrupt change of direction in the airstream to drop out the particulate. This change of direction is caused by fixed baffle plates within the unit.

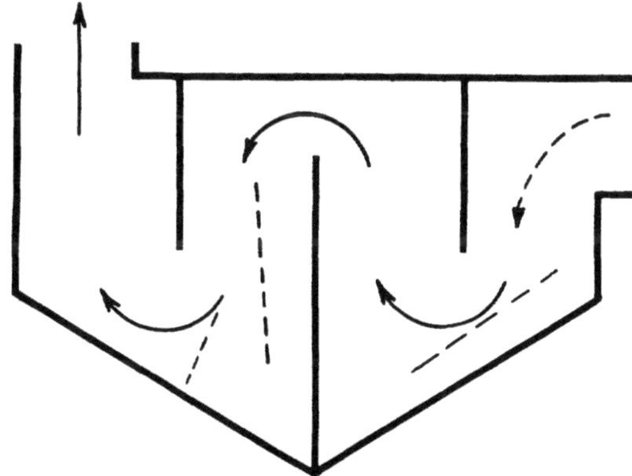

FIGURE 7-3 Multiple baffle chamber.

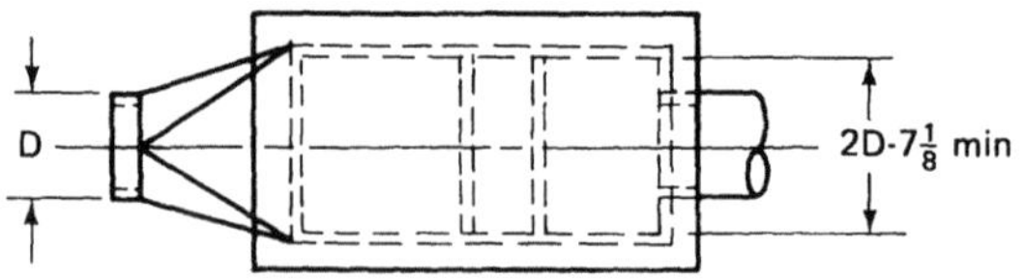

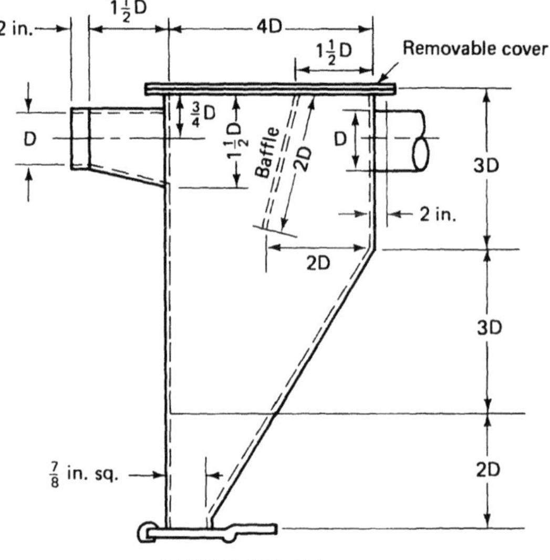

FIGURE 7-4 Chip trap.

The baffle-type collector is used primarily as roughing collectors or as pre-cleaners to more efficient collectors. It can be used on the same applications as the settling chamber. Since velocities are low, the space requirements are large.

The operating principle of a dust louver type dust separator can easily be understood by examining a cross-sectional view (Figure 7-5). The air enters through the total air inlet and as the air passes the louver, the direction of flow is reversed. About 90% of the air is drawn off on the clean-air side, with the remaining 10% being drawn off with the dust in the secondary air or dust circuit. The dust is concentrated in a relatively small air volume and can be discharged into the atmosphere, settled into a hopper or into a more efficient secondary collector.

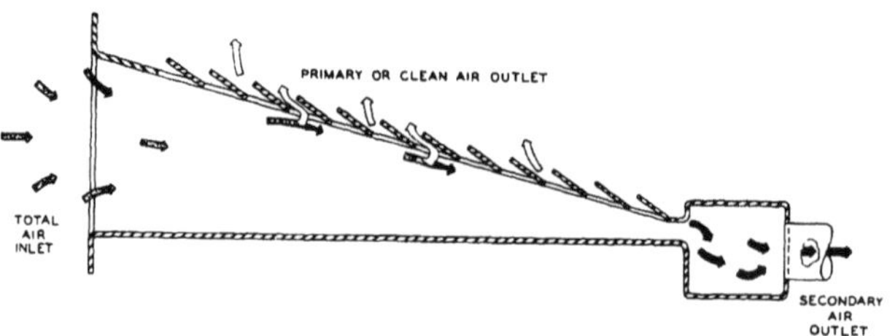

FIGURE 7-5 Cross-sectional view of dust louver.

FIGURE 7-6 V-pocket dust louver.

The design of louvered separators is governed by angle of the element, shape of the blade, pitch of louvered blade, gauge and type of blade material, ratio of total inlet air to secondary air outlet, and percentage of secondary air volume. The efficiency is a function of louver spacing, with closer spacings yielding higher efficiencies. The louver separator is available in several basic designs, as shown in Figures 7-6 and 7-7 [1].

Cyclones

The most prevalent type of all mechanical collectors, especially the dry-inertial type of collector, is the cyclone. Although this type of inertial collector will rarely meet air pollution control codes when used as a primary cyclone collector, the dust collector more than justifies its popularity if applied correctly. It can increase the

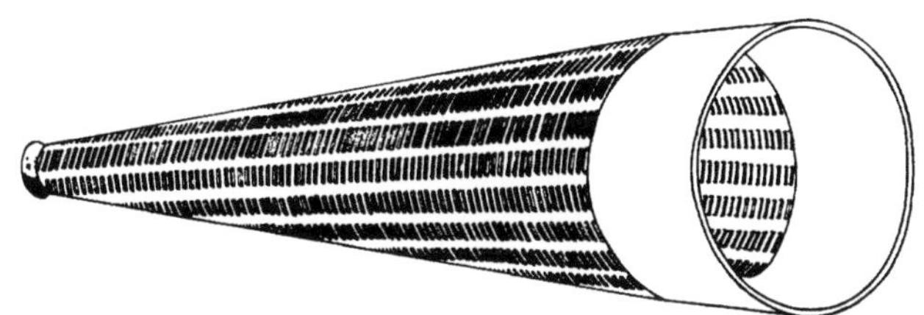

FIGURE 7-7 Conical pocket dust louver.

reliability of a secondary collector by removing excessive dust loadings from the inlet of scrubbers or electrostatic precipitators. It will act as a spark arrestor when installed upstream from a fabric filter. The cyclone can be designed in numerous arrangements and can be fabricated out of a variety of materials.

7.2   DESIGN PROCEDURES

Structurally, the design of a cyclone must have an axial gas outlet, a dust discharge, and an inertial forces for gas inlet that will produce the gas rotation necessary to create the dust separation. These three elements can be combined in a number of ways.

Cyclones are usually classified in four categories (Figure 7-8), depending on how the gas stream enters and leaves the unit: (1) by tangential inlet and axial dust discharge, (2) tangential inlet and peripheral dust discharge, (3) axial inlet and axial dust discharge, and (4) axial inlet and peripheral dust discharge. The types appearing in Figure 7-8a and 7-8c are most widely used. Large-diameter cyclones with body diameters three or more times the volume of the inlet duct are useful where large gas-handling capacities are required and moderate particle collection efficiencies are acceptable. Higher-efficiency cyclones are attained by decreasing the body diameter of the cyclone. This is due to increased separation forces caused by the smaller vortex radius. Individual high-efficiency, small-diameter cyclones have a small capacity and can be operated in parallel to handle nominal gas volumes. They have a common gas inlet, dust hopper, and gas outlet, and are arranged in banks with several hundred separating elements.

Rotation can be produced by tangential gas entrance or by axial gas entrance through a set of contoured twisted blades. Separated dust can be removed either axially or tangentially from the periphery. Dust is usually removed from the end

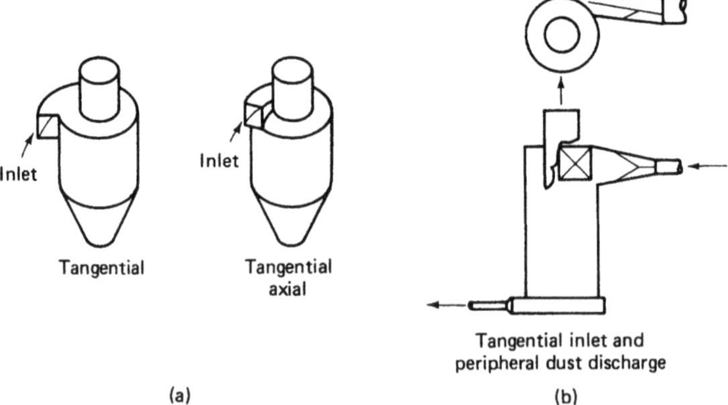

FIGURE 7-8 (Courtesy of A. J. Buonicore and L. Theodore, *Industrial Air Pollution Control Equipment for Precipitators,* CRC Press, Inc., West Palm Beach, Fla. 1975.)

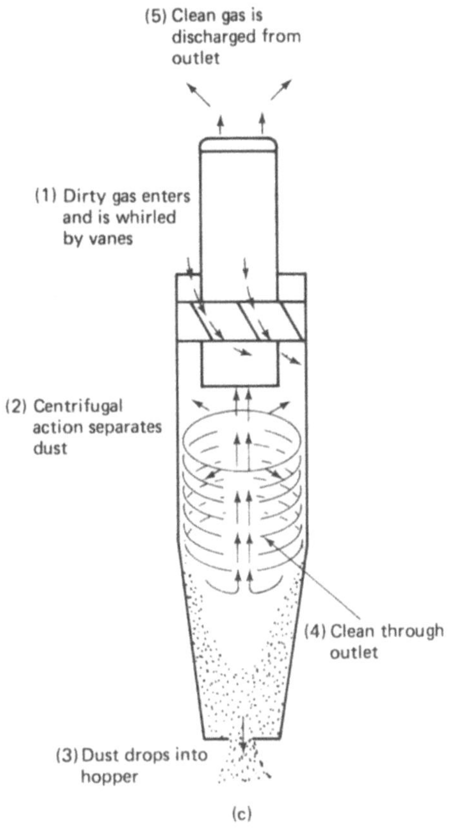

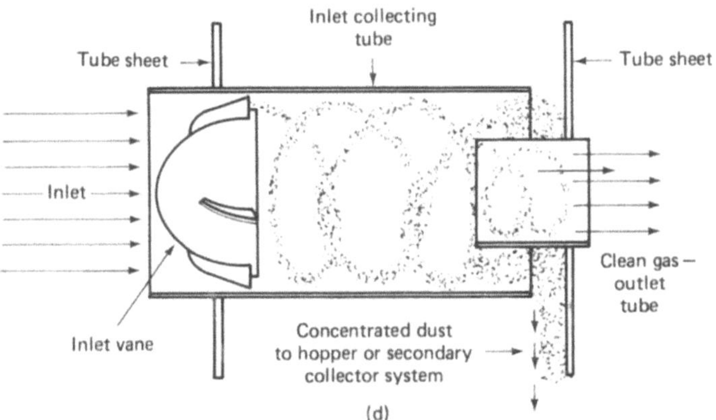

FIGURE 7-8 (con't.)

opposite the axial gas outlet. There can be either one or a multiplicity of tangential inlets. The cyclone body can be completely cylindrical, completely conical, or made of both cylinders and cones. The gas outlet can be either cylindrical or conical.

Figure 7-9 is a simple schematic of the basic cyclone collector. The entire mass of the gas stream with the entrained particulates is forced into a vortex action in the body portion of the cyclone.

Rotation of the carrier gas around the axis of the tube imparts a tangential velocity (V_t) to the particles. The higher density of the particles with respect to the gas imparts a radial velocity which forces the particles toward the tube wall. Here they are carried by gravity or secondary eddies toward the dust outlet at the bottom of the tube. The flow vortex is reversed in the lower portion of the tube, leaving most of the entrained particulates behind. The cleaned gas then passes through the central or exit tube and out of the collector.

There are many variations to the basic cyclone. For an axial entry, the vortex is achieved with a vane cascade rather than a simple tangential entry. These are usually called multitube-type collectors (Figure 7-8c). Many multitube types use an energy-recovery device in the gas outlet tube. After separation of entrained particulates, this device converts some of the rotational kinetic energy back into static pressure and thus reduces the total energy input required to establish the desired separation velocities.

Performance of the cyclone collectors is governed by the following equations:

$$V_t = K\sqrt{\Delta P/\rho} \tag{7-3}$$

$$V_r = \frac{2V_t^2}{9\mu R}(\rho_p - \rho)(R_p^2) \tag{7-4}$$

$$E_{ff} = f(V_r) \tag{7-7}$$

where
$\quad V_r =$ radial velocity of particle, ft/sec

$\quad V_t =$ tangential velocity of particle, ft/sec

$\quad \mu =$ gas viscosity, lb/ft sec

$\quad R =$ average radius of rotation, ft

$\quad R_p =$ particle radius, ft

$\quad \rho =$ gas density lb/ft^3

$\quad R_p =$ particle radius, ft

$\quad K =$ constant dependent on design of cyclone

$\quad \Delta P =$ pressure drop, in. w.g.

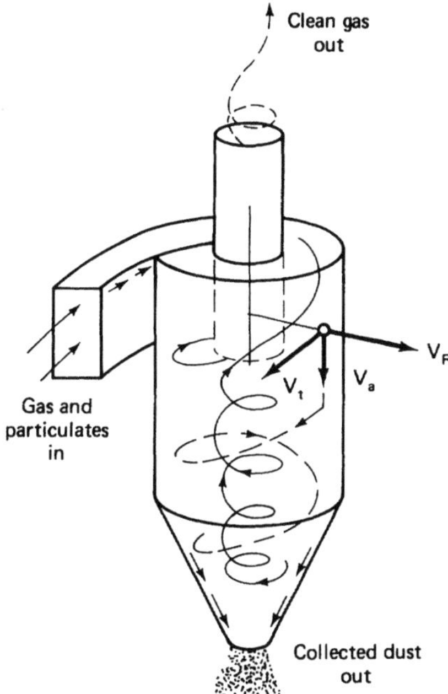

FIGURE 7-9 Basic cyclone collector.

The first requirement for dust collection is to accelerate the gas in the cyclone to attain the required tangential velocity V_t. The relationship between the operating variables in cyclone design and tangential velocity is shown in equation (7-3), which is derived from Bernoulli's equation. The object is to supply energy in the form of total pressure drop (ΔP) and accelerate the gas and entrained particles to V_t. The amount of energy depends on operating variables such as gas pressure, temperature, and molecular weight plus the design of the collecting element, which essentially determines how efficiently static pressure is converted to velocity. Rotation of the particles around the tube axis at V_t results in centrifugal force, which causes the particles to migrate outward at a velocity relative to the gas of V_r.

The variables affecting V_r are defined in equation (7-4). These variables consist of particle radius, particle density, and tangential velocity. Tangential velocity is a function of power input through pressure drop, gas viscosity, and tube radius. Generally, the efficiency is a function of V_r (7-5) [2].

Although mechanical collectors have been used for many years to collect suspended materials, a satisfactory theory relating the many operating variables has never been universally accepted. The performance of cyclone collectors is based almost entirely on experience and empirical data, probably because of the complexity of the aerodynamics involved in cyclone collector flow. However, the foregoing equations at least identify those factors that can either increase or decrease effi-

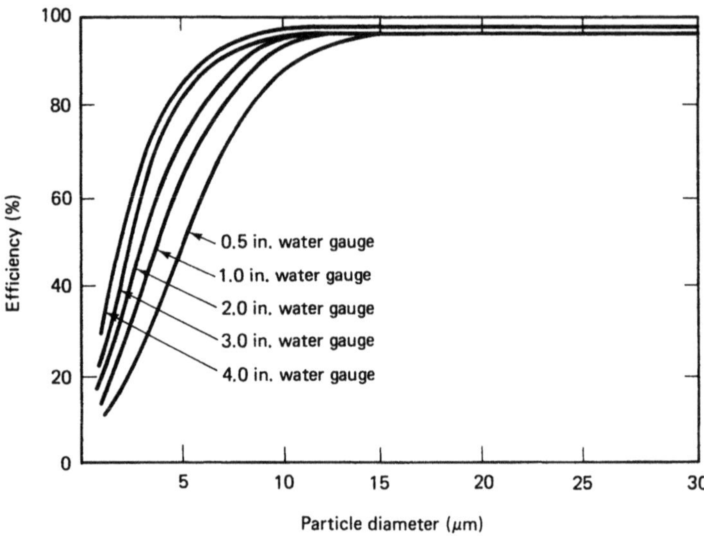

FIGURE 7-10 Basic performance characteristics of a multitube mechanical collector.

ciency. For example, as particle radius (A) increases, V_r will appreciably increase and thus efficiency should increase. This is illustrated in Figure 7-10, which shows the basic performance characteristics of a typical multitube mechanical collector. This is known as the collector fractional efficiency curve and is usually established by the supplier of the cyclones. By knowing the designed pressure drop across the cyclone, the particle-size distribution, and specific gravity of the dust to be collected, the overall efficiency can be calculated by simple numerical calculation.

The most prevalent factor affecting the collection efficiency is the size and nature of the particulate. As the particle size becomes smaller, the separating force decreases, so that for any design and gas volume, there is a limiting particle size below which collection ceases. The nature of the particles also affects the collection efficiency since some types of small particles tend to agglomerate and behave like heavier particles in the separate field.

Factor	*Effect*
Inlet loading	Efficiency decreases as inlet loading decreases
Specific gravity	Efficiency decreases at lower specific gravity
Particle-size distribution	Efficiency decreases as percent of dust particle size less than 10 μm increases
Pressure drop	Efficiency decreases as pressure drop decreases

Additional information on some of these effects is discussed in the next section.

7.3 INSTALLATION PROCEDURES

The prerequisite for good collection performance is to follow proper installation procedures. To eliminate reentrainment, the collector *must* be airtight. Thus, when installing the equipment, access doors, inlet and outlet plenums, and dust disposal areas must be completely sealed. The same criteria as above hold true for multitube collectors, with emphasis on additional critical areas. Header sheets must be welded tight to walls of shell. Gasketing of tubes must seal any opening between tube and header sheets. Outlet tubes must be tightly sealed to the top of header sheets by welds or gaskets. It should be noted that any leakage in the collector can cause a loss of over 25% in collection efficiency. One bit of caution is advised in the installation of multitube collection elements fabricated of cast-iron material. Generally, to attain the hardness required today, the tubes are cast of gray or white cast iron with a Brinell hardness of 350 up to ni-hard. This causes the tube to be very brittle. Any knock or even overtorquing when attaching the tubes to the underside of a header sheet could cause breakage around the edges of the tube. This permits leakage of dust fines to escape through the opening between the header sheets and tubes and pass through the collector.

Because of high field labor costs, the present trend is to factory-assemble as much (and/or as many) of the mechanical collector components as possible within shipping limitations.

7.4 OPERATION

The operating variables in cyclone performance are the gas temperature, pressure, and composition, and dust characteristics, including size and distribution, shape, density, and concentration.

Increasing the gas temperature decreases its density and increases its viscosity. The direct effect on efficiency by changes in gas density can be neglected since the gas density is minute compared to the dust density. If the collector pressure is held constant, the tube capacity will increase because of the lower gas density. Higher temperature increases the inlet velocity, and this increases the particle velocity toward the wall. However, because of the temperature rise, the increasing gas viscosity tends to decrease the particle velocity toward the wall. The net effect of these factors caused by the temperature rise is essentially negligible in the normal operating range of 40 to 700°F. Thus, at constant pressure drop, efficiency is essentially constant with temperature. As the temperature extends above 1000°F, the viscosity effects become more predominant and there is gradual decline in efficiency. Gas composition can also affect both gas viscosity and density.

Typical performance of conventional and high-efficiency cyclones is shown in Table 7-1 [3].

Some procedures can be instituted to increase efficiency. However, tests show that any attempt to increase efficiency by means of blades, spiral ducts, or any

TABLE 7-1 Performance of conventional and high-efficiency cyclones.

Efficiency vs. Particle Size	Percent of Particles Below 10 μm in Size	Efficiency Range
Fly ash (power)		
Spreader stoker-fired boilers	20	80–85
PC-fired boilers	42	75–90
Cyclone-fired boilers	65	55–65
Nonmetallic minerals (when collector is part of process and collector catch is reusable)		
Cement (kilns and process)	40	70–85
Asphalt plant	10	80–95
Lightweight aggregate (kiln)	30–40	80–90
Refractory clays (kiln)	40–50	70–80
Lime (kiln)	40–50	75–80
Fertilizer plant (process equipment)	40	80–85
Steel (ore benefication)		
Pelletizing (vertical shaft and rotary kiln)	10–40	80–95
Foundry (general)	10–40	80–95
Chemical process (drying, calcining)	10–40	80–95
Incinerators (municipal)	20–40	65–75
Coal processing (thermal drying)	10	90–97
Petroleum (catalytic cracking process)	0.6	99+
General industrial application (in plant)	10–60	65–95

Variable	Efficiency
Dust size, increases	Increases
Dust concentration, increases	Increases
Dust density, increases	Increases
Gas molecular weight, increases	Decreases

appliance for the purpose of directing the rotating gases is useless, as it only results in the checking of the free rotation of gases in the body of the cyclone. This rotation, which causes the separation of the dust, is continued in the surge hopper or some sort of dust bunker. Even in these, situated outside the cyclone body, the checking of the rotating gases has an unfavorable influence on the efficiency.

A cyclone with a large dust bunker or hopper will have a higher efficiency than one whose dust outlet is closed by means of a rotating valve. Similarly, the placing of vanes in the gas exhaust pipe in order to decrease the pressure drop in the cyclone will influence the efficiency unfavorably. It is most desirable, therefore, to provide the cyclone with a dust hopper and an outlet vortex tube in which the gases can continue their rotation.

The shape of a cyclone is determined by the following principal dimensions (see Figure 7-11): (1) inlet angle of gases; (2) diameter of cyclone, D; (3) gas exhaust diameter; (4) length of outlet vortex pipe, S; (5) height of cyclone body, h; (6) overall height of cyclone, H; and (7) area of gas inlet, $A \times B$. Changes in these dimensions will affect the efficiency of the cyclone. For example, a gradual entry of the gases into the involute part of the cyclone is essential to obtain a

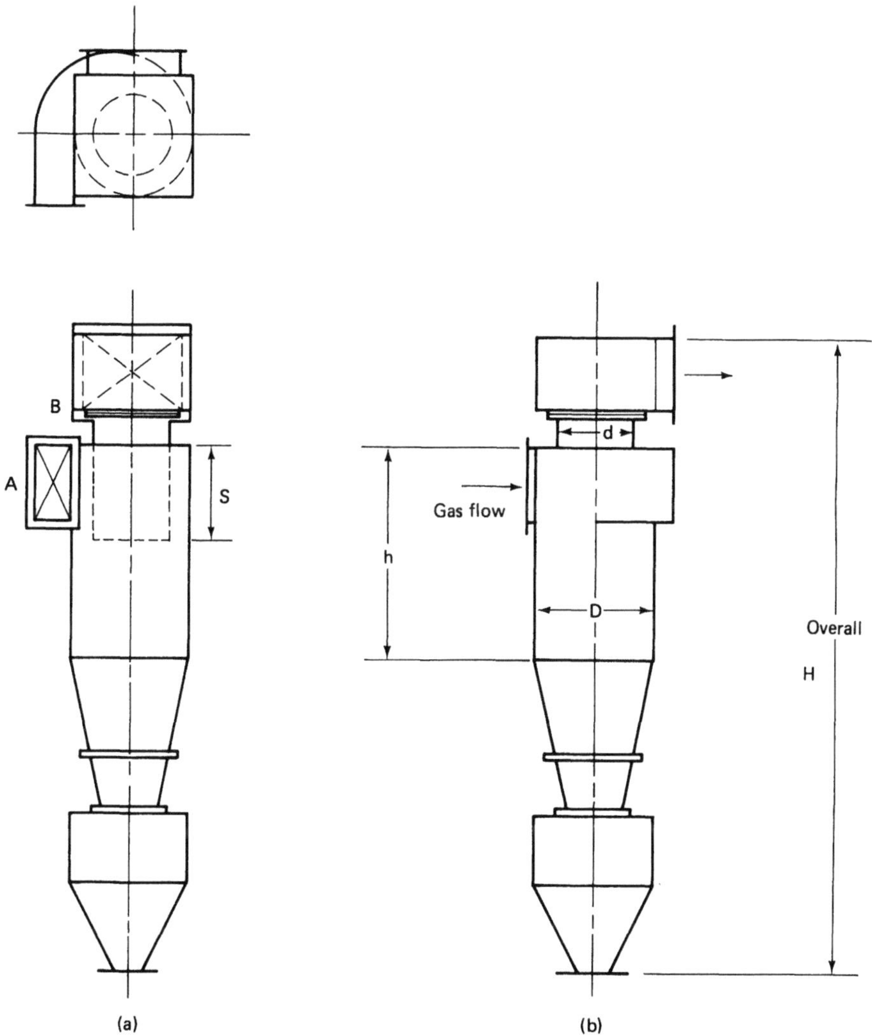

FIGURE 7-11 Dimensions of a cyclone collector: (a) end elevation; (b) side elevation.

reasonable efficiency. The best results are obtained when the inlet angle from the involute is 180° (see Figure 7-11). Further increase of this inlet angle does not raise the efficiency and usually makes manufacture unnecessarily expensive.

The influence of the length of the outlet vortex pipe, S, on the efficiency is shown in Figure 7-12. The efficiency decreases quickly when the length of the vortex pipe becomes too short, the maximum efficiency being reached when S approximates d, or S approximates B when B is greater than d. Increase of the overall height of the cyclone improves the efficiency, although it probably will increase the initial cost and space requirements of the cyclone. Tests show that an overall height greater than $3D$ is recommended (see Figure 7-13). Decreasing the diameter d of the vortex pipe in relation to the body diameter of the cyclone increases the pressure, which in turn increases efficiency (Figure 7-14). However, to fabricate a practical cyclone, the diameter of the vortex should be approximately 0.4 times the cyclone body diameter and the length should be $^1/_6$ or $^1/_7$ times the overall cyclone height [4].

The entry area of the cyclone will also influence the efficiency to some extent. In general, a higher-than-wider entry, where $A > B$, will increase the efficiency, but a more practical shape is a square inlet, where $A = B$. To decrease pressure drop in the cyclone, it is desirable to make the cross-sectional area of the inlet duct, $A \times B$, not much smaller than that of the exhaust.

The influence of gravity on the dust separation in a cyclone is slight, so that efficiency is almost independent of the position of the cyclone. Experiments have shown that separation remains satisfactory not only in a horizontal position, but also even with the cyclone upside down, with the exhaust gases directed vertically downward and the dust carried off upward from the cyclone. A good cyclone separates the dust satisfactorily in any position. Difficulties may arise, however, with coarse particles of dust which keep rotating in the conical part of the cyclone

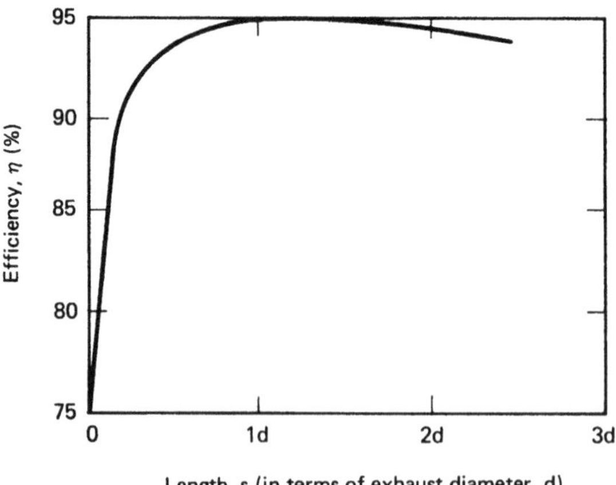

FIGURE 7-12 Influence of length of exhaust pipe inside cyclone on efficiency.

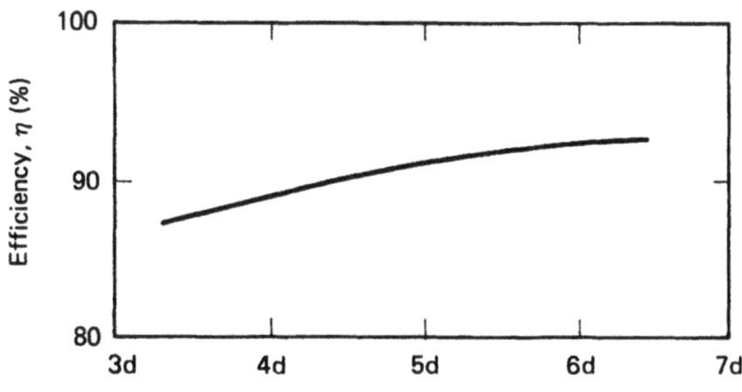

Height, h (in terms of exhaust dimater, d)

FIGURE 7-13 Influence of height H on efficiency.

and cannot reach the dust outlet. This difficulty may arise with a cyclone in the normal vertical position, but to a lesser extent.

To understand the operation of a cyclone, it is necessary to trace the velocities and pressures within the unit and thereby understand their affect on efficiency. The gas flow is three-dimensional, so that at any point within a cyclone, the velocity can be resolved into three components: tangential component of gas velocity, radial component of gas velocity, and vertical component of gas velocity. Figures 7-15 and 7-16 show the variations of the tangential, radial, and vertical gas velocities inside the cyclone. With the exception of the highly turbulent area in the center, the tangential component V_t predominates, so that the total velocity of the gas deviates only slightly from V_t. From the circumference to the center, the tangential component increases and attains its maximum at a radius of approximate two-thirds of the cyclone radius [5].

In the cylindrical part of the cyclone, the fluctuation of V_t, is generally affected by the velocity at the circumference and the radius of the point at which the gas velocity V_t is estimated.

In the conical part of the cyclone, V_t increases considerably as the cone becomes narrower, and at the same distance from the axis, V_t is considerably greater in the conical part than in the cylindrical part.

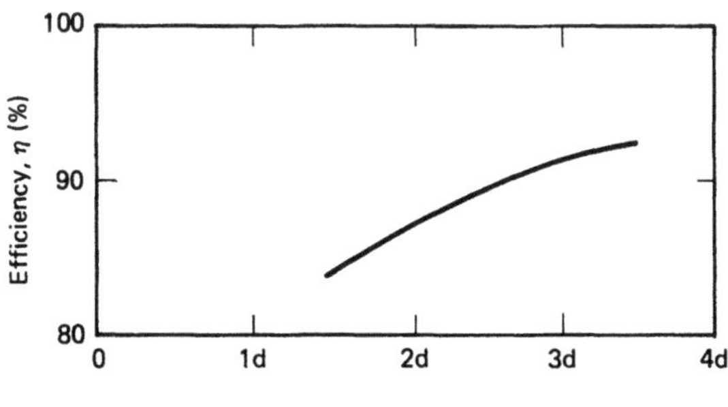

Diameter, D (in terms exhaust diameter, d)

FIGURE 7-14 Influence of diameter D on efficiency.

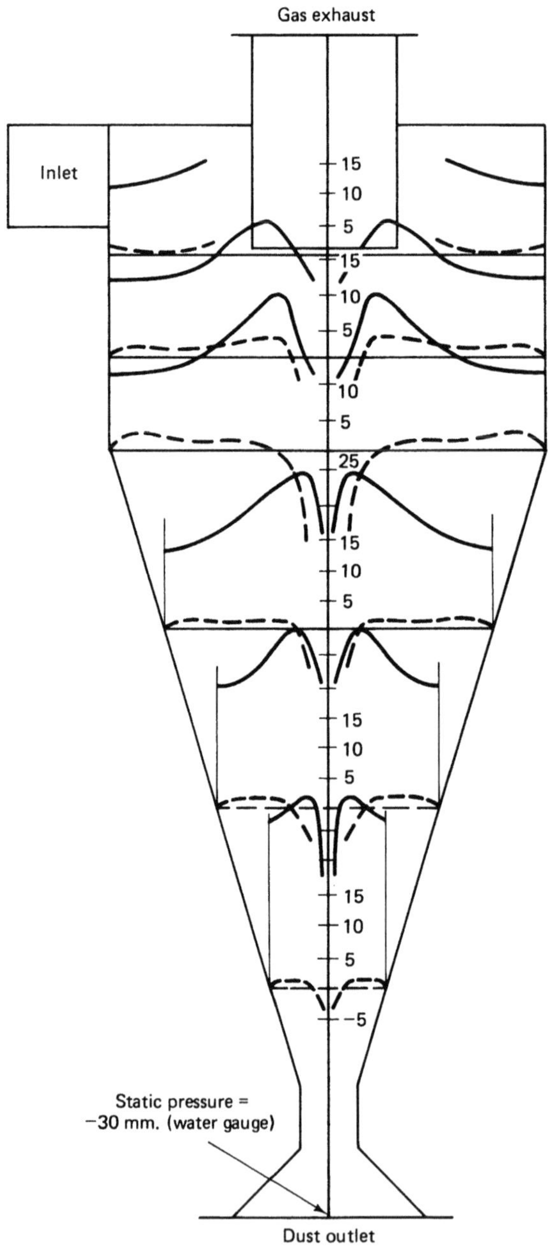

FIGURE 7-15 Variation of tangential velocity V_t and radial velocity V_r at different points in a cyclone: solid lines, tangential velocity; dashed lines, radial velocity.

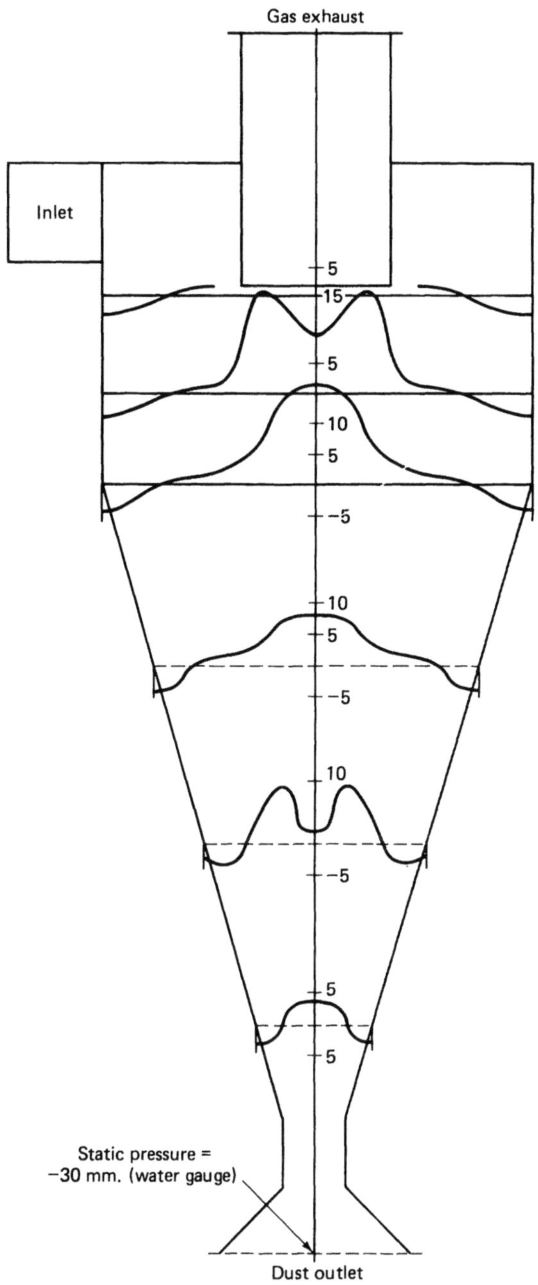

FIGURE 7-16 Variation of vertical velocity V_h at different points in a cyclone.

In the turbulent and disturbed center of the cyclone, the total velocity deviates considerably from the tangential. As shown in Figure 7-16, the vertical component V_h is directed downward on the outer walls of the cyclone. The downward flow of gas along the walls ensures the carry-off to the dust outlet and bunker of the particles that have been thrown out to the circumference.

The radial velocity V_r is directed toward the center throughout most of the cyclone. This velocity tends to carry the particles to the center against the action of centrifugal force. In the center, the radial velocity is directed outward.

From the variations of velocities, it appears that the area in the center of the cyclone does not lend itself to favorable particle separation. A particle in this area will be carried by the vertical gas currents to the gas exhaust and will not drop to the dust outlet. To attain a high level of efficiency, it is desirable to keep the particles outside the center of the cyclone and to increase the tangential velocity V_t as much as possible in proportion to the radial and vertical components V_r and V_h respectively.

Collection efficiency in a cyclone is largely dependent on particle size distribution. Efficiency can only be guaranteed if the settling rate, or alternatively, the particle size distribution is known. Generally, guarantees are made when the dust composition is constant or contains only a small fraction of particles less than 10 microns in diameter.

7.5 MAINTENANCE

An adequate scheduled maintenance program for cyclones will repay itself many times over by reduction in production time losses and costly major repairs. Effective maintenance is easily accomplished if the following basic rules are followed:

1. At initial installation, provide for easily accessible access for cleaning and removal of any part of the system. Provision for convenient maintenance begins on the drawing board. The designer should allow for the eventual removal of parts that may wear out. Special material should be designated for those parts normally exposed to the greatest wear or possible abuse.

2. Establish an information file for history of all collection equipment. The file should include actual operating data, such as gas volumes, operating temperatures, inlet and outlet velocities, pressure differentials, dust loadings, and any and all other data affecting operation.

3. Maintain data as designated on nameplate.

4. Maintain copies of arrangement or construction drawings, including parts lists and a spare-parts reference drawing. These contain commercial descriptions of such items as gaskets, bolts, nuts, and washers.

At least one copy of the information file should always be available to maintenance personnel, and a master copy should always be kept in the plant engineering department.

Erosion, pluggage, and corrosion cause the most problems for cyclones. Under erosion conditions, metal generally fails by continuous gouging and channeling of dust particles. Once started, it is self-accelerating. The metal begins to wear through in a small area. Erosion will increase with dust loading, specific gravity, hardness of dust particles, velocity of the gas, and impingement angle. At light dust loadings, erosion tends to be concentrated in the cone. As the loading increases, the entire cone is affected.

Erosion can be minimized by the wise choice of cyclone diameter size as well as by special construction features. Maintenance features for erosion control should include the use of heavier-gauge metal for cones, use of abrasion-resistant removable wear plates at the impingement zone, avoidance of welds and joints that cannot be ground smooth, and the use of larger cyclones for prevention of excess gas velocities.

Two types of cyclone fouling usually occur: dust outlet pluggage and wall buildup. Outlet pluggage is usually induced by large chunks of material lodging in the outlet and forming an obstruction about which smaller particles can build up. This buildup of dust on the walls and in the hopper leads to reentrainment. Once a hopper has filled sufficiently to plug dust outlets and has been emptied, it is not uncommon to find dust plugs remaining in the outlets after the hopper has been emptied. In large-diameter cyclones, one maintenance procedure to minimize pluggage is to provide an axial poke hole with a bolted cover plate in the top of the outlet pipe or scroll in which a rod can be inserted to break up an outlet dust plug.

Wall buildup is primarily a function of the dust. As a general rule, the finer the dust, the greater the tendency for wall buildup. Another reason for buildup of particles below 3 μm is their inherently greater cohesive and adhesive forces. Also, it is possible for smaller particles to remain in the stagnant boundary layer. A second major cause of wall buildup is moisture, picked up by hygroscopic particles at the walls of the cyclone. Buildup of these particles can be minimized by continuously flushing the dust to the outlet, but if this wet operation is used, corrosion-resistant liners or special alloys may be needed.

Sticky-material problems may be minimized by keeping the tangential velocity above the transport velocity. Also, smoothness of the wall will greatly minimize the tenacity with which the particles stick.

Corrosion is usually caused by chlorides and sulfides in contact with the metal walls. Corrosion can be minimized by using the proper material—alloys, or in some application plastics. Corrosion control can be provided only in the initial stage design.

In most processes, the exhaust gas temperature is usually maintained as low as possible to obtain the best thermal efficiencies, as well as for operating economics. Precautions should be taken to avoid reaching the dewpoint. Some methods of avoiding this include: preheating the gases after shutdown until all the dust has been evacuated, insulating the collector to retain heat, or providing artificial heating units. These steps will prevent condensation of moisture and minimize the danger of corrosion.

For proper maintenance, two general rules should be followed: a regular schedule for removing the collected material from the dust collection hoppers, and inspection of all parts of the dust collector at definite intervals and cleaning or repairing of the dust collector when inspection indicates. To operate efficiently, all cyclone dust collectors should be kept air tight. Gaskets must close the gap between flanges. In axial-entry collectors, header sheets must be sealed without any breaks in the welds. Doors, ports, and poke holes must be sealed to prevent reentrainment. A certain number of fine dust particles are always suspended in the air, and any type of leakage—no matter how slight—causes reentrainment and escaping of dust particles from the collector.

Detection of leaks in a cyclone is not very complicated. One method for checking leakage in a collector is in the use of bright floodlights. By shining a light up from the bottom of the cyclone, erosion holes, gasket leakage, and weld breaks can be detected when viewed from the outside. When directing the light upward from the hopper, for axial-entry collectors, breaks in header sheets, decomposed gaskets, and cracks in material can be detected.

Much emphasis is made on prevention of leaks. The main reason for this concern is that a cyclone operates on an inertial principle and leakage disrupts the flow pattern of the gas stream, thereby forcing the fine particles to reentrain through the outlet vortex. It is most important to seal the cyclones, as it has been estimated that over 25% collection efficiency can be lost due to dust reentrainment.

7.6 IMPROVING OPERATION AND PERFORMANCE

The pressure drop within the cyclone determines the collection efficiency. The pressure drop can be increased or decreased by varying the diameter of the cyclone body—the smaller the diameter, the higher the efficiency, or by increasing the volume per tube, thereby causing increased velocity and increased dropout of dust particles. These features must be designed into the system. Should it be known that the system will be turned down with less volume quite frequently, the collector should incorporate dampers so that gas velocity going to the cyclones will be increased.

Fine tuning a high-efficiency tangential cyclone is very difficult, as there are no moving parts. With axial-entry cyclones, some manufacturers use a spiral vane that permits some control of volume by moving a vane in and out of a constricted opening of the collecting element. When fully in, maximum rotation is induced, thereby creating greater centrifical action. When fully out, only a partial amount of gas is rotated; the rest bypasses the vanes.

Turn-down of gas flow can be controlled by use of dampers at the inlet. This can be done if the collector is sectionalized, thus allowing the designed pressure drop to be maintained. Should the reduction in gas volume be permanent, the collector can be modified by capping off the excess tubes. This again permits the remaining tubes to handle the reduced gas volume at the designed pressure drop.

The best control of cyclone operation and performance is to initially design the collector to meet the criteria. It is very difficult—in most cases impossible—to modify a cyclone once it has been improperly designed or if the operating conditions have changed appreciably.

7.7 CONCLUSIONS

Because of their simplicity of construction and lack of moving parts, cyclones can be fabricated of a variety of materials, including ceramic, or lined with various materials, including ceramics. They are applicable over a broad range of temperatures from below ambient to above 2000°F. Cyclones can also be used at high gas pressures. Generally, the only criterion for good performance is the requirement for proportionate increase in power input to maintain separation velocities. The collector is insensitive to inlet concentration, and in fact, the efficiency can increase with increasing particle concentration because of particle interactions. Cyclones find most use when operating under the following situations: where dust is coarse, where dust loading is above 3 g/ft^3, where high efficiency is not a critical factor, and where stack opacity is not a criterion.

Cyclones are the most economical of the four generic classes of particulate collectors. They are also the most versatile. As unsophisticated as this type of collection device is, it does the job, and that is the only requirement for success.

7.8 NOMENCLATURE

R_p	particle radius
D_p	minimum particle size that can be completely removed (particle diameter), actual or aerodynamic equivalent
ρ	gas density, lb/ft^3
ρ_p	particle density, lb/ft^3
E_{ff}	fraction collected, wt %
g	acceleration due to gravity, 32.2 ft/(sec^2)
H	chamber height
K	constant dependent on design of cyclone
L	chamber length, ft
μ	gas viscosity, lb/ft/(sec)
ΔP	pressure drop, in w.g.
R	average radius of rotation ft
μ_t	settling velocity of dust ft/sc
μ	gas viscosity, lb/ft^2
V	uniform gas velocity, ft/sec

V_r radial velocity of particle ft/sec

V_t tangential velocity of particle ft/sec

7.9 REFERENCES

[1] IGCI Gravity, *Louver and Dynamic Mechanical Collector*. Publication M-3, February 1970.

[2] WALKER, A. B., "Cyclone Theory and Applications." Lecture for Technical Staff Personnel, Research-Cottrell, Somerville, N.J., 1957.

[3] BUONICORE, A. J., AND THEODORE, L., "Stdy of Industrial Air Quality Control Methods for Particulates," Proceedings and Am. Environ. Eng. and Sci. Conference, Louisville, Ky., 1972.

[4] TER LINDEN, A. J., "Investigations into Cyclone Dust Collectors," Proceedings of the Institute Mechanical Collectors, Amsterdam, Holland, 1949.

[5] VAN DER HEGGE ZIGNEU, B. G., Proceedings of the Royal Academy, Amsterdam, Netherlands, 1929.

8

Baghouses

John D. McKenna

President
ETS, Inc.
Pollution Control Consultants
Roanoke, Virginia

and

Gary P. Greiner

Executive Vice President
ETS, Inc.
Pollution Control Consultants
Roanoke, Virginia

8.1 DESCRIPTION OF CONTROL DEVICE

Fabric filter dust collectors, commonly referred to as baghouses, are among the oldest and most widely applied particulate emission control devices. It is reported that there are over 200,000 currently (1980) in operation in the United States. The universally employed principle of operation, simply stated, is the removal of dust from dust-laden gas by passing the gas through a filtration medium, normally a fabric. The cleaned gas emerges from one side of the medium while the dust is collected on the other side. Subsequently the collected dust is removed from the fabric. Different collector designs, fabric types, configurations, and a variety of cleaning methods provide numerous hardware combinations. A number of the basic designs emerged dominant in the marketplace. These designs and their maintenance features will be discussed subsequently.

There are roughly 100 U.S. fabric-filter manufacturers. The larger companies usually offer more than one fabric-filter design type. The market is extremely competitive, and therefore designs are constantly changed by the more aggressive companies. With so many manufacturers, product design possibilities, and a dynamic market, one realizes that hard and fast operating and maintenance rules are difficult to come by, and it is important for the user to establish where the fabric-filter dust collector fits into the design and application picture.

Categorizing Fabric Filters

To better understand one's fabric-filter dust collector (baghouse), it is helpful to determine where it fits among the various types of fabric filters. When one tries to group fabric filters into a number of categories, it soon becomes obvious that the task is not simple. There appears to be an exception to each of the rules. The creation of certain categories, though they are not rigid, is yet very helpful.

One such approach is to group fabric-filter designs by *cleaning method,* as shown in Table 8-1. There are three major cleaning methods: shakers, reverse-air, and pulse jets. In addition to these three dominant cleaning methods, there exist a large number of other cleaning methods which are less often applied. Recently, combinations of the three primary methods have been occasionally employed. For example, reverse-air and shake have been used in combination, and reverse-air with a pulse assist.

Another approach to grouping fabric filters is by *capacity* (see Table 8-2). Generally, the groupings are small volumes (i.e., below 10,000 acfm), medium volumes (i.e., 10,000 to 100,000 acfm), and large volumes (i.e., >100,000 to the multimillion acfm level).

The *filter-media type* and temperature *capabilities* provide two other ways to categorize and view fabric filters. Woven vs. felted media are the media-type categories and high-temperature (>400°F), medium-temperature (200 to 400°F), and low-temperature applications (<200°F) are useful temperature groupings. Two additional ways to group filters are *intermittent vs. continuous* service and *inside vs. outside* bag collection.

One should be aware of these distinctions and attempt to find where the collector in question fits; thus, when considering operating or troubleshooting recommendations, one only applies recommendations that are suited to the type of collector being used.

Cleaning Methods

Shakers Simply stated, a shaker cleaning system removes the collected dust from the surface of the bag by mechanically shaking the bag. This can be done manually in the case of small dust collectors. Generally, however, it is accomplished

TABLE 8-1 Design groupings for fabric filter dust collectors: grouping by cleaning method.

Cleaning Method	General Characteristics Associated with Design Type
Shaker	Inside collection
	Manual and automatic cleaning
	Woven filter media
	Cake filtration
	Generally intermittent—small operation
	Large and small volumes
	Often no compartments
Reverse-air	Inside collection
	Automatic cleaning
	Woven-filter media
	Cake filtration
	Continuous operation
	Medium and large volumes
	Compartments
Pulse-jet	Outside collection
	Automatic cleaning
	Felt-filter media
	Media filtration
	Continuous operation
	Medium and small volumes
	Often no compartments

TABLE 8-2 Design groupings for fabric-filter dust collectors: grouping by capacity.

Capacity Range	General Characteristics Associated with Design Type
Low capacity	Off-the-shelf, small units, handling hundreds to a few thousand cfm; little or no assembly; prebagged; manual cleaning systems available with some types
	Pulse-jet, shake, cartridge
Medium capacity	Modular designs: a few thousand to less than 100,000 cfm; little field assembly required; generally prebagged
	Pulse jet; high G/C
High capacity	Shippable modular design: fifty thousand to one million CFM; moderate field assembly; field bagged
	Reverse-air and shake
	Structural: 100,000 to 1 million cfm; field-assembled and field-bagged
	Reverse-air and shake; low G/C

by the use of a motor driving an eccentric (Figure 8-1), which in turn has a rod connected to the bags. It is probably the oldest bag cleaning technique and is currently employed over the full range of baghouse capacities from very small off-the-shelf units to the extremely large structural design units (Figure 8-2).

The bag is generally open at the bottom and closed at the top, being fixed in the tube sheet at the bottom and attached to the shaking mechanism at the top (Figure 8-3). With this configuration the dust is collected on the inside of the bag. The bag normally contains no rings or cages. The flow of dirty gas to the bags is stopped during the cleaning process.

In the United States, shakers normally employ woven cloth at gas/cloth ratios below 4:1. Attempts thus far to use shake cleaning in combination with domestic felts have led either to ineffective cleaning and thus high pressure drop or to the filter medium being shaken apart.

The important parameters affecting the efficiency of cleaning are frequency, oscillation, and amplitude. These are shown in Table 8-3.

To achieve a continuous shake-clean baghouse, the unit needs to be subdivided

FIGURE 8-1 AAF's rugged shaker mechanism—used on Amertube shaker-type fabric collectors. (Courtesy of American Air Filter Co., Inc.)

FIGURE 8-2 An AAF Amertube shaker-type fabric collector. (Courtesy of American Air Filter Co., Inc.)

into a number of isolatable compartments, such that one compartment at a time may be taken off-stream for cleaning.

The normal procedure for the intermittent-shake cleaning process is to first damper off the inlet flow and, if necessary, the outlet flow of dirty gas. This should almost immediately result in zero pressure drop across the bags. Once this condition

TABLE 8-3 Shake cleaning—parameters.

Frequency	Usually several cycles/second; adjustable
Motion type	Simple harmonic or sinusoidal
Peak acceleration	$1-10g$
Amplitude	Fraction of an inch to few inches
Mode	Off-stream
Duration	10–100 cycles, 30 sec to few minutes
Common bag diameters	5, 8, 12 in.

FIGURE 8-3 AAF's shaker support logs. (Courtesy of American Air Filter Co., Inc.)

is achieved, shaking of the bags can be initiated. Shaking continues for as many as 100 shake cycles. The cleaning is then stopped and the compartment is momentarily kept dormant until the dust settles. The dampers are then opened and the filtration process resumed. If reverse-air is employed to assist the cleaning process, the cleaning cycle may consist of a series of repeated shake, then reverse-air cycles before the unit is brought on-stream.

Reverse-flow cleaning Reverse-air cleaning is perhaps the gentlest cleaning method. In this method the dust is removed from the bags by back-flushing with low-pressure (a few inches water gauge) reversed flow. In the case of high-temperature applications, the just-cleaned hot gas is employed to back-flush rather than ambient-temperature air. Woven-filter media are generally employed in conjunction with reverse-air cleaning. The cleaning is motivated by a separate cleaning fan which is normally much smaller than the main system fan since only a fraction of the total system is cleaned at any one time. In the case of a negative pressure system, one often can clean without a reverse flow fan. The flow rate of cleaning gas is normally about equal to that of the dirty gas (see Table 8-4).

Most often, reverse-flow systems are comprised of a number of isolatable compartments. The gas/cloth ratios usually employed are less than 4:1. Normally, the dust is collected on the inside of the bag, the bag being open at the bottom and closed at the top. The bag will contain rings to keep it from collapsing completely

TABLE 8-4 Reverse-air cleaning—parameters.

Frequency	Cleaned one compartment at a time, sequencing one compartment after another; can be continuous or initiated by a maximum-pressure-drop switch
Motion	Gentle collapse of bag (concave inward) upon deflation; slowly re-pressurize a compartment after completion of a back-flush
Mode	Off-stream
Duration	1–2 min, including valve opening and closing and dust settling periods; reverse-air flow itself normally 10–30 sec
Common bag diameter	8, 12 in.; length 22, 30 ft
Bag tension	50–75 lbs typical, optimum varies; adjusted after on-stream

during flow reversal. Complete collapse would, of course, prohibit cleaning since the dust particles would be unable to fall down within the bag to the hopper. Cleaning is accomplished both with and without flexing (partial collapsing) of the bag. Reverse flow without flexing can be employed when the dust is very easily dislodged from the bag surface. Figure 8-4 shows a typical cleaning cycle. In the on-stream-gas filtering mode, the compartment and outlet dampers are open and the dirty gas enters the bag at the bottom. The dust is collected on the inside of the bag and the cleaned gas exits the outlet damper. During the bag cleaning cycle the

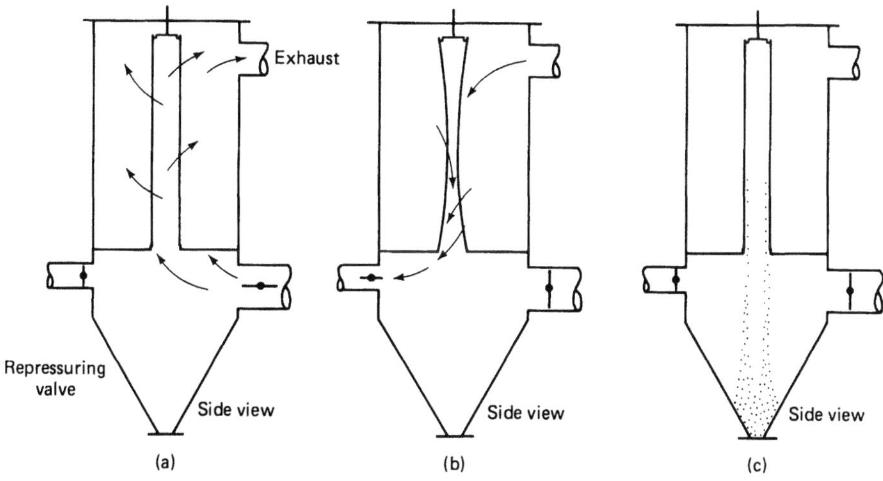

FIGURE 8-4 Reverse air cleaning: (a) filtering; (b) collapsing; (c) cleaning.

flow is reversed by closing the outlet plenum and opening a third damper, which allows cleaned gas to enter the compartment on the clean side of the bags, thus back-flushing the bags and exiting the compartment through the inlet damper. This now-dirty gas progresses to the balance of compartments which are on-stream. It should be noted that this process increases the system gas/cloth ratio by adding to the total gas volume, the volume of gas employed in the back-flushing process. The system described here, although generally applicable, is normally associated with large gas volumes. Low- and medium-volume reverse-air systems can be found; however, these often employ felted media with a moving reverse-flow plenum.

Pulse-jet cleaning   Pulse-jet cleaning is a method that employes high-pressure (60 to 120 psi) compressed air, with or without a venturi, to back-flush the bags vigorously. This method creates a shock wave that travels down the bag, knocking the dust away from the filter medium. Normally, this method is employed in conjunction with felted-filter media and the gas/cloth ratio is generally higher than that of the shake and reverse-air cleaning methods. The duration of cleaning is lower than that of the other two methods; generally, the pulse lasts only a fraction of a second. The house is often not subdivided when pulse-jet cleaning is employed.

The usual configuration has the bag closed at the bottom and open at the top, as shown in Figure 8-5. A metal cage is employed within the bag to keep the bag from collapsing. In the normal mode of operation, the dirty gas enters the hopper and proceeds to the bags. The dust is collected on the outside of the bags and the cleaned gas exits through the top of the bags and baghouse. Usually, a row of bags is cleaned simultaneously by introducing compressed air briefly at the top of each bag. The shock wave thereby created drives the dust off the outside of the bag and down into the hopper. Continuous discharge of the dust from the hopper is often employed. The importance of the venturi to provide an onrush of secondary air for cleaning is now being questioned, and some pulse systems operate with only the compressed-air manifold and without venturies over each bag. Cleaning parameters are given in Table 8-5. Notable is that this system has no internal moving parts and allows for removal of the bags from the clean side of the house, since the bags are usually connected only at the top.

Miscellaneous cleaning methods   In addition to the three most common cleaning methods, a number of other methods have been employed. These methods were either made obsolete by one or more of the three major methods or else never achieved the commerical significance of the three major cleaning methods.

One method that achieved some degree of commercial success is "reverse jet" cleaning, sometimes referred to as "blow ring" cleaning. This method employed a jet ring surrounding each filter tube; the rings were carried up and down the bags on a carriage driven by chains and sprockets or a cable system. Today, this method is found only in very limited application.

In addition to the reverse jet, high-frequency agitation has been employed. One

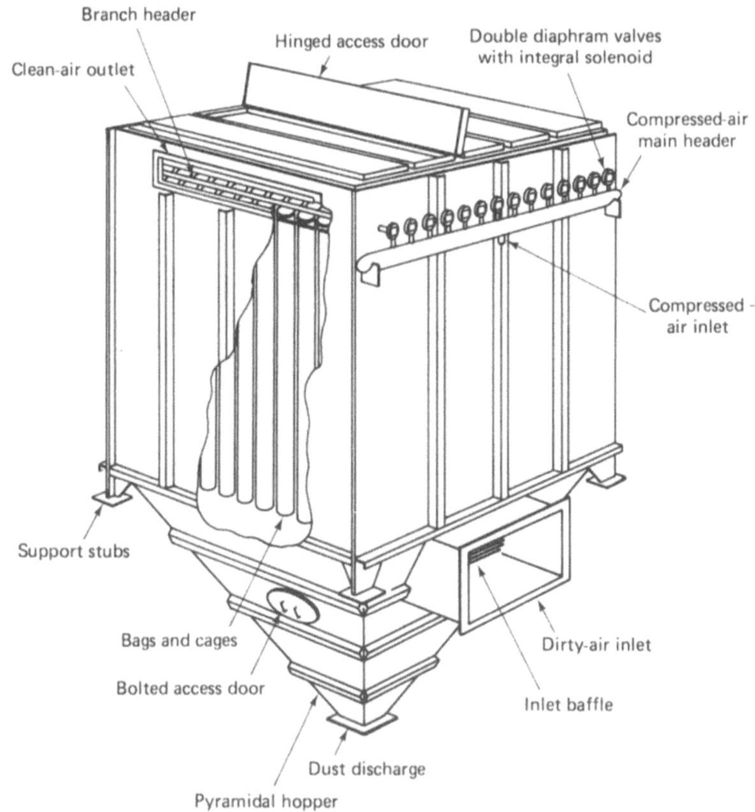

FIGURE 8-5 (Courtesy of Buell Emission Control Division, Envirotech Corporation.)

TABLE 8-5 Pulse-jet cleaning—parameters.

Frequency	Usually, a row of bags at a time, sequenced one row after another; can sequence such that no adjacent rows clean one after another; initiation of cleaning can be triggered by maximum-pressure-drop switch or may be continuous
Motion	Shock wave passes down bag; bag distends from cage momentarily
Mode	On-stream: in difficult-to-clean applications such as coal-fired boilers, off-stream compartment cleaning being studied
Duration	Compressed-air (100 psi) pulse duration 0.1 sec; bag row effectively off-line
Common bag diameter	5–6 in.

such approach is to achieve fluttering of the cloth by vibration and rapping. This is normally associated with envelope filters. In the case of tubular bags, fluttering has been achieved by the use of an "air shake."

Even-higher-frequency agitation is achieved by the use of sonic and ultrasonic cleaning. Sonic cleaning has also been employed in combination with reverse flow.

One of the simplest cleaning methods used in the case of very small collectors is manual cleaning, where the bags are struck, shaken, or even removed and turned inside out.

A variation of the pulse jet which is commerical today but not as common as the pulse jet is the plenum pulse method of cleaning. Like the pulse jet, this method employes compressed air, but instead of delivering the air to each bag via a manifold, the compressed air is delivered to a compartment of bags via a compressed air plenum and damper system, cleaning the entire compartment simultaneously.

Cleaning Methods Versus Capacity

As shown in Tables 8-1 and 8-2, fabric filters can be grouped by both cleaning method and gas capacity. There is a loose correlation between the two groupings. The extremely small collectors are often found to be shakers. This is especially true if an intermittent collector is suitable. Recently, compact cartridge filters (similar to truck air filters) employing pulse cleaning have been finding success in the small-volume marketplace. The small and especially the medium-size collectors, where continuous operation is required, are most often found to be pulse jet collectors. The large continuous collectors of recent vintage are most often reverse-air collectors, although large shakers are found as well. In both cases, continuous operation is achieved by subdividing the unit into a number of compartments so that when one compartment is off-stream for bag cleaning, the total process gas volume can be handled by the balance of the baghouse system.

Bags and Cages

As previously mentioned, each fabric-filter dust collector consists of a number of fabric-filter elements. These elements normally consist of textile material and are constructed as tubes, envelopes, or most recently, cartridges (see Figure 8-6). The textile can be either felted or woven, and if woven, there are numerous constructions that can be employed, such as sateen, 2×2 twill, or crowfoot (see Figure 8-7). In addition to various yarn constructions, the filter element can be varied further by the finishing process applied. The most commonly employed finishing steps include glazed, silicone graphite, Teflon, napping, calendaring, and singeing. Details of these finishes are provided in Table 8-6. Finally, when reviewing bag construction, one must also consider stitching and seam types.

The most commonly employed filter element design is that of a tubular shaped "bag." This is employed to collect dust either on the outside or the inside of the tube. The envelope-constructed filter element is not as common in the United States

FIGURE 8-6 Cartridge baghouse. (Courtesy of Torit Division, Donaldson Corporation.)

as the tubular-shaped. The flat-enveloped baghouse normally collects on the outside of the filter element and provides more filtering area per volume of housing. Finally, the cartridge design, which is a pleated construction, provides maximum filtering area, but at this time seems to have temperature and filtering velocities limitations more restrictive than the other two constructions.

When reviewing the filter element and its construction, one must also consider the filter element support structure. In the case of shakers, as mentioned, the filter element normally has no support other than attachment at the top and bottom. Reverse-air filter elements normally have support rings sewn into the bag. In the case of pulse-jet cleaning systems, the bag is usually supported by a cage placed within the bag. The standard construction has been 6 to 20 vertical wires (stringers)

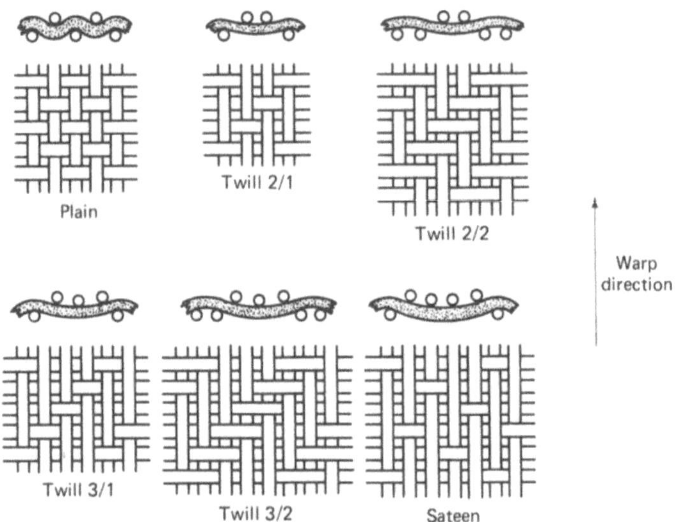

FIGURE 8-7 Filter-cloth weaves.

spot-welded to a number of rings, the rings being located about 6 to 12 in. apart. The recent use of glass bags in conjunction with pulse-jet cleaning has led to the use of close-mesh cages with either a 2×1 in. or a $\frac{1}{2} \times \frac{1}{2}$ in. mesh to provide greater fabric support (Figure 8-8). Variations in cage design beyond mesh considerations include differences in the bottom cage pan, top collar, and top flanges. Usually, these variations are related to bag–cage attachment schemes.

TABLE 8-6 Filter element finishes.

Calendaring	High pressure pressing of the fabric: pushes the surface fibers down onto the body of the filter medium
Napping	A scraping of the filter-medium surface that raises the surface fibers
Singeing	Passing of the filter medium over an open flame, thereby removing the independent surface fibers
Glazing	High-pressure pressing of the filter medium at elevated temperatures: fuses surface fibers to the body of the filter medium
Coating	Immersing the filter medium in a solution to provide the fibers with a coating that will lubricate and thereby reduce self-abrasion; in the case of woven-glass bags, the most common coatings have been Teflon and silicone-graphite

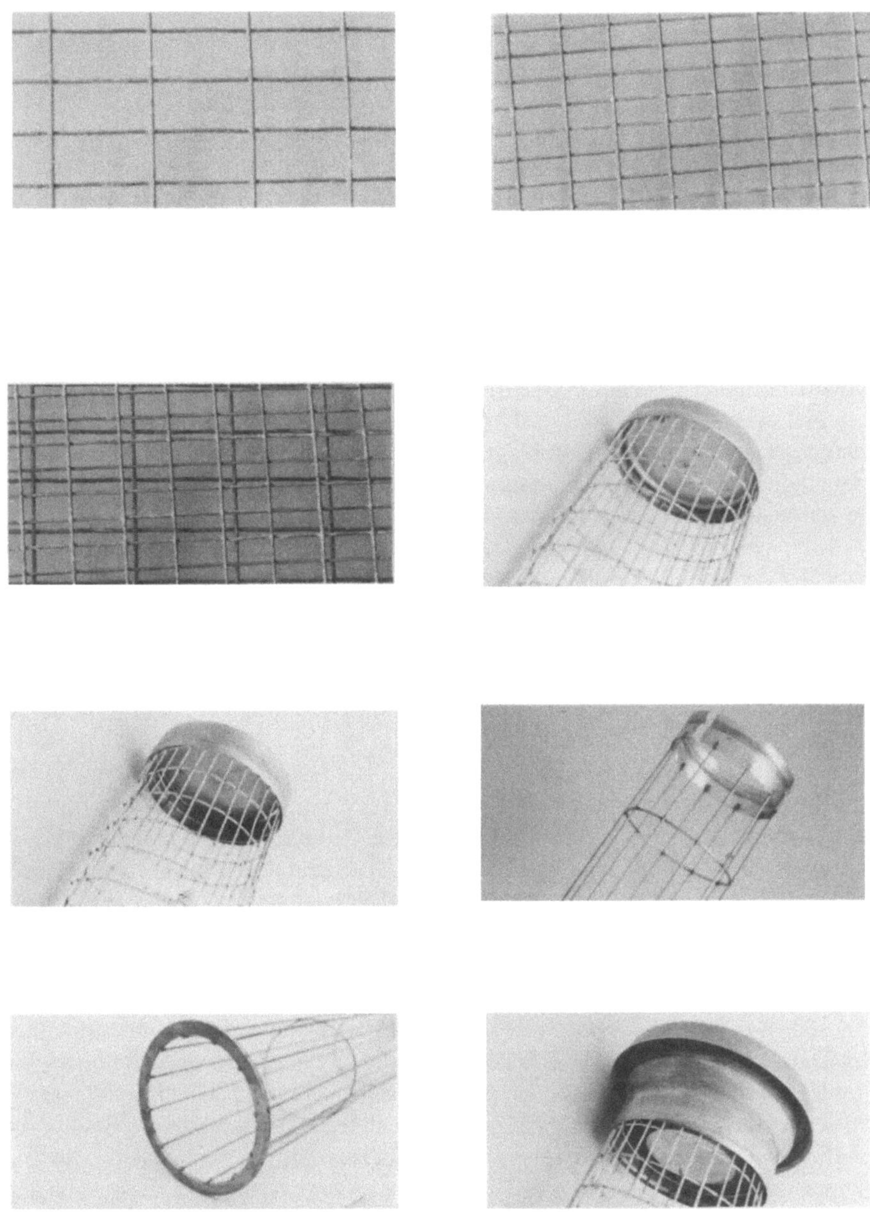

FIGURE 8-8 Cage types. (Courtesy of BHA Accessories Co., A Division of Standard Havens, Inc.)

8.2 DESIGN PROCEDURES

When selecting a baghouse, two major areas of design need to be considered in conjunction with each other. The first is that of the total dust collection system design. This includes all ductwork, fans, heater, dampers, dust conveying equipment, and controls that will directly affect and are intimately associated with the dust collector itself. The second area of concentration is that of the flange-to-flange dust collector, its cleaning method, bag type, materials selection, and so on. The flange-to-flange collector should not be selected without concurrent specification of the total dust collection system, and this in turn requires full consideration of the dust source operation (i.e., normally the production process). It can be that such an overview will lead to some modification of the production process or process controls in order to attain trouble-free baghouse operation.

The specification of the dust collector and its auxiliaries must begin with a definition of the emission problem. The better and more complete the problem definition, the better are the chances that the dust collection system will function in a trouble-free, reliable manner. The dust properties of primary interest are the density, particle size, and dust loading. The gas properties essential to proper problem definition are volume, temperature, moisture content, and acid gas concentration. A general description of the source or production operation is helpful, particularly in defining whether the collector is to be continuous or intermittent. Additional information, as requested in Figure 8-9 can be of great help. It is not uncommon to find a dust collector selected simply on the basis of estimated gas volume, temperature, and a brief description of the production operation; however, one probably enhances the risk of selecting a baghouse system requiring constant maintenance and bag replacement.

The size of the fabric-filter dust collector is tied directly to a parameter known as the air/cloth (A/C) or gas/cloth ratio (G/C). This parameter is the apparent velocity of the dust-laden gas stream as it approaches the filter elements. It can be calculated by dividing the total gas volume, usually in ft^3/min, by the total area in ft^2 of cloth or filter medium available for filtration. Variations of the gas/cloth ratio definition, adjusted for cloth off-stream, or for including cleaning air in the total gas volume, are shown in Table 8-7. Baghouse design is still very much an art form rather than an exact science, and nowhere is this more evident than in the selection of the gas/cloth ratio. Factors influencing the gas/cloth selection include the cleaning method, filter media, dust size, dust density, dust loading, and other factors. Quantifying the general relationship between these parameters has been attempted with very limited success [1]. Models specific to a given source operation and dust collector type have been provided more accurate G/C selection for this more limited application [2].

Most commonly, two design approaches have been used successfully. The first has been to collect all of the empirical data available for the source in question. If there is none in one's industry, one goes to an analogous industry that is using baghouses and determines the G/C range successfully employed in that industry and

CUSTOMER INQUIRY DATA SHEET

Firm name _______________________________ Telephone ( ) _____________

Street __________________________________ Telephone extension ________

City ___________________________________ Per ____________________

State ___________________ Zip __________ Title __________________

ESSENTIAL INFORMATION REQUIRED

1. Description of application ____

____

2. Gas inlet volume _______________________ ACFM 3. Gas inlet temp. ________ °F

4. Description of dust ____

OTHER INFORMATION DESIRED

1. Space available for equipment (sketch or dwgs) _______________________

____

2. Equipment auxiliaries wanted _______________________________________

____

3. Particulate pollutants: Type of dust: _______________ Particle size ______________

   Abrasive? ____________ Explosive? ____________ Inflammable? ______________

   Sticky? ________________ Moisture content (wt %) ____________________

   Bulk density ___________________ Specific gravity ____________________

   Inlet loading ___________________________ grains/SCF of lb/min

   Allowable outlet loading ___________________________ grains/SCF or lb/min

4. Gaseous pollutants: Type(s) and concentration(s) _____________________

____

5. Additional comments (i.e., location, etc.) ___________________________

____

6. Type of duty: Intermittent ____________ Continuous ____________

7. Operating pressure ____________ in. water gauge

8. Existing electrical and compressed air capabilities ___________________

____

FIGURE 8-9 Customer-inquiry data sheet.

conservatively applies it. The risk here is that there is a subtle but essential difference
between the two industries which prohibits the G/C translation. The alternative to
this approach is to operate a pilot unit on a slipstream and vary the gas/cloth ratio.
Then one designs the full-scale unit around the G/C, which appeared to provide
trouble-free operation, at emission and pressure drop levels within predetermined
ranges. The pilot plant usually needs to be run for a few months minimum in the

TABLE 8-7 Gas/cloth ratio.

Gas/cloth ratio:
$$\text{gas/cloth ratio} = \frac{\text{gas volume}}{\text{cloth area}}$$

Gross gas/cloth ratio:
$$\text{gross gas/cloth ratio} = \frac{\text{total inlet gas volume}}{\text{total filter cloth in collector}}$$

Net gas/cloth ratio:
$$\text{net gas/cloth ratio} = \frac{\text{total inlet gas volume} + \text{cleaning volume}}{\text{on-stream cloth}}$$

Units usually employed:
$$\text{gas/cloth ratio} = \frac{\text{gas volume}}{\text{cloth area}} = \frac{\text{ft}^3/\text{min}}{\text{ft}^2} = \text{ft/min}$$

same manner as the full-scale unit to obtain data representative of long-term operation. Once the G/C has been selected, the size of the baghouse is nearly defined. Variations in number of walkways between bags, hopper slope, and so on, have some further influence on the overall dimension, but none have as key a role on size as does the G/C. Examples of typical gas/cloth ratios are given in Table 8-8. The G/C and cleaning method are tied to each other; thus, what has been stated regarding G/C selection is also true for the cleaning method selection. Once the G/C and the cleaning method have been selected, the next major decision is the selection of the filter medium type. Here the basic selection is more of a science than an art. As shown in Table 8-9, temperature limitations and chemical resistance are determining factors in the filter medium selection. As one goes up in temperature, the choices become fewer and fewer. The maximum temperature for which economical and commerically available filter media are provided is about 550°F. When confronted with temperatures above this level, the normal approach has been to cool the gas down to this level. This, of course, has the added advantage of allowing the use of a smaller baghouse since the gas volume has been reduced.

Once the G/C ratio, cleaning method, and filter medium have been selected, the essence of the flange-to-flange design-selection process is complete. The only major consideration left is the baghouse material of construction. Generally, this is mild steel, although in certain specialty applications stainless steel units are employed. Somewhat thicker house and hopper gauge are employed in applications where acid gases are present, such as boiler-fly-ash control systems.

The overall system component selection and system design can be straightforward and relatively simple, as in the case of a single point pickup for an ambient application, thus involving a single hood, a small amount of ductwork, and a system fan on stack, or it can be fairly complex, as in the case of a high-temperature, acid gas, dust control system involving system bypass, auxiliary heat, numerous dampers, temperature and pressure alarms, automatic shutdown, extra modules for maintenance, and off-line cleaning. The system design must start with a definition of system objectives. Consideration must be given to local ordinances (e.g., bypass restrictions, height restrictions, etc.). The production system requirements must be

TABLE 8-8 Gas/cloth ratios.[1]

	Shaker/Woven Reverse-Air/Woven	Pulse-Jet/Felt Reverse-Air/Felt
Alumina	2.5	8
Asbestos	3.0	10
Bauxite	2.5	8
Carbon black	1.5	5
Cement	2.0	8
Clay	2.5	9
Coal	2.5	8
Cocoa, chocolate	2.8	12
Cosmetics	1.5	10
Enamel frit	2.5	9
Feeds, grain	3.5	14
Feldspar	2.2	9
Fertilizer	3.0	8
Flour	3.0	12
Fly ash	2.5	5
Graphite	2.0	5
Gypsum	2.0	10
Iron ore	3.0	11
Iron oxide	2.5	7
Iron sulfate	2.0	6
Lead oxide	2.0	6
Leather dust	3.5	12
Lime	2.5	10
Limestone	2.7	8
Mica	2.7	9
Paint pigments	2.5	7
Paper	3.5	10
Plastics	2.5	7
Quartz	2.8	9
Rock dust	3.0	9
Sand	2.5	10
Sawdust (wood)	3.5	12
Silica	2.5	7
Slate	3.5	12
Soap, detergents	2.0	5
Spices	2.7	10
Starch	3.0	8
Sugar	2.0	7
Talc	2.5	10
Tobacco	3.5	13
Zinc oxide	2.0	5

[1] Generally safe design values; application requires consideration of particle size and grain loading.

TABLE 8-9 Fiber selection chart.

Fiber	Generic Name	FIBER PROPERTIES					RECOMMENDED OPERATING TEMPERATURES (°F)	
		Tensile Strength	Abrasion Resistance	Chemical Acids	Resistance Alkalies	Supports Combustion	Continuous	Surges
Cotton	Natural fiber cellulose	Good	Average	Poor	Excellent	Yes	+180	+225
		Comment: Excellent selection in ventilation-type collector.						
Polypropylene	Polyolefin	Excellent	Good	Excellent	Excellent	Yes	+190	+190
		Comment: Strong fiber, low moisture absorption and possesses excellent chemical resistance.						
Glass	Glass	Excellent	Poor	Good	Poor	No	+500	+550
		Comment: All properties are highly dependent on fabric treatment. Can be used at high temperature and has high tensile strength						
Nylon	Polyamide	Excellent	Excellent	Poor	Excellent	Yes	+200	+250
		Comment: Rugged fiber with excellent resistance to abrasion and alkalies.						

Dacron[1]	Polyester	Excellent	Excellent	Good	Fair	Yes	$+275^2$	$+325^2$
		Comment: High tensile strength, good dimensional stability, excellent heat resistance. Susceptible to moist-heat hydrolysis.						
Orlon[1]	Acrylic	Average	Average	Very Good	Fair	Yes	+240	+260
Microtain[3]	Acrylic	Average	Average	Very Good	Fair	Yes	+260	+280
		Comment: Good at elevated temperatures and in acid conditions. Microtain fabrics possess excellent dimensional stability.						
Wool	Natural fiber protein	Poor	Average	Fair	Poor	No	+200	+250
		Comment: Good filterability.						
Nomex[1]	Aromatic polyamide	Very good	Very good	Fair	Very good	No	+400	+425
		Comment: Outstanding heat resistance and good resistance to abrasion.						
Teflon[1]	Fluorocarbon	Average	Below average	Excellent	Excellent	No	+450	+500
		Comment: Can be used at elevated temperatures and possesses excellent chemical resistance.						

[1] E.I. du Pont registered trademark.
[2] Dry heat.
[3] Registered trademark for Globe Albany 100% homopolymer acrylic fabrics.
Courtesy of Globe Albany.

entered into the analysis (e.g., impact of back pressure and baghouse upset conditions). Last, but certainly of equal importance, are the flange-to-flange baghouse requirements (e.g., temperature limitations and volume and dew-point restrictions). Full consideration of the interplay of baghouse and production operations must be made since the baghouse can restrict production and the production process can restrict or shut down the baghouse. The designer must not only consider typical operating conditions but must also review and consider the extreme ranges of the production process and their impact on the dust control system.

8.3 INSTALLATION PROCEDURES

Depending on the type of system chosen, installation of baghouse systems can take a few days to install or several months to erect. In either case, proper planning and installation procedures will save time and money, during both the installation and future operation and maintenance periods.

In this section we discuss recommended installation practices, typical installation errors and oversights, as well as their potential immediate and long-term consequences. We also review a typical pre-startup inspection checklist.

The installation stage of a project plays an important role in the future operation and maintenance of a dust collection system. If poorly done, it can cause operation and maintenance (O&M) problems for years or require large expenditures to correct. If properly executed, it has the potential to ensure initiation and continuation of a successful O&M program. How? This stage allows for the first time, on-site assessment of the compatability of the system design and O&M procedures and requirements. Usually, and unfortunately, it is the first meeting of the system designers and the O&M personnel. Not enough emphasis is generally placed on a complete combined designer/O&M engineering review and inspection during the installation phase. This is understandable because of the usual separation of responsibilities and company affiliation; the design engineer's role terminates with completion of paper design and analysis; the erection engineer is primarily concerned with meeting a schedule and often never understands the dust collector's function; plant O&M personnel, overburdened with existing production equipment O&M, have not yet found time to understand this new monster that normally will provide no additional operating capacity and is generally regarded as a "pain in the butt."

If, however, these obstacles are overcome and effective coordination of the system designer, installer, and O&M personnel is accomplished, the payoff in future dollars saved will be substantial. Some key areas to reevaluate during the installation period are:

Is there easy access to all potential maintenance areas (i.e., fans, motors, conveyors, discharge valves, dampers, damper operators, pressure and temperature monitor ports, mechanical linkage)?

Is there easy access to all potential maintenance areas (i.e., fans, motors, conveyors, discharge valves, dampers, damper operators, pressure and temperature monitor ports, mechanical linkage)?

Is there easy access to all inspection and test areas?

How will weather affect the above?

Are manual operators available with respect to location and access? If a worker has to climb a 50-ft ladder in winter to open a bypass damper to ensure plant equipment or personnel safety or continued operation—better rethink.

Can the system design tolerate all anticipated operating conditions?

Although most of the above should have been considered in the original system design, often they are not. This is primarily because the designer and real O&M personnel have never gotten together or O&M procedures have changed.

Equally important is that the erection engineer and/or O&M personnel not make field changes without the approval of the system designer. Most items of a system are interrelated, and modifying one part without regard for the whole is often costly.

Typical installation errors and their effect on O&M are shown in Table 8-10.

TABLE 8-10 Typical installation errors and their effect on O&M.

Item	Immediate Potential Effect	Long Term
Baffle plates and turning vanes improper installation or left out	Uneven dust distribution; uneven hopper loading; higher pressure loss	Bag wear, duct wear, hopper fires
Poor seal of flanges and access areas	In-leakage resulting in: Reduced inlet volume Higher fan volume Higher operating costs Lower baghouse temperature	Localized cold spots resulting in: Component corrosion Bag degradation
Poor seal at dust discharge flanges	Incomplete discharge, reentrainment; hopper fires	Reentrainment; creeping ΔP
Cracked or chipped paint and coatings	Esthetics	Corrosion
Improper bag tensioning	Ineffective cleaning; bag collapse	Bag wear; high ΔP
Improper bag seating	Stack emission	Compliance failure; bag wear; high ΔP
Incomplete insulation	Cold spots	Corrosion
Seal between dirty and clean air compartments	Stack emission; dirtying of clean side of plenum	Compliance failure
Duct damper alignment	Loss of flow control	Poor maintenance ambient
Screw conveyor direction reversed	No discharge	Bent screw; full hopper; fires
Fan mount	Noise—vibration	Broken components
Fan belt alignment	Noise—improper fan volume	Broken belts
Exposed compressed-air lines without dryer	Freeze-up—condensation	Damaged downstream components
Lack of inspection access	Lack of early warning signs	Major problems
Lack of maintenance access	Lack of regular preventative maintenance	Major breakdowns

A Properly Designed and Installed System Will Provide:

1. *Uniform air and dust distribution to all filters*. Duct design, turning vanes, and deflection plates all assist in obtaining this. Often, they arrive loose and are field-installed. If improperly installed, they can induce high-airflow regions that will abrade the duct or bag filters or cause reentrainment, high-dust-concentration regions that can produce uneven hopper loading and uneven filter bag dust cake.

2. *Total seal of system from dust pickup to stack outlet*. In-leakage at flanges or collector access points either adds additional airflow to be processed or short-circuits the process gases. In-leakage to a high-temperature system is extremely damaging, as it creates cold spots and can lead to dew-point excursions and corrosion. If severe, it can cause the entire process gas temperature to pass through the dewpoint and result in condensate on the bags. Early bag failure and high-pressure drop will generally result. The best check for leaks is to inspect the walls from inside the system during daylight. Light penetration from outside isolates the problem areas. It is particularly important to seal the dust discharge points in negative systems. In-leakage here will result in incomplete or no discharge, which can lead to reentrainment problems, yielding high-pressure drop and hopper fires.

3. *Effective coatings and paint*. Most systems are painted on the exterior surfaces only. Take extra care to touch up damaged areas with a good primer and if equipment is not delivered finish-painted, apply it as soon as possible following erection. Unprotected primers will soon allow corrosion and require sandblasting plus many dollars to repair. If your system has been *internally* protected with a coating, thoroughly inspect for cracks or chips, particularly in corners, and repair before operating. A poor interior coating can be worse than none at all because it will trap corrosive elements between the coating and the surface it was intended to protect.

4. *Properly installed filter bags*. The filter bags are the heart of any fabric filter collection system. Improper installation can result in early bag failure, loss of cleaning effectiveness, and thus high pressure drop and operating costs or increased stack emission. Each manufacturer provides instructions on the proper filter bag installation and tensioning (where required). These must be explicitly followed. Very often, early bag failures can be traced to improper installation. Remember, it is much easier to check and recheck bag connections, tensioning, locations, and so on, in a clean, cool, dry collector than it will be one day after startup. Bag maintenance usually accounts for 70% of annual maintenance time and money. Extra efforts in this area during installation can have a significant effect.

5. *Proper insulation installation*. High-temperature collector systems require special consideration in order to prevent O&M problems. When handling high-temperature gases, it is important to maintain the temperature of the gas and all collector components coming in contact with it above the gas dew point. Insulation is usually employed. Much of the time, all or a part of the insulation is field-installed. Check to see that all surfaces and areas of potential heat loss are adequately

covered. In particular, check to see that field flashing also has insulation beneath it. Cold spots cause local corrosion. Gross heat loss may cause excessive warm-up time or lower the gas temperature below the dew point.

6. *Total seal between dirty side and clean side of collector.* Remember, the primary purpose of the dust collector is to separate the particulate from the gas by means of fabric filtration. This means that *all* the gas must pass through the fabric. Any leaks bypassing the fabric filters will directly emit dust to the stack and therefore reduce the collection efficiency of the system. The time to inspect "bypass leaks" is *before* startup, when everything is clean and accessible. The best technique is to utilize a bright light on one side of the plenum and visually observe for light penetration on the other. This is the most effective in total darkness. Take that extra time to check this important area. Tracking down stack emissions not associated with bag failures can be extremely difficult after startup.

7. *Properly installed and operating dampers.* Most systems employ several dampers to isolate areas of the system or control the volume of air flow. Proper alignment is important both in the case of internal blades and the operating linkage. In high-temperature applications, special care must be taken to allow for proper operation and sealing at the operating temperatures. Some dampers may require readjusting after reaching high-temperature operation. In modular systems, single modules are normally isolated for bag cleaning and maintenance. Leakage through these isolation dampers can cause improper bag cleaning. It will also create a very poor ambient condition for maintenance workers to work in. This, in some applications, can pose a health hazard, and in all applications results in lower-quality workmanship or incomplete maintenance.

8. *Properly operating mechanical components.* Most mechanical components are designed with a normal operating direction. Cylinder rod location, motor rotation, and so on, must be checked. Remember, when hooking up an ac motor, one has a 50% chance of being correct on the first try. Not only will a backward-moving conveyor produce no discharge, but it can pack material so that it bends the screw. Left uncorrected, a reversed screw conveyor will result in a full hopper. The industry abounds with horror stories where full hoppers have led to burned bags, or dust set up, requiring jackhammers to remove it.

9. *Smoothly running fans.* Fans must be checked for proper rotation, drive component alignments, and vibration. Fans should be securely mounted to sufficient mass to eliminate excessive vibration.

10. *Clean, dry compressed air.* Most systems employ compressed air to operate dampers, controls, instruments, and so on. Probably more systems suffer shutdowns and maintenance problems due to poor-quality compressed air than for any other reason. Clean, dry air is necessary to maintain proper operation of the pneumatic components. In installations where the ambient temperature drops below 32°F, a desiccant dryer system is generally employed. Sometimes, insulation of air lines and pneumatic components will be required. Often, these considerations are not included in the dust collector system, with "clean, dry compressed air to be supplied by owner." Remember, the air must be clean and dry when it reaches the pneumatic component.

1. Visually inspect:	
Structural connections for tightness	
Duct flanges for proper seal	
Filter bags for proper seating in tube sheet	
Dampers for operation and sequence	
System fan, reverse-air fan, and conveyors—check for proper rotation	
Electrical controls for proper operation	
Rotary valves or slide gates for operation	
2. Remove inspection door and check conveyor for loose items or obstructions	
3. Adjust ductwork dampers—open or at proper setting	
4. Remove any temporary baffles	
5. Test horn alarm system, if included, by jumping connected sensors	
6. Start screw conveyors and check for proper operation	
7. Start reverse air fan, if included	
8. Start system fan	
9. Log manometer and temperature (if appropriate) readings at 15-minute intervals; log readings	
10. Check to see that reverse air dampers are cycling	
11. Adjust ΔP switch, if included	
12. Determine system air volume and adjust dampers, as required	
13. Check cells for dust leaks	
14. Check to see that dust is being discharged from hopper	

FIGURE 8-10 Inspection and startup checklist.

The Check List

Each installation will have its own checklist reflecting its unique components. The important thing is to prepare one *before* beginning your final inspection and initial startup. Startup is normally a hectic time, so "be prepared." Figure 8-10 shows a typical checklist for a fairly simple system.

The Formula for a Successful Startup

Implementation of the following steps will go a long way toward providing a successful startup:

Advance planning—create your checklist

Good communication with operating personnel both before and during startup

Be certain of the system's initial state through a thorough inspection

Define and communicate to all personnel involved the operating conditions requiring an *abort* and the emergency steps required to accomplish it

Activate each part of the system carefully and in proper sequence

Observe and log all pertinent data at frequent intervals until the system has "settled" out

Reinspect key areas under operating conditions

Identify any special action required of operating personnel

8.4 OPERATION

The operation phase of a fabric filter collection system should start long before the actual startup. Prior to bringing the collector system on-stream there should be familiarization and training sessions for all O&M personnel. In these sessions the following subject areas should be covered:

System design

Component functions

Control system

Critical limits and alarms

Key parameters to monitor

Good operating practices

Specific *no-no's*

Preventive maintenance program

Startup and shutdown procedures

Emergency shutdown procedures

Safety considerations

This training phase is too often left to on-the-job training. Such an approach can be both dangerous and costly. It should be the plant manager's and/or plant engineer's responsibility to see that all responsible O&M personnel are both qualified and trained in the operation and maintenance of the system prior to startup.

Figure 8-11 shows a system schematic for a typical high-temperature system. Its operational-mode chart shows the condition of all critical components for each mode. A schematic and chart of this type should be prominently displayed in the control area. Startup and shutdown procedures, including emergency shutdown for the system, should also be displayed.

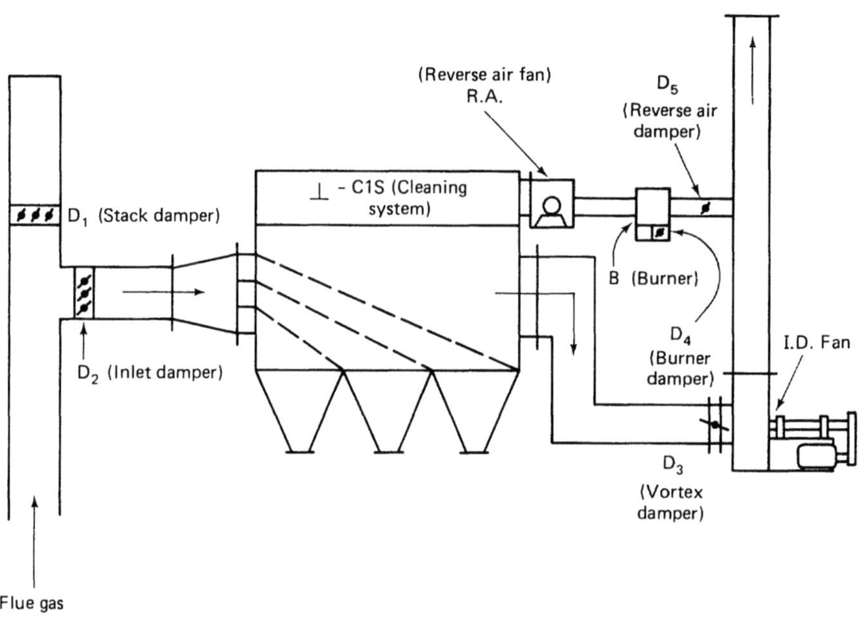

Mode	D₁	D₂	D₃	D₄	D₅		I.D.	R.A.	B.	C1S.
Preheat	O	C	O	O	O		OFF	ON	ON	ON
Normal	C	O	Auto	C	O		ON	ON	Auto	Auto
Purge	O	C	O	O	C		OFF	ON	ON	ON
Bypass	O	C	O	C	C		OFF	OFF	OFF	OFF
Supl. heat	C	O	Auto	O	O		ON	ON	ON	ON
Supl. cool	C	O	Auto	O	O		ON	ON	OFF	ON

Figure 8-11 Typical high temperature baghouse system.

Startup and Shutdown

Operating a baghouse is like flying a plane—the most critical moments are usually at the beginning and the end of a run.

Improper startup can lead to long-term negative effects. For example, partial or total blinding of the bags and residual pressure drop can result from allowing too high a filtering velocity or dew-point excursions. Improper shutdown can result in dewpoint excursions, over- or undercleaned bags, inadequate hopper discharge, and so on. High-temperature applications are particularly sensitive to startup and shutdown periods. Each application has its special concerns and the baghouse manufacturer should be consulted as to the proper do's and don'ts for your system.

Initial startup   The initial startup is potentially the most dangerous time for permanently damaging the fabric. The clean filters have not developed a protective dust cake and are particularly sensitive to dust abrasion and penetration of fine particles that can lead to permanent residual pressure drop. For some applications, the bags are precoated with a protective dust prior to the introduction of flue gas. In all cases, the filtering velocity should be kept very low until a sufficient dust cake is developed. This will be indicated by a pressure drop of 1 to 2 in. water gauge. The gas flow can then be slowly increased until design volume is achieved.

A system flow control damper is usually available to regulate the gas flow. If there is none, the pickup lines should be dampered back by using the balancing blast gates or restricting the pickup hoods with a temporary baffle.

Some general rules for routine startup and shutdown are:

Startup

Make sure all collector components are in working order and in proper mode.

Do not allow higher-than-design filtering velocities or air flow.

Avoid passing through (below) the dew-point within the baghouse when dirty gases are present. The system should be preheated to above the dew point with clean, hot air before the introduction of flue gas. During normal operation, maintain the temperature above the dew-point level.

Operate the bypass system to assure its readiness in an emergency situation.

Check all indicating and monitoring devices for proper operation.

Shutdown

Purge the collector with clean (hot when necessary) dry air before allowing temperature to descend below the dewpoint. Even ambient applications will pass through the dewpoint when the collector cools at night.

Do not store dust in the collector. Many maintenance workers have resigned after spending a day with pick and shovel inside a dust collector hopper.

Allow bags to clean down after dust flow ends, but do not overclean.

Check to see that all components are in the proper shutdown mode.

All O&M personnel should become familiar with each of the system components, the control system, and monitoring devices. The operator should monitor and keep records so that system or component deviations can be identified. These will serve as early warning signals for maintenance and assist in troubleshooting.

Normally, the principal parameters of concern are pressure drop, temperature, outlet emissions, and hopper discharge rates. Other intelligence related to specific equipment may provide early warning signs of problems that are not easily detected or understood through the major parameter indicators. The object is to identify and correct problems before they reach the magnitude requiring shutdown. It is also recommended that alarms be included within the system sufficiently below shutdown levels to provide necessary lead time in correcting the problem before shutdown.

After "living" with the new system for a while, the operating and/or maintenance personnel may define additional monitoring or control equipment to accomplish this task.

Typical monitoring and indicating devices include:

Pressure-drop indicators, magnehelics, photohelics, or manometers to define pressure drop across the baghouse or at other points of the system. Recorders of this parameter are very useful and often employed.

Opacity meters which monitor the outlet and indicate the percent opacity.

Gas flowmeters to indicate actual flow or percent of design flow rate.

Temperature measuring and recording devices measuring inlet, outlet, and key temperature locations.

Indicating lights defining operation condition of all motors, solenoids, dampers, and so on.

Adjustable timers to control the frequency and duration of the cleaning mechanism.

Fan-current indicator, fan-bearing-temperature indicator, fan-vibration detector, all aimed at identifying early warning signs of maintenance needs.

Heat-sensitive materials placed in key areas such a bag housings, cylinder heads, and motor housings. These change color when exposed to temperature.

Corrosion chips placed inside the ductwork, collector areas, and so on, to monitor the degree of internal hardware corrosion.

Safety

All O&M personnel must know the safety limits of the system and when to abort the system. Normally, there will be an emergency shutdown button for this purpose and often the abort system is automatic. Usually, the important thing to do in such an emergency situation is to stop the flue gas from entering the collector and that means shutting off the fan, opening the bypass damper, and shutting any inlet or outlet dampers, where available.

Whenever personnel enter a collector hopper or bag plenum, protection from dust, temperature, and flue gas contaminents and entrapment must be provided. All entry doors should include an electrical interlock to ensure compartment isolation and a positive mechanical method for ensuring that the door cannot close accidentally. When performing work inside a collector, it is advisable that one member of the team always remain outside. Reports of personnel becoming trapped inside collectors have been alarmingly frequent. If adequate ventilation cannot be achieved, it will be necessary for maintenance and inspection personnel to wear life support systems and proper protective clothing.

Personnel should be trained as to emergency procedures in case of a fire, explosion, or electrical hazard in or around the collector. A dust collector is a relatively safe piece of equipment, but caution and preparedness are still required.

8.5 MAINTENANCE

The award of the year for *least*-maintained equipment again goes to the *fabric-filter*!!! This unfortunate lack of care not only costs industry millions of dollars each year, but also reduces the potential impact of air pollution control on air quality levels.

Normally, a dust collector has been or will be installed at the encouragement of the local, state, or federal regulatory agency or union, not a production department. Even though it is often tied directly in series with your production process, it is often treated like a stepchild. The first step toward efficient, trouble-free operation of a fabric filter system is to treat it the same as you would any other piece of *production* equipment. It is a machine, and machines need "tender loving care." It requires routine inspection, preventative maintenance, and quick response to malfunctions.

Routine inspection of key components and operating parameters are the key ingredients to normal maintenance. Table 8-11 shows a typical periodic maintenance checklist.

A preventative maintenance program is a must for systems requiring continuous duty. The preventative maintenance procedures must be continually updated to reflect your actual experience, not some typical guidelines.

The extent of preventative maintenance during scheduled shutdowns must be evolved, starting with the equipment supplier's recommendations, supplemented by your own routine inspection and maintenance records.

Ah, records—the heart of any maintenance system! No one wants to keep them, but everyone wants to see them. Log keeping of actual inspections, observations, maintenance performed, and so on, is invaluable for:

Updating the maintenance program as a whole, modifying the preventive maintenance schedule or procedures

TABLE 8-11 Routine maintenance schedule.

Item	Check	Daily	Weekly	Monthly	Quarterly	Annually
Dust pickup areas	Dust capture effectiveness	×				
	Monitoring flue gas volume	×				
	Test hood face velocity				×	
	Rebalance system					×
System dampers	Proper operation		×			
	Proper valve seating				×	
	Wear or corrosion					×
Bags	Observe stack (visually or through opacity meter)	×				
	Spot-check bag tensioning (inside collectors)		×			
	Spot-check bag condition and seating			×		
	Thoroughly inspect bags				×	
Cleaning system	Monitor cleaning cycle	×				
	Check compressed air	×				
	Inspect mechanical components for wear			×		
	Replace high-wear parts					×
Control system	Observe all indicators on panel	×				
	Log ΔP	×				
	Blow out manometer lines		×			
	Check compressed air system, including filters		×			
	Activate key shutdown or bypass controls		×			
	Verify accuracy of temperature-indicating equipment		×			
	Check accuracy of all other indicating equipment			×		
System fan	Check drive components		×			
	Inspect for corrosion and material buildup			×		
	Check for vibration		×			
Outlet stack	Check emission visually or monitor opacity meter	×				
	Calibrate opacity meter				×	
Discharge system	Monitor discharge rate	×				
	Check all moving parts for wear and alignment			×		
General	Walk through entire system to check for normal and ab-normal visual and audible conditions	×				
	Inspect system for corrosion			×		

TABLE 8-11 Continued

		FREQUENCY				
Item	*Check*	*Daily*	*Weekly*	*Monthly*	*Quar-terly*	*Annually*
	Inspect door gaskets				×	
	Check for dust buildup in ducts				×	
	Inspect paint				×	
	Inspect baffles, hopper duct, etc., for wear					×
	Inspect general structural integrity of system					×

Defining problem areas to be corrected

Defining maintenance costs and manpower needs

Maintaining spare-parts control

Each plant normally has its own method of maintenance record keeping, but the following items should be included:

Activity date and time of day

Type of activity (i.e., routine inspection, emergency maintenance, preventative maintenance, etc.)

Specific action(s) taken (i.e., replacement modification, adjustment, etc.)

Components affected—by part number and manufacturer

Problem cause, if known

Manhours expended

Materials used

Spare-parts usage

Any recommendations to prevent reoccurrence

The time used to record the details listed above will be well worth it. The maintenance section that recognizes the value of these items, is diligent in obtaining and utilizing it, will be under control and not be known as the local firefighters.

Troubleshooting

Detecting and correcting small problems before they snowball into large and expensive crises is the goal of any maintenance program.

Trouble symptoms A major malfunction in your collector system will result in one or more of the following identifying symptoms:

TABLE 8-12 Troubleshooting guide.

Symptom	Possible Cause	Remedy
High collector pressure drop	Malfunction of bag-cleaning system	Check all cleaning-system components
	Ineffective cleaning	Modify cleaning cycle Review with designer
	Reentrainment of dust in collector due to low-density material or in-leakage at discharge	Check discharge valves Lower A/C ratio
	Wetting of bags	Control dew-point excursions Dry bags with clean air Clean bags with vacuum or wet wash
	Too high A/C ratio either through added capacity or improper original design	Verify gas volume Reduce inlet volume if possible Review with designer
	Change in inlet leading or particle distribution	Test Review with designer
Abnormally low pressure drop	Manometer line(s) plugged	Blow back through lines Protect sensing point from dust or water buildup Incorporate autopurging system in sensing lines
	Manometer line(s) broken or uncoupled	Verify with local manometer Inspect and repair
	Overcleaning of bags	Reduce cleaning energy and/or cycle time
Stack emission	Broken bag	*See* Bag maintenance section
	Bag permeability increase	Test bag Check cleaning energy/cycle and reduce if possible
	Clean-to-dirty plenum leakage	Inspect and repair
	Change of inlet conditions	Test and review
Puffing	High pressure drop across baghouse	*See above*
	Low system fan speed	Check drive system Increase speed
	Improper duct balancing	Rebalance system
	Plugged duct lines	Clean out
	Poor hood design	Evaluate temporary modifications and implement
	Improper system fan damper position	Check and adjust
Low dust discharge	Inleakage at discharge points	Inspect and repair seals or valves
	Malfunction of discharge valve, screw conveyor or material transfer equipment	Inspect and repair
	Reentrainment of dust within collector	Lower A/C ratio

226

TABLE 8-12 Continued

Symptom	Possible Cause	Remedy
	Retainment of dust on filter bags	Increase cleaning
Loud or	Vibrations	
unusual	Banging of moving parts	
noises	Squealing of belt drives	
Corrosion	Improper paint material or application	
	Improper insulation	
	Emission of nearby equipment	
	Dew-point excursions	
	Improper shutdowns	

1. Abnormally *high pressure drop* across the collector
2. Visible *emission* of dust in the exhaust stack
3. Inadequate face velocity or *"puffing"* at the pickup points and hoods
4. *Lower*-than-normal *dust discharge*
5. *Loud* or *unusual noises*
6. Severe *corrosion* of *material*

A general guide is shown in Table 8-12. If unable to identify and correct a problem in a reasonable period of time, do not hesitate to contact the system designer or the equipment manufacturer. Often, a phone call to the right person is all that is required. Remember, the manufacturer has the advantage of experience and feedback from many sources to assist in solving your problem.

Bag Maintenance

The highest source of maintenance and cost are generally the filter bags. All bag sets have a finite lifetime which will vary by application, installation, operating parameters, fabric type, and so on. Typical causes of bag failure are:

High G/C abrasion

Metal-to-cloth abrasion

Chemical attack

Bag-to-bag abrasion

Inlet velocity abrasion (on inside-out cleaning)

Accidents

Upset conditions (e.g., temperature)

Thread mismatched

Cuff mismatched

In addition, each bag in a set may have a different life as a result of fabric quality, bag manufacturing tolerances, location in the collector, variation in the bag cleaning mechanism, and so on. Any one or a combination of these factors can cause bags to fail. This means that your baghouse will experience a series of intermittent bag failures until the failure rate requires total bag replacement. Typically, a few bags will fail initially or after a short period of operation due to installation damage or manufacturing defects. The failure rate should then remain very low until the operating life of the bags is reached, unless a unique failure mode is present within your system.

The importance of when you correct a broken bag will depend on the type of collector you have and the resultant effect on outlet emissions. In "inside bag collection" types of collectors, it is very important that dust leaks be stopped as quickly as possible to prevent adjacent bags from being abraded by jet streams of dust emitting from the broken bag. This is called the "domino effect" of bag failure. "Outside bag collection" systems do not have this problem, and the speed of repair is determined by whether the outlet opacity has exceeded its limits. Often, it will take several broken bags to create an opacity problem and a convenient maintenance schedule can be employed instead of emergency maintenance.

In either type of collector, the location of the broken bag or bags has to be determined and corrective action taken. In a noncompartmentized unit, this requires system shutdown and visual inspection. In inside collectors, bags often fail close to the bottoms, near the tube sheet. Look for accumulation of dust on the tube sheets, the holes themselves, or unusual dust patterns on the outside of the bags. Other probable bag failure locations in reverse-air bags are near anticollapse rings or below the top cuff. In shaker bags, also inspect the area below the top attachment. Improper tensioning can cause early failure.

In outside collectors, which are normally top-access systems, inspection of the bag itself is difficult; however, location of the broken bag or bags can normally be found by looking for dust accumulation on top of the tube sheet, on the underside of the top-access door or on a blow pipe.

If the system is compartmentized, the search for broken bags can be narrowed by monitoring the stack while isolating one known compartment at a time. Through an opacity monitor, and sometimes visual observation of the stack, the compartment containing broken bags can be identified, since the emissions will be reduced when a compartment containing broken bags is isolated from the system. Once the compartment is identified, find the specific broken bag as detailed above.

One of the latest techniques for locating failed bags is the use of fluorescent or phosphorescent powder and an ultraviolet light. The powder is injected into the inlet gas stream and the ultraviolet light used to scan the clean-air side of the collector. Very small leaks can be detected from the glow of the powder under the ultraviolet light where it has penetrated the clean air plenum.

Once you have found a broken bag, you must correct it. Individual bags within a compartment should not be replaced with clean bags, as they will be forced to filter a much higher volume of air than the older dust-laden bags, due to its lower pressure drop. This higher filtering velocity could cause permanent "blinding" and high pressure drop or early failure through dust abrasion. Instead, the broken bag should be tied off or plugged up. In inside collection bags where the failure has not occurred too close to the tube sheet, the bag should be cut, tied, and stuffed into its hole. For outside collection bags and inside collection bags with holes too close to tie, a hole plug should be used. Your collector manufacturer can probably supply you with hole plugs or they can be made from steel plate with proper gasketing or heavy sand bags made from the same material as the filter bags.

Further action will be dictated by increased pressure drop. As the number of bags taken out of service rises, so will the pressure drop. At some point you will need to replace the broken bags. When this occurs, it is recommended that you replace all broken bags with used bags from one designated compartment. Change out all bags in the designated compartment with new bags and damper the flow into that compartment until a sufficient dust cake has been developed. Store extra "used" bags for future replacement.

When inspecting for plugging or changing out broken bags, care must be taken for the safety of the maintenance personnel. Protection from dust, temperature, and flue gas contaminents must be provided. This is especially important for inside collectors, where the maintenance personnel must enter the bag plenum.

Care should also be taken not to damage adjacent bags and to properly reinstall the new bag. This operation should not be rushed and as a good working environment as possible should be provided. Errors here will only create more problems tomorrow.

8.6 IMPROVING OPERATION AND PERFORMANCE

In this section our comments are directed toward improving the operation and performance of fabric filter dust collection systems. It is assumed for the sake of this discussion that the user is at the point where the baghouse system has no major problems, the dust collector is operating smoothly; however, there is a desire to improve operational reliability and reduce operating costs. As one orients to optimizing costs, it becomes clear that general cost rules can be very misleading because the gas volume capacity can vary greatly from one unit to another, and therefore operating, maintenance, and capital cost relationships can change dramatically. The approach to optimization can be uniform, however, and may include trading off additional capital investment or increased maintenance costs to reduce overall operating costs.

The initial cost objective when purchasing the system was to achieve a technically reliable system capable of meeting the outlet emission goals and achieving all of this at the minimum annualized cost. Once the system has been purchased

The first and basic requirement for improved operation and operating costs is that of record keeping. Specifically, a daily log of any breakdowns, parts replacement, and operational problems needs to be kept. In addition, monitoring and recording of operational cost parameters is needed. For example, records of specific utility component costs need to be separately kept (e.g., main fan, reverse-air fan, conveyors, etc.). Examples of their cost parameters are shown in Table 8-13.

After about six months of operation these data should be reviewed and operational changes made based upon minimum operating cost considerations. It should be noted that data obtained in the early weeks may not be representative of long-term operational data, and therefore one should evaluate where the data appear to level out and more weight should be given to these data.

Analysis of the data obtained falls into two categories, the first being a review for any recurring operational problems and/or component failures. If this review identified any problems of a repetitious nature, corrective actions need to be defined and executed. These remedial actions can include component changes, modifications, additions, relocations, increased or modified preventive maintenance procedures, and additional monitoring equipment and controls. The second area of data review is a cost study to determine whether there are hardware or operational mode changes that can result in operation-cost savings. This analysis should be done in a rigorous fashion and probably put in graphical form. Points to be considered here are discussed below.

Can the system be run at a higher gas/cloth ratio and allow for increased gas volume, and therefore increased production or the addition of another source operation?

TABLE 8-13 Installed and operating cost components.

Installed cost components
   Design
   F.O.B. baghouse
   Fans and motors
   Ducting
   Dust conveyors
   Instrumentation
   Freight
   Foundations
   Structural steel
   Insulation
   Installation labor
   Start-up
Operating cost components
   Electric power (fan motors, conveyor drives, compressors, lights)
   Auxiliary heat
   Bag replacements
   Maintenance (fans, motors, valves, cylinders, filters, labor)
   Dust disposal
   Overhead

Can the cleaning system be modified to allow for either pressure-drop reduction or longer bag life? Bag cost vs. utility costs need to be weighed here. It is at this point that the user recognizes that the original selection of the G/C and cleaning cycle was purely an empirical selection and now, based on the actual operation, this can be modified to arrive at lower-cost operation. Bag type and finishes are evolving rapidly in the current market, and it may be worthwhile to initiate one's own bag screening program. This can be done by installing one or more test bags and monitoring their performance. Such a program may be simply oriented to monitoring bag life or if conditions allow, provide pressure drop and emission data as well. In cases of very poor bag life (i.e., less than 1 year), excessive pressure drops (e.g., greater than 6 to 8 in. water gauge), or very costly bags, it is recommended that a bag screening program be instituted.

8.7 CONCLUSIONS

One final note is directed toward the total production process. A review of operating cost will indicate that utilities are of ever-growing concern and the dust collection system fan is usually the major utility component of the dust collection system costs. Reduction of the system fan operating cost requires reduction of the system pressure drop. This can be accomplished by optimization of the duct run design, inlet and outlet damper designs, the bag cleaning system, filter media type, or reduction of the gas volume processed. Often, the fan speed can be adjusted to an improved fan curve operating point. It is important that constant vigilance be maintained to ensure that the process volume is no greater than required (e.g., boiler excess air control) and that process gas inleaks are eliminated.

The manufacturer's operating manual is at best only a guide to proper operation. The optimization of operation and minimization of operating costs are left in the user's hands. Awareness and sensitivity to the operational cost parameters and a semiannual review of the operational data collected are the route to improved operation, performance, and costs.

8.8 REFERENCES

[1] BILLINGS, C. E., AND WILDER, J., *Fabric Filter Systems Study: Handbook of Fabric Filter Technology*, Vol. 2. U.S. Dept. of Health, Education and Welfare, Washington, D.C., 1970, App. 5.1.

[2] Dennis, Richard, et al., *Filtration Model for Coal Fly Ash with Glass Fabrics*. EPA-600/7-77-084, 1977.

9

9
Wet Scrubbers

Anthony J. Buonicore, P.E.

President
Buonicore-Cashman Associates, Inc.
Bridgeport, Connecticut

9.1 DESCRIPTION OF CONTROL DEVICE

Wet scrubbers have found widespread use in cleaning contaminated gas streams because of their ability to remove effectively both particulate and gaseous pollutants. Specifically, wet scrubbing describes the technique of bringing a contaminated gas stream into intimate contact with a liquid.

The particulate collection mechanisms involved in the wet scrubbing operation may include some or all of the following.

Inertial impaction

Direct interception

Diffusion (Brownian movement)

Electrostatic forces

Gravitational forces

Condensation

Thermal gradients

Inertial impaction occurs when an object (the droplet), placed in the path of a particulate-laden gas stream, causes the gas to diverge and flow around it. Larger particles, however, tend to continue in a straight path because of their inertia; they may impinge on the obstacle and be collected (as in Figure 9-1a). Direct interception also depends on inertia and is merely a secondary form of impaction. A collision occurs due to direct interception if the dust particle's center misses the target object by some dimension less than the particle's radius (see Figure 9-1b). Direct interception is, therefore, not a separate principle, but only an extension of inertial impaction. Diffusion is another extension of the impaction principle. Very small particles (submicron) suspended in a gas stream have an individual oscillatory motion known as Brownian movement (see Figure 9-1c). In this case, particle and target collide as a result of relative motion within limited space. As in all diffusional processes, the rate of diffusion is favored by large areas for diffusion, thereby necessitating small liquid droplets with high surface-to-volume ratios for high collection efficiencies. While collision or impaction may be the result of either inertia or Brownian movement, the results are the same. Gravitation forces can also cause a particle, as it passes an obstacle, to fall from the streamline and settle on the surface of the obstacle (see Figure 9-1d). Such forces come into play, however,

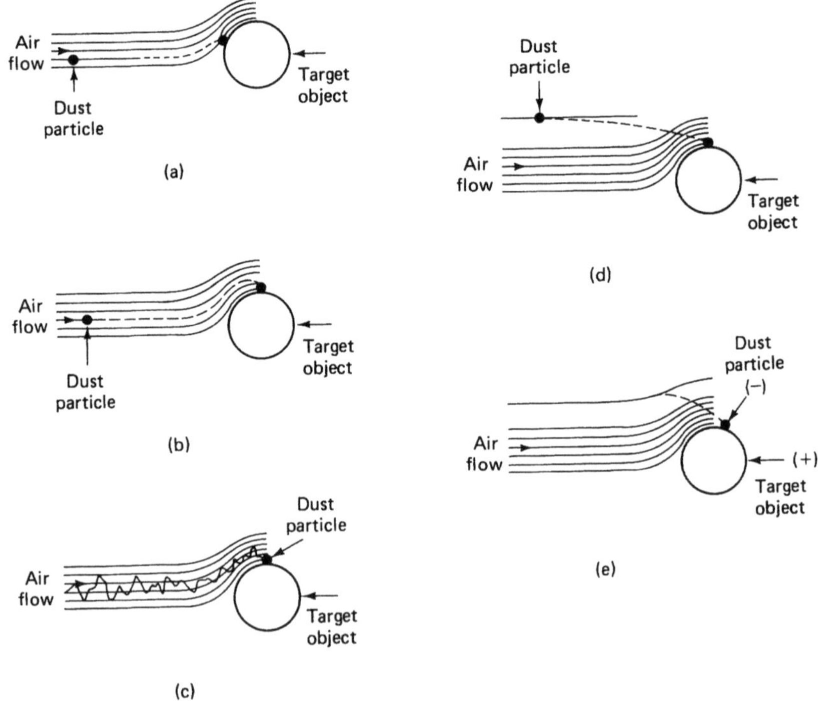

FIGURE 9-1 Principal collection mechanisms in wet scrubbers: (a) intertial impaction; (b) direct interception; (c) diffusion; (d) gravitational forces; (e) electrostatic forces.

when dealing with the larger particles (usually greater than 40 μm). Electrostatic forces result when particles and liquid droplets become electrically charged (see Figure 9-1e). In addition, when only the particle or obstacle is charged, a charge may be induced on the uncharged component, resulting in a polarization force that can also effect particle removal. The electrostatic charge may be acquired, for example, by liquid droplets during their formation. An electrical charge may also be induced by flame ionization, friction, or the presence of charged matter. The effect of the electrostatic mechanism may be significant when the charge on the particle or obstacle is high and when gas velocity is low. Condensation effects may also come into play. Condensation occurs if the gas or air is rapidly cooled below its dew-point. When moisture is condensed out of the gas stream, fogging occurs, and the dust particles can serve as condensation nuclei. The dust particles can become larger as a result of the condensed liquid, and the probability of removal by impaction is increased. Particle collection may also be affected by thermal gradients. Such forces can drive particles from hotter to colder regions. The motion can be caused by unequal gas molecular collision energy on the surfaces of the hot and cold sides of the particle; it is directly proportional to the temperature gradient.

9.2 DESIGN PROCEDURES

Wet scrubbers are constructed with such a multiplicity of designs that no single type can be considered representative of the category as a whole. Some units simply consist of an existing dry-type collector modified by the introduction of a liquid phase to assist in particulate removal and to prevent particulate reentrainment; other units are specifically designed to operate as wet collectors. It is difficult to generalize about relationships among operating parameters such as pressure drop and liquid flow rate. It is also difficult to classify wet scrubbers according to particle collection mechanisms since the usual case finds many of the collection mechanisms working simultaneously in each type of scrubber. However, in a very general sense, wet scrubbers may be loosely categorized by pressure drop (or energy consumption). Low-energy scrubbers are those with typical pressure drops less than 5 in. of water; medium-energy scrubbers are those with typical pressure drops from 5 to 15 in. of water; and high-energy scrubbers are those with typical pressure drops greater than 15 in. of water. Spray chambers and spray towers, for example, provide the lowest pressure drop and, correspondingly, the lowest collection efficiencies; they would be classified as low-energy scrubbers. The medium-pressure-drop group could include centrifugal fan wet scrubbers, atomizing impingement collectors, and certain packed-bed scrubbers. The most familiar of the high-pressure-drop (high-energy) group is the venturi-type collector. Since many scrubbers can conceivably be listed in more than one of the categories listed above, each type of collector will be considered individually in this section, with its eventual application determining whether it will be classified as a low-, medium-, or high-energy scrubber. Only conventional scrubber systems will be considered.

Spray Towers

The most common low-energy scrubbers are gravity spray towers in which liquid droplets are made to fall through rising exhaust gases and are drained at the bottom of the chamber (see Figure 9-2). The droplets are usually formed by liquid atomized in spray nozzles. The spray is directed into a chamber shaped to conduct the gas through the finely divided liquid. In a vertical tower the relative velocity between the droplets and the gas is eventually the terminal settling velocity of the droplets. However, to avoid spray droplet reentrainment, the terminal settling velocity of the droplets must be greater than the velocity of the rising gas stream. In practice, the vertical gas velocity typically ranges from 2 to 5 ft/sec. For higher velocities, a mist eliminator must be used in the top of the tower.

Spray towers are suited for both particle collection and mass transfer (gas absorption), as are all wet scrubbers. Operating characteristics include low pressure

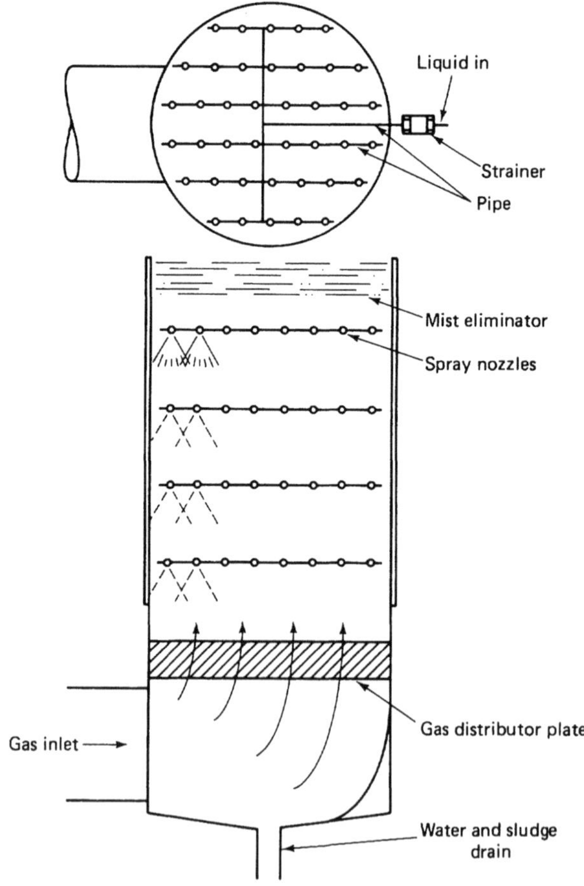

FIGURE 9-2 Gravity spray tower.

drop (typically less than 1 to 2 in. of water, exclusive of mist eliminator and gas distribution plate), ability to handle spray liquids having a high solids content, and liquid requirements ranging from 3 to 20 + gal/1000 ft^3 of gas treated. Spray towers are capable of handling large gas volumes and are often used as precoolers to reduce gas stream temperatures. Gas rates from 800 to 2500 lb/hr ft^2 are typical. Gas retention times within the tower typically range from 20 to 30 sec. The chief disadvantage of spray towers is their relatively low scrubbing efficiency for dust particles in the 0- to 5-μm range. Particles larger than 10 μm can typically be removed with efficiencies up to about 70%. The efficiency may be improved by the addition of high-pressure sprays, ranging from 100 to 400 psig. Depending on the spray pressure utilized, efficiencies can be increased by as much as 20%. The use of baffles can also increase collection efficiency. An important advantage of spray towers is their capability to treat relatively high dust concentrations without fear of any plugging. Although the units have a relatively large space requirement, they are inexpensive and are used in practice primarily for the collection of coarse dusts (greater than approximately 25 μm) and as a precooler.

Centrifugal Collectors

Particulate collection efficiency is improved by increasing the relative velocity between the droplets and the gas stream. The particulates are assumed to travel at the same velocity as the gas stream. An increase in this relative velocity may be achieved by utilizing the centrifugal force of a spinning gas. The spin may come from introducing gases to the scrubber tangentially or by directing the gas stream against stationary swirl vanes. Although normally designed for specific installations, cyclonic spray scrubbers can accommodate gas entrance velocities up to 100 + ft/ sec. In the centrifugal systems shown in Figures 9-3 to 9-5, the principal benefit is derived from the wetted walls hindering reentrainment of collected particulates. Water rates of 2 to 5 gal/1000 ft^3 of gas and pressure drops from 2 to 6 in. of water are typical. In Figure 9-6 the liquid spray is directed outward from sprays set in a central pipe manifold. Sprays in this fashion enhance collection efficiency by increasing particle impact on the spray droplets. Sprays directed in from the walls, as in Figure 9-7, are more easily serviced. However, since they can be made accessible from the outside of the scrubber, an unsprayed section is usually provided to give liquid droplets containing collected particles sufficient time to reach the walls of the collector and prevent them from exiting the unit in the cleaned gas stream. Figure 9-8 is an example of the rotating motion given to the gas stream by fixed vanes and impellers, with the scrubbing liquid introduced centrally either as a spray or a liquid stream.

Dynamic Collectors

Dynamic collectors, as depicted in Figure 9-9, serve primarily as air movers (centrifugal blowers). Spray nozzles at the inlet offer an opportunity for impaction. Liquid droplets from the nozzles and uncaptured dust must pass through the many

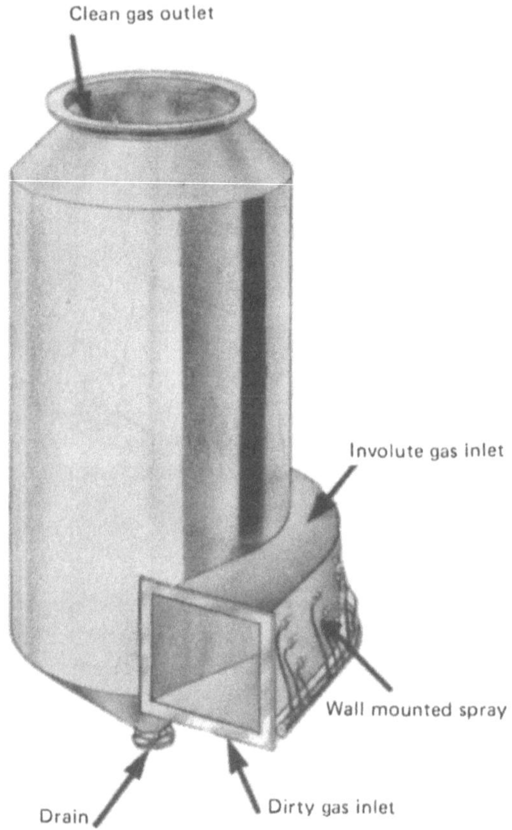

FIGURE 9-3 Polycon cyclone spray scrubber. (Courtesy of Polycon Corporation.)

specially shaped blades of the rotating fan wheel. Centrifugal force and impingement on the blades are utilized for further collection and water separation. Problems have arisen, however, due to instability attributed to sludge buildup on the fan wheel. Slurry is removed and discharged separately from the cleaned air. Design pressure drop is about 6 in. of water with a maximum pressure drop of 9 in. Water requirements typically range from 0.5 to 1.5 gal/1000 ft^3 of gas, and power requirements range from 1 to 2 hp/1000 acfm of gas treated. The collection efficiency is similar to other scrubbers with a 6-in. pressure drop. The chief advantages are compactness, moderate power requirements, and low water consumption.

Packed-Bed Scrubbers

Packed-bed scrubbers find normal application in the removal of pollutant gases and vapors. However, such systems have also been considered for the removal of particulate matter. There are four general ways in which these beds may be operated

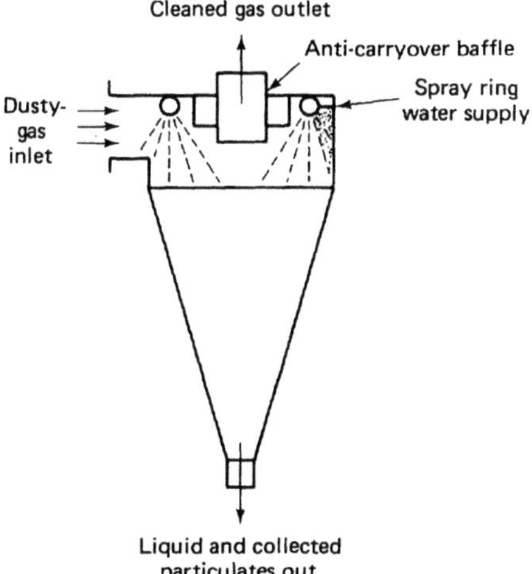

FIGURE 9-4 Large-diameter irrigated cyclone.

FIGURE 9-5 Tubular-type centrifugal wet collector. (Courtesy of American Filter Company, Inc.)

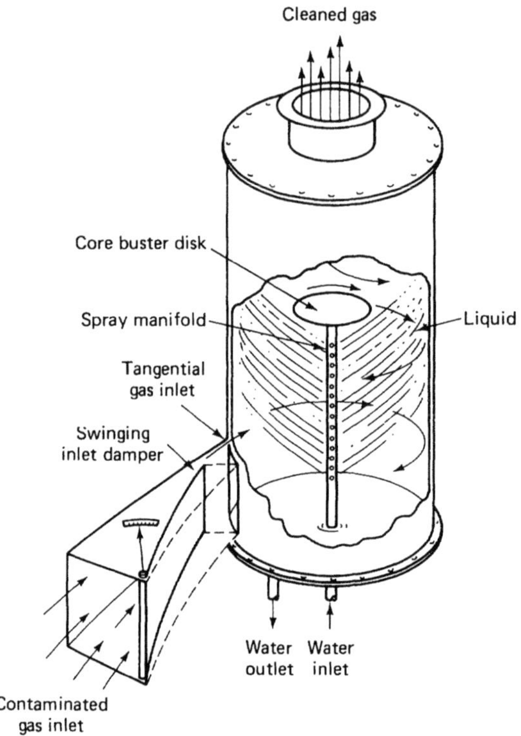

FIGURE 9-6 Cyclonic scrubber.

to effect particulate removal (see Figure 9-10). Cross-flow scrubbers will efficiently remove the relatively large particulates; they are also used for the removal of liquid particulates. The cocurrent flow scrubber is somewhat more efficient than the cross-flow and is effective in removing both solid and liquid particulates. For the collection of particulates as small as 3 to 5 μm, packed-countercurrent scrubbers have been found to be most efficient. These scrubbers operate generally at much higher liquid rates than either the cross-flow or the cocurrent-flow types. Bed depth in packed countercurrent scrubbers is typically 2 to 6 ft. Gas velocities typically range between 3 and 6 ft/sec. Cross-flow beds are usually less than 2 ft in depth.

The difficulty of removal of particulate material increases rapidly with a decrease in particulate size. A 1-μm particle is only 10 times larger in diameter than a 0.1 μm particle, but it will have 100 times more surface area and 1000 times more weight. As particle size decreases, the surface area/mass ratio increases so rapidly that eventually the surface properties can dominate over the mass properties. When this happens, higher velocities and more acute changes in direction are required to permit the particle to be separated from the gas stream by impingement. Cross-flow, cocurrent-flow, and countercurrent-flow packed scrubbers are not capable of achieving a high-enough gas velocity to remove effectively particulates smaller than

3 to 5 μm in diameter. To remove particles in this small size range, much more energy must be used to attain the velocities needed, and to avoid flooding, the liquid should be introduced in the same direction as the gas (i.e., the bed is operated in cocurrent flow).

Liquid requirements for cross-flow operation typically range from 1 to 4 gal/1000 ft³ of gas, with pressure drops from 0.1 to 0.25 in. of water per foot of bed depth. Liquid requirements for countercurrent-packed scrubbers usually range from

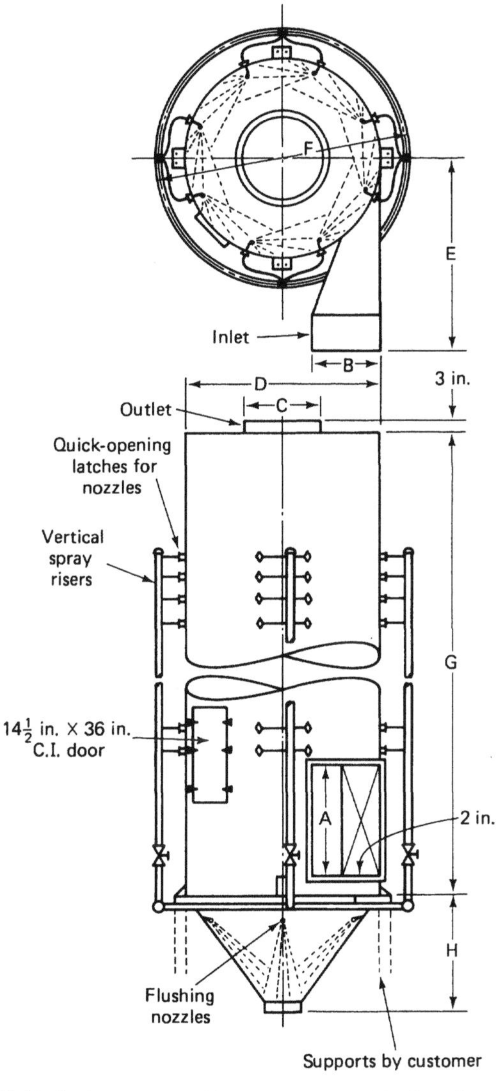

FIGURE 9-7 Cyclonic spray scrubber. (Courtesy of Buffalo Forge Company.)

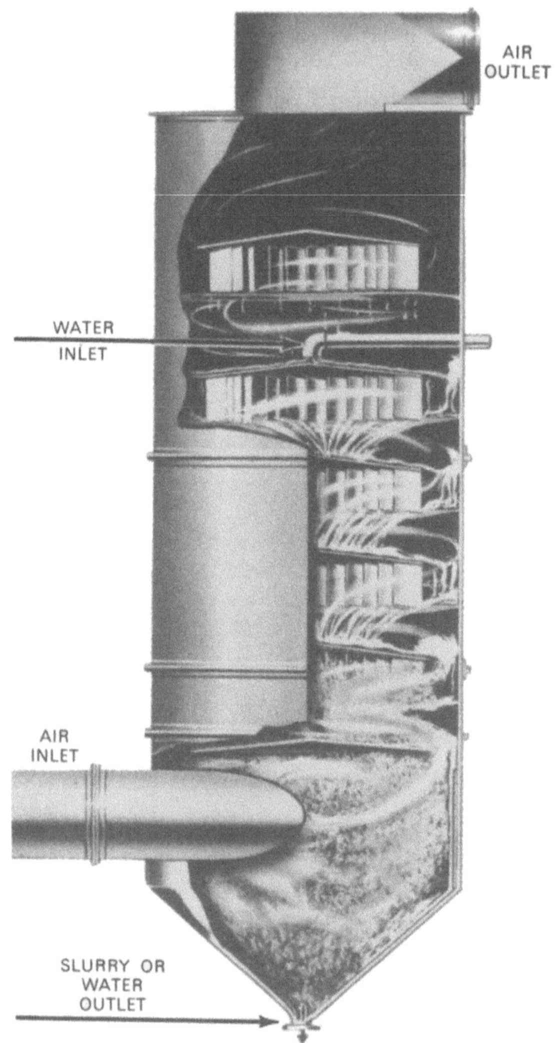

FIGURE 9-8 Multiwash scrubber. (Courtesy of Claude B. Schneible Company.)

10 to 20 gal/1000 ft^3 of gas, with pressure drops typically 0.2 to 1.5 in. of water per foot of bed depth. However, some adaptations, which make use of a highly efficient turbulent layer above the bed (see Figure 9-11), require considerably less scrubbing liquid, typically 2 to 2.5 gal/1000 ft^3 of gas at a pressure drop ranging from 4 to 6 in. of water. In cocurrent flow a liquid requirement of 7 to 15 gal/1000 ft^3 of gas is fairly common.

Coarsely packed beds are used for removing coarse dusts and mists 10 μm and larger, with velocities through the bed on the order of 400 ft/min. Finely packed beds may be used for removing contaminants in the smaller size range, but because of pressure drop considerations the velocity through the bed must be kept relatively low, preferably below 50 ft/min. Finely packed beds have a greater tendency to plug, and their applications are generally limited to gas streams with relatively low grain loadings.

Moving-Bed Scrubbers

Moving-bed (fluid-bed) scrubbers incorporate a zone of movable packing where gas and liquid can mix intimately. The system shown in Figure 9-12 uses packing consisting of low-density polyethylene or polypropylene spheres about 1½ in. in diameter; these are kept in continuous motion between the upper and lower retaining grids. Such action keeps the spheres continually cleaned and considerably reduces any tendency for the bed to plug. Pressure drops typically range from 3 to 5 in. of water (per stage), and collection efficiencies can be in excess of 99% for particles down to 2 μm. Particle collection may be enhanced by using several moving-bed stages in series.

Atomizing-Inertial Scrubbers

The atomizing-inertial category of wet scrubbers probably represents the largest grouping of commercial dust collectors. Available designs vary more than in any other category. For the sake of convenience, this category will be further divided into those that utilize the kinetic energy of the gas stream to fragment or atomize

FIGURE 9-9 Centrifugal fan wet scrubber. (Courtesy of American Air Filter Company, Inc.)

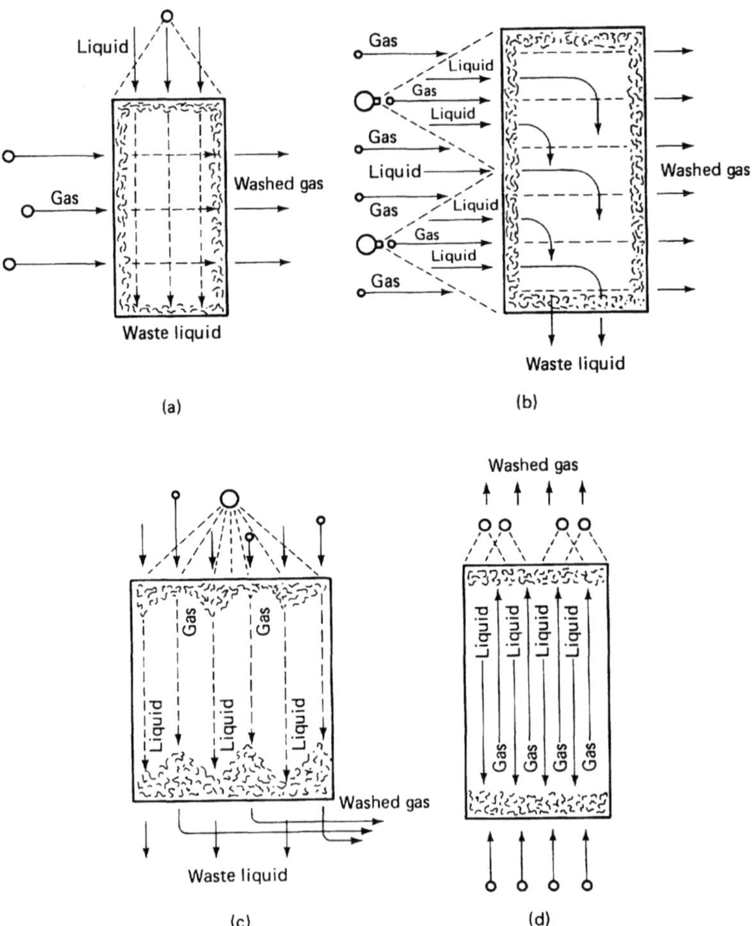

FIGURE 9-10 Packed-bed scrubbers: (a) cross-flow; (b) horizontal cocurrent flow; (c) cocurrent flow; (d) countercurrent flow.

the liquid on which the particle impacts (orifice and venturi scrubbers) and those that utilize a perforated plate with an impingement baffle over each perforation.

Orifice-Type Wet Scrubbers

In orifice-type wet scrubbers (sometimes referred to as self-induced spray scrubbers), the gas stream comes into contact with a pool of liquid at the entrance to a constriction (submerged orifice). Liquid is entrained and carried into the restriction, where greater liquid–particulate interaction occurs, resulting in a high frequency of particulate impaction on the droplets (see Figure 9-13). Upon leaving the restriction, most of the water droplets (those large enough) are separated by gravity since the gas velocity is reduced from what it was in the restriction. Smaller droplets are subsequently removed by centrifugal force and impingement on baffles located in the upper part of the unit. Water rate in motion through the restriction ranges from 10 to 25 gal/1000 ft³ of gas. At a typical gas velocity of 50 ft/sec, droplets in the range 300 to 400 μm can be created. Most of the water is recirculated from the pool (hence this does not contribute to the scrubbers' water requirements, which typically range from 1 to 3 gal/1000 ft³ of gas). Pressure drop typically ranges from 3 to 10 in. of water, with collection efficiencies often in the 90 to 95% range. The chief advantage of these systems is their ability to handle high dust concentration (no fine clearances to cause plugging) and high-solids-content slurries.

A slight modification of this system is the combined venturi-impingement scrubber. The scrubber section contains a liquid reservoir just below the throat section of a venturi. An adjustable weir is provided to maintain a constant liquid level. The

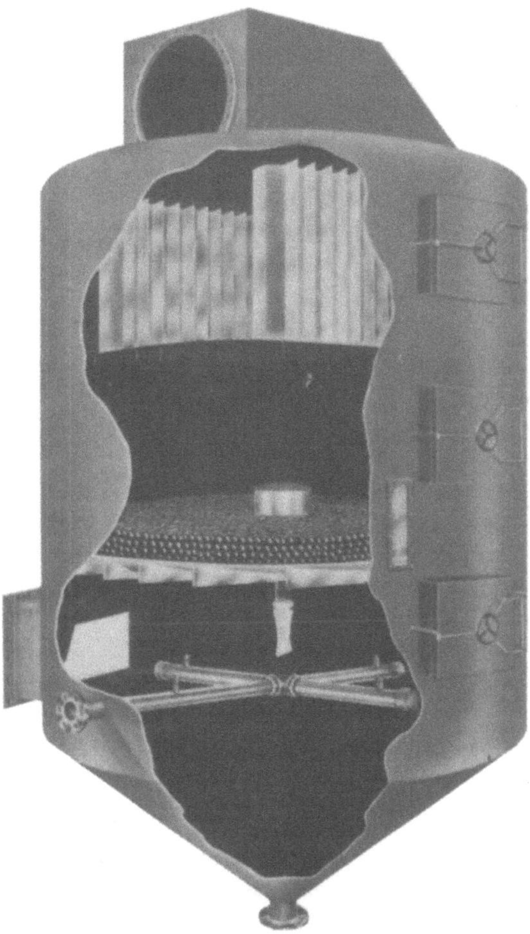

FIGURE 9-11 Flooded-bed wet scrubber with turbulent contact layer. (Courtesy of Riley-Environeering, Inc.)

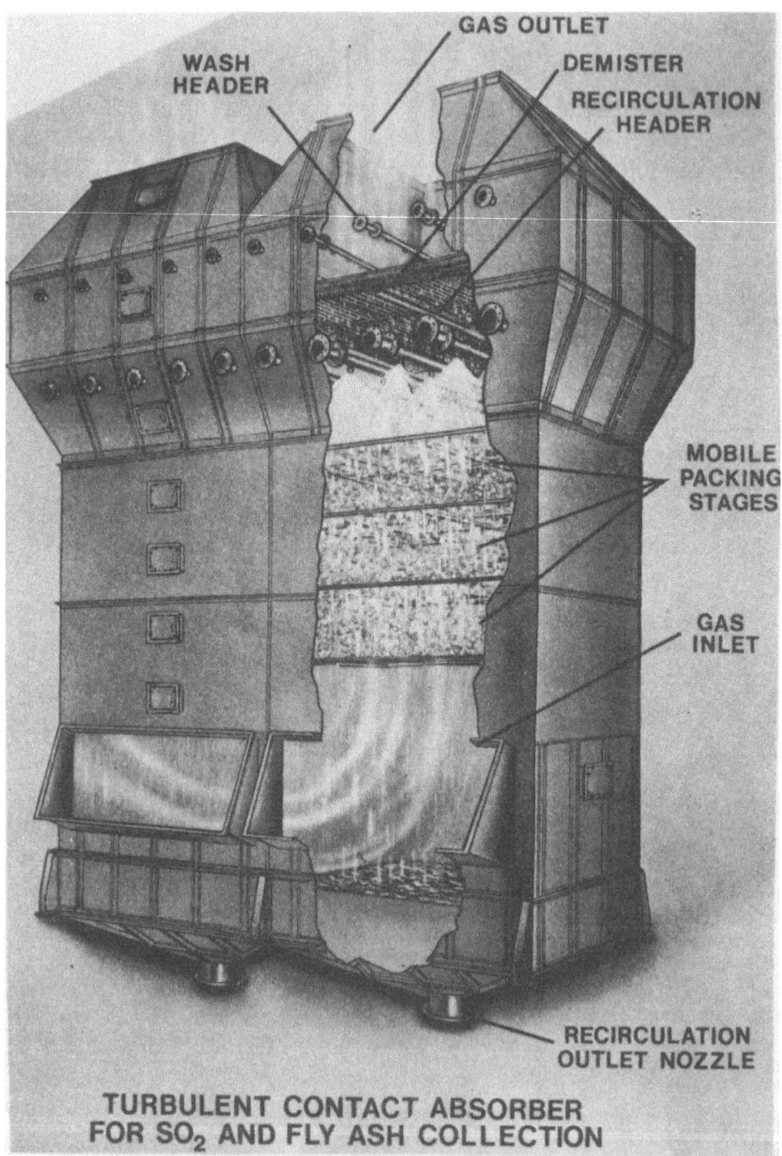

FIGURE 9-12 Moving-bed scrubber. (Courtesy of UOP Air Correction Division.)

scrubbing liquid is recycled (recirculation is normally 3 to 7 gal/1000 ft^3 of saturated gas) through nozzles near the top of the scrubber tube. The scrubbing liquid flows down the inside of the tube and into the throat section of the venturi, where it is deflected and atomized by the high-velocity gases. The particulates impact with the droplets formed by the high-velocity gas stream. A second scrubbing occurs when the gases impinge on the liquid level immediately below the throat section. Water requirements are limited to make up for evaporation loss and slurry bleed-off; for most conditions, this amounts to about 0.5 gal/1000 ft^3 of gas. Efficiency is relatively high for particulates as small as 0.5 μm in diameter. Pressure drop usually ranges from 8 to 15 in. of water.

Venturi Scrubbers

To achieve high collection efficiency for particulates by impaction, a small droplet diameter and high relative velocity between the particle and droplet are required. In a venturi scrubber this is accomplished by introducing the scrubbing liquid at an angle to a high-velocity gas flow in the venturi throat (vena contracta). Very small water droplets are formed, and high relative velocities are maintained until the droplets are accelerated to their terminal velocity. Gas velocities through the venturi throat typically range from 12,000 to 24,000 ft/min. The velocity of the gases alone causes the atomization of the liquid. The energy expended in the scrubber (except for the small amount used in the sprays and mist eliminators) is accounted for by the gas stream pressure drop through the scrubber. Another factor

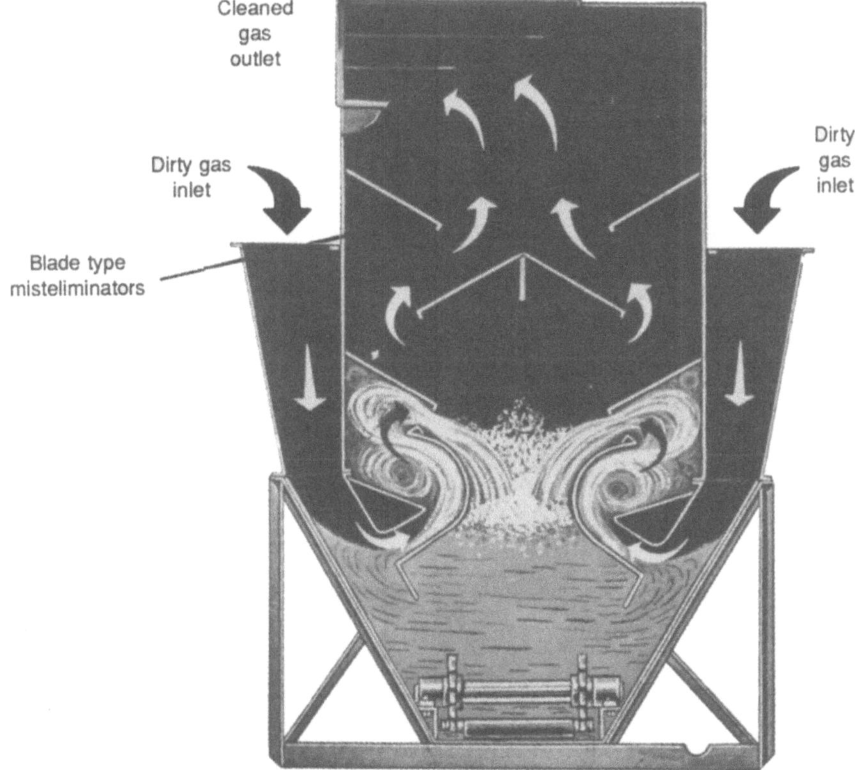

FIGURE 9-13 Swirl-orifice scrubber.

important to the effectiveness of the venturi scrubber is the conditioning of the particulates by condensation. If the gas in the reduced-pressure region in the throat is fully saturated, or supersaturated (preferably), some condensation will occur on the particulates in the throat due to the Joule–Thompson effect. Because of the cooling effect of the scrubbing liquid, condensation will be more pronounced if the gas is hot.

The venturi itself is only a gas conditioner and must be followed by a separating section for the elimination of entrained droplets (see Figures 9-14 and 9-15). Water is injected into the venturi in quantities ranging from 6 to 15 gal/1000 ft^3 of gas. Very high collection efficiencies are achievable with operating pressure drops ranging from 6 to 70 in. of water. Pressure drops of 25 to 30 in. are not at all uncommon. For example, a 10-in. pressure-drop venturi can typically remove particles as small as a few micrometers with virtually 100% efficiency, while a 60-in. pressure-drop venturi is often required to remove particles as small as 0.3 to 0.4 μm. Since collection efficiency is directly related to pressure drop, variable-throat venturi scrubbers have been introduced (see Figure 9-16) to maintain constant pressure

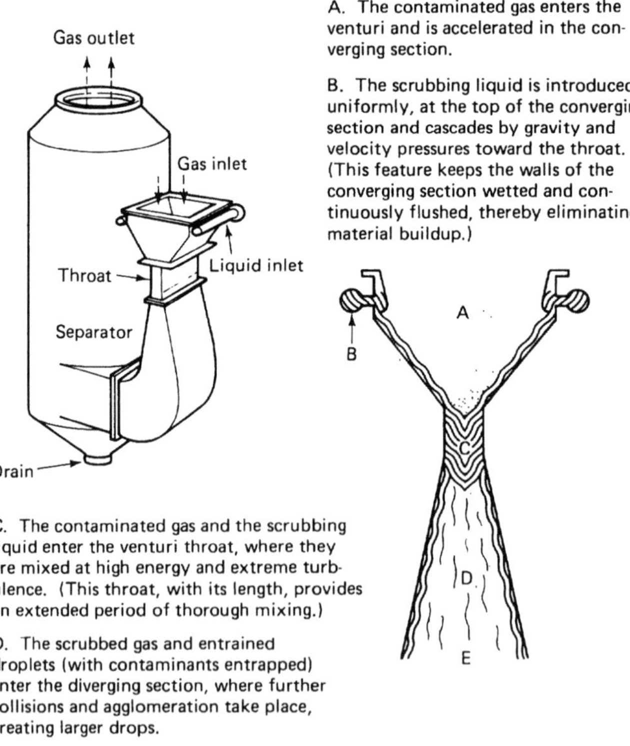

A. The contaminated gas enters the venturi and is accelerated in the converging section.

B. The scrubbing liquid is introduced, uniformly, at the top of the converging section and cascades by gravity and velocity pressures toward the throat. (This feature keeps the walls of the converging section wetted and continuously flushed, thereby eliminating material buildup.)

C. The contaminated gas and the scrubbing liquid enter the venturi throat, where they are mixed at high energy and extreme turbulence. (This throat, with its length, provides an extended period of thorough mixing.)

D. The scrubbed gas and entrained droplets (with contaminants entrapped) enter the diverging section, where further collisions and agglomeration take place, creating larger drops.

E. The gases then proceed to the separator, where liquid drops are easily removed from the gas stream and collected.

FIGURE 9-14 Venturi scrubber with cyclone separator: principles of operation.

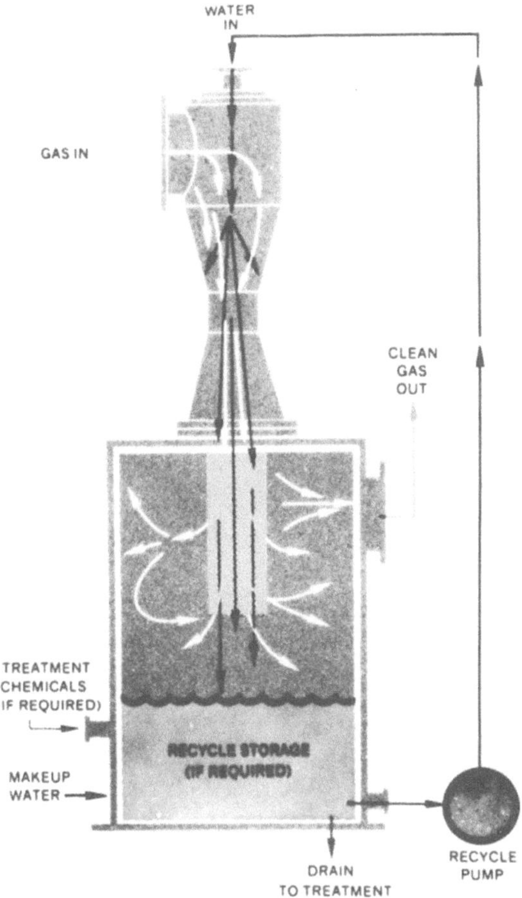

FIGURE 9-15 Jet-venturi scrubber. (Courtesy of Croll-Reynolds Company, Inc.)

drop with varying gas flows. In these systems the scrubbing efficiency and pressure drop may be adjusted by changing the position of a disk located in the venturi throat.

Impingement Plate Scrubbers

Impingement plate scrubbers (see Figure 9-17) utilize perforated plates with an impingement baffle over each perforation. The intention here is to expand the surface area of the liquid through use of the gas stream's kinetic energy. Gas flowing upward is divided into thousands of jets by the orifices. Gas velocities of 15 to 20 ft/sec through the orifices are common. Each jet aspirates liquid from the blanket and creates a wetted surface on the baffle, located at the point of maximum jet velocity. The directed impingement on a wetted target dynamically precipitates particles and entraps them in the scrubbing liquid. On impingement, each jet forms

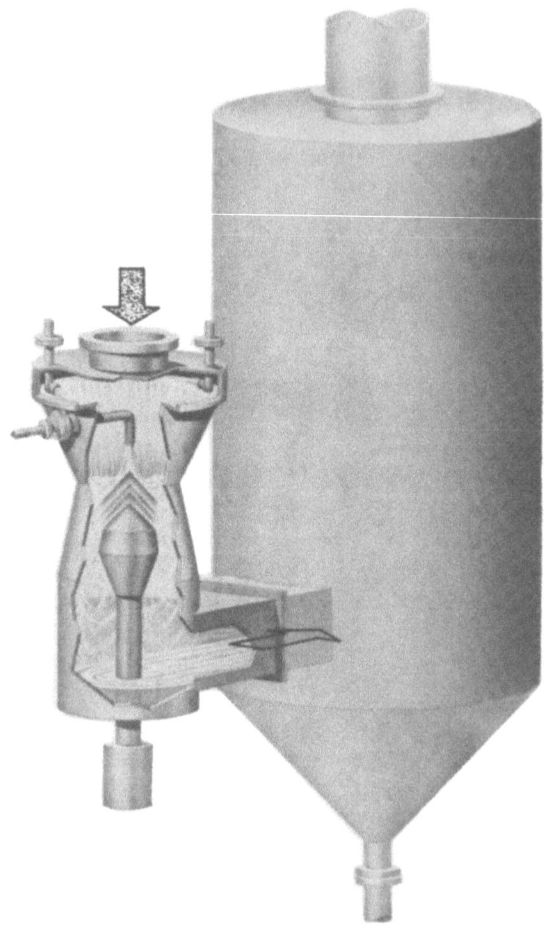

FIGURE 9-16 Variable-throat venturi scrubber. (Courtesy of Koch Engineering Company, Inc.)

minute gas bubbles which rise through and create turbulence in the liquid blanket. This provides extremely close gas–liquid contact for maximum cleaning. Continuous violent agitation of the blanket by the bubbles prevents settling of entrapped particles and flushes them away in the scrubbing liquid.

Overall collection efficiencies for a single plate may range from 90 to 98% for 1-μm particles; pressure drops from 1 to 4 in. of water per plate are typical. Water requirements usually range from 3 to 5 gal/1000 ft^3 of gas.

Atomizing-Mechanical Scrubbers

In mechanically induced scrubbers, high-velocity sprays are generated at right angles to the direction of gas flow by a partially submerged rotor (see Figure 9-18). The dirty gas stream passes through the area of the collector that contains the mechanically produced droplets. Scrubbing is achieved by impaction because of both high radial-droplet velocity and vertical gas velocity. Liquid atomization occurs at the rotor and the outer wall.

Power and liquid requirements range from 3 to 10 hp and from 4 to 5 gal/1000 ft³ of gas, respectively, for the high-velocity design (see Figure 9-18), depending on particle size and the desired collection efficiency. The chief advantages of these scrubbers are relatively low liquid and small space requirements, high scrubbing efficiency, and high dust-load capacity. The rotor, however, is susceptible to erosion from large particles and abrasive dusts. In addition, high-energy scrubbing applications usually require a mist eliminator.

Collection Efficiency and Pressure Drop

The design of wet scrubbers usually focuses on those parameters affecting collection efficiency and pressure drop. In most cases the scrubber must be designed to guarantee a specified collection efficiency, which, in turn, is strongly dependent upon pressure drop (among other parameters). The system pressure drop also dictates the power requirements and the size of auxiliary equipment such as the fans.

Collection efficiency equations as a function of particle diameter typically are of the form originally suggested by Johnstone et al. [1,2]:

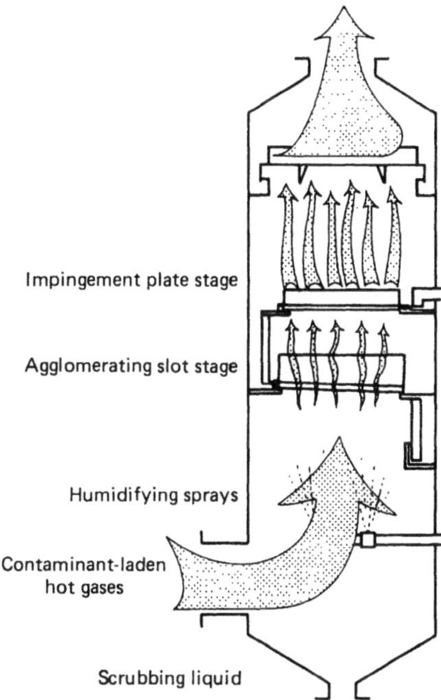

FIGURE 9-17 Impingement plate scrubbers. (Courtesy of Peabody Engineering Corporation.)

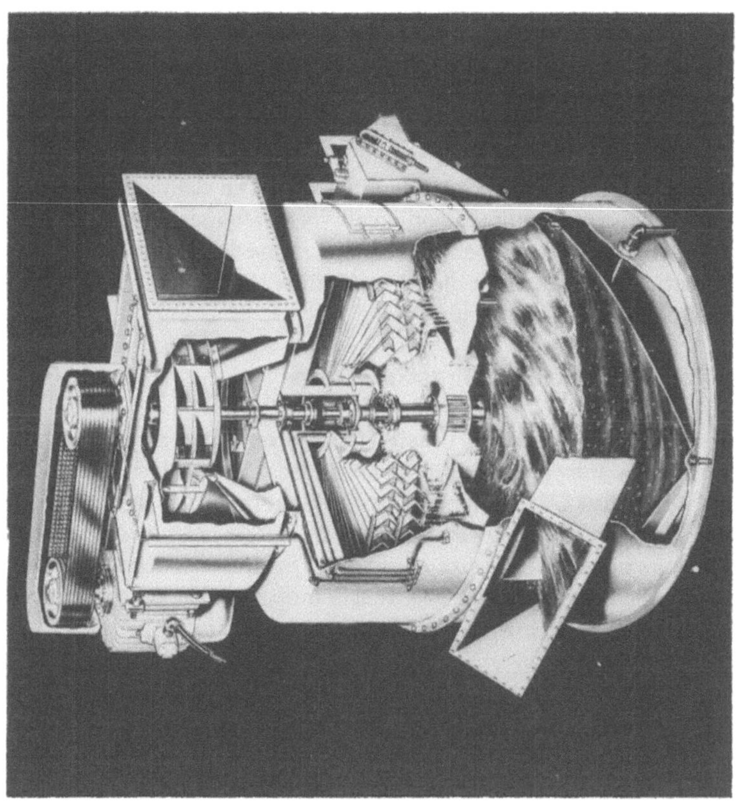

FIGURE 9-18 Center spray high-velocity scrubber. (Courtesy of Centri-Spray Corporation.)

$$\eta = 1 - \exp\left[-k(L/G)\sqrt{\psi_I}\right] \tag{9-1}$$

where η = collection efficiency for particle of diameter d_p

     L/G = liquid/gas ratio, gal/1000 acfm

     ψ_I = inertial impaction parameter

        $= C\rho_p V_G d_p^2 / 18 d_0 \mu_g$

     C = Cunningham correction factor [see equation (9-2)]

     ρ_p = particle density, lb/ft^3

     V_G = gas velocity at venturi throat, ft/sec

     d_p = particle diameter, ft

$$d_0 = \text{droplet diameter, ft}$$

$$= (16{,}400/V_G) + 1.45(L/G)^{1.5} \text{ for air–water system}$$
in a venturi scrubber, μm

$$\mu_G = \text{gas viscosity, lb/(ft) (sec)}$$

$$k = \text{correlation coefficient whose value depends upon}$$
system geometry and operating conditions, typically
0.1 to 0.2

The Cunningham correction factor is given by

$$C = 1 + \frac{2A\lambda}{d_p} \tag{9-2}$$

where
$$A = 1.257 + 0.40 \exp(-1.10d_p/2\lambda)$$

$$\lambda = \text{mean free path of gas molecules } (2.14 \times 10^{-7}/\text{ft in}$$
ambient air)

Pressure drop equations typically take the form [3]

$$\Delta P = k'V_G^2(L/G) \tag{9-3}$$

where
$$\Delta P = \text{pressure drop, in. water}$$

$$k' = \text{correlation coefficient for particular scrubber}$$
design, typically in the range of 0.00005

The practical design of a scrubber system can also be developed from the contact power theory. Contact power theory relates particulate collection efficiency in scrubbers to the pressure drop for the gas plus any power expended in atomizing the liquid. The total pressure loss, P_T, is assumed to be composed of two parts: the pressure drop of the gas passing through the scrubber, P_G, and the pressure drop of the spray liquid during atomization, P_L. These two terms can be estimated by

$$P_G = 0.157\Delta P \tag{9-4}$$

where
$$P_G = \text{contacting power based on gas stream energy}$$
input, hp/1000 acfm

$$\Delta P = \text{pressure drop across the scrubber, in. water}$$

and

$$P_L = 0.583p_L(L/G) \tag{9-5}$$

where P_L = contacting power based on liquid stream energy
                input, hp/1000 acfm

          p_L = liquid inlet pressure, psi

          L = liquid feed rate, gal/min

          G = gas flow rate, ft³/min

Then,

$$P_T = P_G + P_L \tag{9-6}$$

To correlate contacting power with scrubber collection efficiency, the efficiency is often expressed as the number of transfer units, defined by

$$N_t = \ln\left(\frac{1}{1 - \eta}\right) \tag{9-7}$$

where N_t is the number of transfer units (dimensionless). The relationship between the number of transfer units and collection efficiency is by no means unique. The number of transfer units for a given value of contacting power (hp/1000 acfm) or vice versa varies over nearly an order of magnitude. For example, at 2.5 transfer units ($\eta = 0.918$), the contacting power can range from approximately 0.8 to 10.0 hp/1000 acfm, depending on the scrubber type and the particulate to be collected.

For a given scrubber and particulate properties, there will usually be a very distinct linear relationship on a log-log plot between the number of transfer units and the contacting power.

$$N_t = \alpha P_T^\beta \tag{9-8}$$

where α and β are characteristic parameters for the type of particulates being collected (see Table 9-1). Although the power-function relationship can represent what has been observed, it should not be used to predict what will happen—except for identical conditions.

Scrubber Selection

Some of the more important conditions which may be indicative of a potential scrubber application are presented below [3].

1. Introduction of liquid to the gas is permissible to the process.
2. The liquid can be purged from the process without causing a water pollution problem. Water quality requirements of the receiving water must be considered, and a satisfactory effluent treatment system must be provided.

TABLE 9-1 Parameters for Equation (9-8).

Aerosol	Scrubber Type	α	β
Raw gas (lime dust and soda fume)	Venturi and cyclonic spray	1.47	1.05
Prewashed gas (soda fume)	Venturi, pipeline, and cyclonic spray	0.915	1.05
Talc dust	Venturi	2.97	0.362
	Orifice and pipeline	2.70	0.362
Black liquor recovery furnace fume			0.620
	Venturi and cyclonic spray	1.75	
Cold scrubbing water humid gases			
Hot fume solution for scrubbing (humid gases)	Venturi, pipeline, and cyclonic spray	0.740	0.861
Hot black liquor for scrubbing (dry gases)	Venturi evaporator	0.522	0.861
Phosphoric acid mist	Venturi	1.33	0.647
Foundry cupola dust	Venturi	1.35	0.621
Open-hearth steel furnace fume	Venturi	1.26	0.569
Talc dust	Cyclone	1.16	0.655
Copper sulfate	Solivore (A) with mechanical spray generator	0.390	1.14
	(B) with hydraulic nozzles	0.562	1.06
Ferrosilicon furnace fume	Venturi and cyclonic spray	0.870	0.459
Odorous mist	Venturi	0.363	1.41

Source: W. Strauss, *Industrial Gas Cleaning*, Pergamon Press, New York, 1966; with permission.

3. The gas must be cooled in any event.
4. Combustible particles or gases must be treated with minimum risk.
5. Vapors or gaseous matter and particulates must be removed from the gas.

Table 9-2 lists some of the advantages and disadvantages of utilizing wet scrubbers.

There are a number of additional factors to consider in selecting a scrubber. In general, they can be grouped into three categories: economic, environmental, and engineering. These are outlined in Table 9-3. Proper selection of the particular type of wet scrubber for a particular application can often be difficult. It is in the best interest of the prospective user to review the literature, request performance information available from the scrubber manufacturers and, if possible, visit an installation(s) with a similar type of application. In the final analysis, one should rely on previous experience. If previous experience does not exist, pilot testing is often necessary.

TABLE 9-2 Advantages and disadvantages of wet scrubbers.

Advantages
1. No secondary dust sources
2. Small space requirements
3. Ability to collect gases as well as particulates (especially "sticky" materials)
4. Ability to handle high-temperature, high-humidity gas streams
5. First cost is low
6. For some processes, the gas stream is already at high pressures (hence high pressure drops to achieve high particulate removal efficiences are not a disadvantage to the scrubber system)

Disadvantages
1. May create water disposal problem
2. Product is collected wet
3. Corrosion problems are more severe than with dry systems
4. Steam plume opacity may be objectionable
5. Pressure drop and horsepower requirements may be high
6. Solids build up at the wet–dry interface may be a problem

9.3 INSTALLATION PROCEDURES

Prior to installing the scrubber system, a careful inspection of the equipment should be made for shipping damage. The packing list should also be checked to ensure that all parts are accounted for.

In the event that the scrubber is not installed immediately, attention must be directed to proper methods of storage. Table 9-4 lists shelf-life requirements, under specific conditions, for typical vendor-supplied equipment. Indoor storage is always preferable. Storage practices are to a large extent common sense. For example, components and equipment should be stored in an upright position, fan belts should be removed, fans should be rotated once a month, and so on.

Installation procedures should follow the vendor specifications precisely. Deviations from vendor instructions may cause later problems and invalidate a warranty.

After the scrubber system is installed, it is important that all of the major items of equipment, connecting pipe, and auxiliaries be inspected, cleaned, tested, and calibrated (if necessary) before startup.

General preoperation practices include checks to ensure that piping is free of debris, oil levels are correct, fans and pumps rotate in the proper direction, and alignments appear proper. Operating instructions should be reviewed with all appropriate plant personnel. A typical checkout might include:

1. Utilities
   a. Power supply
   b. Instrument air
   c. Process air

TABLE 9-3 Factors involved in scrubber selection.

Environmental Factors	Engineering Factors	Economic Factors

1. Equipment location
2. Space available
3. Ambient conditions
4. Availability of adequate water and power utilities and sludge disposal facilities
5. Maximum allowable emission (air pollution codes)
6. Visible water vapor or steam plume
7. Equipment noise levels (i.e., in case of high-energy scrubbers, the noise level of high-tip-speed fans may be objectionable)
	1. Characteristics of the dust, fume, mist, or fog to be collected
	  a. Particulate-size distribution
	  b. Concentration or loading
	  c. Chemical reactivity
	  d. Physical and chemical properties (e.g., density, solubility in scrubbing liquid, agglomeration tendencies, shape, explosiveness, stickiness, etc.)
	  e. Corrosiveness and abrasiveness
	  f. Toxicity
	2. Characteristics of the gas stream (e.g., temperature, pressure, humidity, volume, composition, etc.)
	3. Characteristics of the scrubbing liquid (e.g., density, viscosity, corrosiveness, foaming tendencies, etc.)
	4. Design characteristics of the scrubber
	  a. Size and weight
	  b. Fractional efficiency curve (i.e., collection efficiency vs. particle size)
	  c. Pressure drop
	  d. Reliability and dependability
	  e. Method of disposal
	  f. Materials of construction
	  g. Effect of air volume changes (efficiency, pressure drop)
	  h. Power requirements other than fan
	  i. Utility requirements
	  j. Temperature limitations
	  k. Maintenance requirements
		1. First cost
		  a. Equipment
		  b. Installation
		2. Operating cost
		  a. Utilities
		  b. Maintenance
		  c. Savings (when recovering valuable products) or disposal costs
		  d. raw materials (if necessary)
		3. Expected equipment lifetime

TABLE 9-4 Storage shelf life.

Equipment	Storage Time (months)		
	0–6	*6–18*	*18–36*
Electric components, control equipment	3	4	4
Gates, mechanical assembly, machine castings	2	3	4
Closed crates and boxes	1	2	3
Vessels and tanks	1	2	3
Rough castings, structural steel	1	2	3

Note: 1, unprotected outdoor storage; 2, protected outdoor storage (elevated and covered); 3, unheated indoor storage; 4, heated indoor storage.

2. Pumps
   a. Belt tensions, pump rotation, pump alignment, lubrication, seal-water operation, and electrical interlocks
   b. Recycle pumps—suction and discharge valves
   c. Flush-water pumps
   d. Pneumatic pumps for scrubber variable volume and liquid level
   e. Spare-pump availability and operation
3. Valves/dampers (stack, isolation, bypass)
   a. Bypass
   b. Density control
   c. Water purge control
   d. pH elements for flush water
   e. Freshwater makeup
4. ID/FD fan
   a. Electrical controls
   b. Fan bearing coolant water
   c. Lubrication
   d. Vibration sensors
   e. Bearing-temperature sensors
5. Process water
   a. Level-detector calibration
   b. Mist-eliminator sump-level alarm calibration
   c. Mist-eliminator sump agitator
   d. Recycle pump

6. System controls/feedback controls
   a. Stack gas flow
   b. Makeup-water control
   c. Slurry pH
   d. Slurry density
   e. Sludge drainoff and disposal
7. Safety system
   a. Interlocks
   b. Alarms for various system components in accordance with design

Routine startup

1. Allow scrubber vessel to fill with liquid to pre-determined level through normal level controls if practical.
2. Start control liquid to all pump glands and fan sprays.
3. Start recycle pumps with liquid bleed closed.
4. Adjust liquid rates to manufacturer's specifications.
5. Check system isolation dampers and place scrubber in series with primary operation.
6. Energize fan and check vibration. If fan has an inlet control damper, it should normally be almost closed until the fan reaches speed.
7. Check the most important operating variables (i.e., gas saturation temperature, liquid flows, liquid levels, fan static pressure differentials, duct pressure drops, and scrubber pressure drop).
8. Slowly open the bleed to drain the system so that the slurry concentration is allowed to build up slowly. Check the final concentration as a cross-check on bleed-rate calculation.
9. Adjust the makeup liquid flow rate to maintain liquid reservice requirements.

Routine shutdown

1. Deenergize the fan and the fan spray water and isolate the scrubbing system from operation.
2. Allow the liquid system to operate for as long as practical. This will cool the scrubber and will reduce scrubbing liquid slurry concentrations.
3. Shut off the makeup water to the system, allowing the system to bleed normally.
4. When levels reach minimum settings (indicated by pump cavitation noise), the pump should be shut down immediately. The pump gland water can then be shut off.

5. Open system manholes, bleeds, and other drains. The system may have to
    be purged first with inert gas if explosive conditions could be created. In
    any event, purge the system with fresh air before entering to inspect it.

9.4 OPERATION

Although wet scrubbers are relatively simple devices, they do require proper care
to ensure long service life and trouble-free operation. Properly designed and installed
units can do an excellent job in helping plants comply with air pollution control
regulations. However, as with most processing equipment, periodic reviews and
inspections must be made to ensure that scrubbers are operating most efficiently.
Improper performance of a unit can usually be traced to any of the following:

Inadequate scrubber system design and/or operation.

Process conditions have changed so that the scrubber system cannot operate
efficiently.

Scrubber system mechanical condition has deteriorated until performance is
below design specifications.

The first step in solving a scrubber-performance problem is to assess and evaluate
the problem. The more thoroughly the problem is defined, the easier it is to find
the proper solution. Problem evaluation can be greatly helped by plant personnel.
Operating experience and performance records can provide valuable clues. Data
that must be studied include production process variables (flow rates, temperatures,
and pressures), stack opacity, and the scrubber's operating conditions.

A number of possible solutions usually suggest themselves for any problem.
The choice can be narrowed only if the situation is thoroughly investigated and
sufficient and accurate data are available.

Often, the solution to the problem can be relatively simple. For example, minor
modifications in the production process may sufficiently upgrade a scrubber's performance and quickly eliminate the problem. However, the overall effect on the
process must be carefully studied, because each process variable can have a critical
influence on an individual operation. Process engineers should always be consulted
when such evaluations are made.

If process flow rates have increased significantly—or the pollutants have changed
in character, size, and quantity—then scrubber upgrading or new equipment may
be necessary.

If equipment modification is justified, efforts must be made to ensure that the
desired results are obtained at minimum cost and operating expense. The job must
also be done with the least disruption to plant layout and operation. The achievement
of these goals requires that detailed tests be conducted. Extensive evaluation of
materials and data is needed before meaningful specifications can be prepared. If
erosion or corrosion is a problem, specimens of different materials and coatings
should be exposed to the exhaust stream and studied.

Rebuilding rather than replacing scrubbers is an approach rapidly gaining in popularity. A properly planned rebuilding program can add years of service to scrubbers and save a plant considerable time and money. Scrubbers operating in parallel may be a wise choice if flow rates are high and maximum operating flexibility is needed.

Assuming that the proper unit was selected for the given application, most scrubber problems usually involve spray nozzle plugging, liquid circuit restrictions, and/or entrainment of droplets from the vessel. Such problems may include:

1. *Wet/dry zone buildup.* Scrubber design may improperly allow dry, dust-laden gas to contact the juncture of the scrubbing liquid and the vessel, causing dust buildup. Proper design prevents this contact by extending ductwork sections sufficiently into the scrubber and thoroughly wetting all scrubber surfaces through reliable means (usually gravity flush and sometimes sprays).

2. *Nozzle plugging.* Nozzles plug through improper selection; too small orifices through which too dense scrubbing liquid must pass; improper header design; drawing off a sump that also settles (and concentrates) solids; erratic pump operation; chemical scaling sealing; and mechanical failure.

3. *Flow imbalance.* The headers external to the scrubber must send the required flow to the proper location at the proper rate. Many problems can be solved through the simple adjustment of flow using existing valves or dampers.

4. *Buildup (scaling).* Scaling is the plating out of deposits on a scrubber surface and is most often related to the chemical composition, solubility, temperature, and pH of the scrubbing liquid. Scaling is usually not operationally significant unless the surface in question is a functional one. Proper control begins with the scrubber design and process control.

5. *Localized corrosion.* Corrosion is a major factor in reducing the operating life of a scrubber, whether properly designed or not. Wells or pockets of liquid should be avoided and points of particulate buildup should be adequately flushed.

6. *Instrumentation fitting blockage.* One problem that can cause major trouble is instrumentation blockage or pluggage. Often, a standard fitting is not adequate in a scrubber. Specially designed fittings and connections are often recommended.

7. *Sump swirling.* Sump swirl problems are most often associated with cyclonic devices. The swirling of the scrubbing liquid can cause severe wear and draining problems unless arrested by antiswirl plates in the scrubber or rapid continuous draining.

8. *Entrainment.* Entrainment occurs when the droplet separator is not functioning properly. Nearly all scrubbers produce entrainment; only the properly designed systems reduce it to acceptable levels prior to discharge.

9. *Reentrainment.* Reentrainment occurs beyond the droplet removal device through improper draining or erratic flow patterns. It can also occur in stacks of very high velocity or where fittings protrude into a high velocity area.

10. *Liquid–gas maldistribution.* The gas and liquid must be properly distributed for the given application. Each affects the other and is aggravated by the influence of baffles (designed or accidental), buildup, mechanical failure, wear, scaling in headers, or improper design.

11. *Thermal shock.* Where hot gases meet cold scrubber liquor, thermal shock may occur. Proper design permits gradual cooling rather than abrupt changes. Thermal shock is a relatively simple problem to prevent (i.e., through the use of multiple cooling zones).

12. *Loss of seal.* The juncture of the scrubber liquid circuit with its surroundings is often a liquid seal. This seal may be at the top of a quencher or from an overflow connection. These lines must have seals able to prevent gas movement to or from the ambient surroundings. Loss of seal can cause entrainment or plugging, and instrumentation malfunction.

13. *Wear.* Wear can be tolerated unless it is localized. Unfortunately, a scrubber's functioning parts are also the wear parts. Expect to replace high wear points at frequent intervals. Excessive wear is often a result of excessively high solids concentration.

14. *Vibration.* This is most common in pumps with the fans associated with wet scrubber systems. Vibrations problems are controlled by monitoring and scheduled preventive maintenance.

Since packed- and moving-bed scrubbers have additional operational problems associated with them by virtue of their packings, these are considered separately.

Packed-Bed Scrubbers

One of the most common difficulties noted with packed-bed scrubbers is high pressure drop on the gas side. As the liquid rate increases, the pressure drop will increase until, at the floodpoint, the liquid will actually begin to form a layer above the packing. Typical packed bed-scrubbers for air pollution control are designed to operate at values well below the floodpoint. Many are equipped with spray nozzles that make it extremely difficult to overload the scrubber with liquid.

The first step in avoiding flooding rates is to measure either the inlet or discharge liquid flows. If a spray nozzle is used, its liquid pressure should be checked as well as its condition. If a trough- or weir-type distributor is used, there should be little restriction on liquid flow through the unit.

A quick check for high liquid rate can be obtained by turning off the liquid temporarily. If the pressure drop immediately falls to the dry value, the liquid rate is in excess of design. If the pressure drop remains high, other problems, such as fouling or plugging, are present.

Assuming that the liquid rate is correct, a check should be made of the gas rate through the unit. As this rate increases at any fixed liquid rate, the pressure drop through the column will increase.

A pitot tube or similar device can be used to obtain stack velocity readings needed for calculating the total volume leaving the column. Exit-gas volume can differ significantly from that of the inlet. Whenever possible, inlet measurements should be taken to ensure that the gas volume is equal to or less than the design volume. High inlet temperatures, or changes in water content, can alter the volume flow rate considerably.

Changes in gas composition from the design values can also alter the pressure drop through the column. As gas density increases, the pressure drop will tend to increase.

Next to improper gas or liquid rates, the presence of the particles themselves becomes the greatest concern. Solids or vapors that precipitate within the scrubber can cause plugging in the mist eliminator, packed bed, or support plates. These items can be visually inspected. Normally, packing plugs at the top or bottom. Plugging results from inlet material accumulating on the bottom of the packing, or from liquor (containing suspended solids) recirculated to the top of the bed.

Since many packed-bed scrubbers use injection-molded plastics, warnings should be posted against steam cleaning. Many problems in scrubber operation can be traced to a maintenance program involving extremely hot solutions used to clean the internals. Although the packing may not actually melt if such solutions are sprayed on the bed, they can cause deformation because of the weight of the packing resting on the bottom layer.

In scrubbers having fiberglass-reinforced plastic (FRP) walls, any incipient corrosion normally attacks the inner resin layer. Exposed fiberglass mat then begins to shred. If such material is picked up in the recycle loop, it can cause plugging at the top of the packed bed. Besides raising the pressure drop, the existence of such material is a sign that more-serious structural damage can be occurring in the scrubber. The cause should be determined immediately. If a particular area is affected, care should be taken during repairs to eliminate the contributing conditions. If these are due to chemicals not previously suspected, additional washdown sprays may prove to be a temporary answer. Serious damage may call for alternative materials of construction.

Previous operational history helps but is not always available. Repeated upsets can shift the arrangement of the internals. When packing is installed randomly, it is normally loose. Gas velocities through such an arrangement are so low that they do not cause problems under most conditions. During startup or upset, the gas flow rate can suddenly rise and fluidize part of the packing, resulting in a nonlevel bed. With a spray- or weir-type distributor, the liquid will then be unevenly distributed. The result is channeling of the gas and liquid, with low particulate collection or gas-absorption efficiencies. A quick visual check of the packing will determine whether leveling is necessary.

Mist carryover usually follows high gas or liquid rates, or a defective mist eliminator. Normally, a high pressure drop will also ensue. In some cases, a temporary upset may force the mist eliminator out of its normal position within the column. Visual inspection to ensure that no gaps appear between the various sections of the separator is advised.

Sometimes, mist carryover can even be experienced at low gas velocities. Mist eliminator systems are normally designed for minimum velocities of 3 ft/sec. If the spray system in the column is not properly designed, it may create a certain amount of fine mist and induce carryover through the mist eliminator at gas velocities below the initial design point. For this reason, a low nozzle pressure is strongly recommended. If an existing scrubber has a high nozzle pressure (over 25 psig), changing the nozzle may eliminate carryover problems.

If high gas velocity is the culprit, it may be possible to adjust the inlet gas rate. If not, substitute an alternative scrubber packing to reduce the overall pressure drop. For example, a more-open-design 2-in. packing may replace existing 2-in. saddles. Although the surface areas may be similar and the overall efficiency the same, pressure drop can be reduced to original design conditions. Care should be taken when considering such substitutions that efficiency is not sacrificed.

Packed-bed scrubbers can operate over a wide range of gas rates. The maximum rate is normally the design condition. As the rate decreases, the effective liquid/gas ratio will increase. This means that the efficiency will remain constant or actually increase slightly. At some point, efficiency will begin to decrease again. This occurs when pressure drop through the column becomes so low that the gas is not forced to distribute itself across the entire packed bed. Such a condition often results at values less than 30% of the design gas rate.

If wide variations in the gas rate are anticipated, the manufacturer of the column should be consulted to assure that it can handle these variations. A spray-type distributor is strongly recommended for such service. In trough- or weir-type distributors, the gas velocity and accompanying turbulence normally ensures good distribution of liquid. But as gas velocity decreases, the distribution of the liquid can be disturbed. In existing columns, a spray-type distributor and a mist-eliminator section can be added where variations in the gas rate are causing problems with efficiencies.

Moving-Bed Scrubbers

Moving-bed scrubbers are often used where packed-bed scrubbers would be expensive or impossible to maintain due to plugging problems. The rolling action of the spheres in the moving-bed scrubbers can prevent scaling and plugging in many instances. In those cases where scaling does occur, it frequently takes place in the bed section. Material accumulates in the grid interstices and between balls. Generally, as the scaling becomes more abundant in one area, the velocity through the open area of the scrubber increases, causing more rapid ball movement in that area. At some point, the ball activity becomes so accelerated that further scaling is prevented. Nozzle plugging may also result in scaling of that portion of the spheres covered by the nozzle.

Sphere movement is an important maintenance characteristic. If the distribution of air entering the bed is uneven, spheres in the high-velocity zone will tend to gravitate to the low-velocity area. This, in turn, creates a greater resistance in the low-velocity area than in the high-velocity area. Since the pressure must equalize, the air flow increases in the lower depth sphere zone and causes liquid carryover. This problem is difficult to correct with baffles underneath the bed.

Nozzles are another potentially high maintenance area in moving-bed scrubbers. This problem is less severe in the plastic ball scrubber because of the more open nozzle design and lower pressure requirement.

Wear and deterioration of the balls also pose a potential problem. When exposed to abnormally high temperatures, the balls may deform. Most applications are designed with safety precautions to prevent loss of water and resultant high-temperature conditions. Nevertheless, occasional malfunctions can occur.

Support grids may also be a source of maintenance. The motion of the spheres tends to wear and vibrate the grids. These problems, however, have been quite minor in most applications.

Entrainment elimination is quite necessary in moving-bed scrubbers because of the quantity of fine droplets created. The horizontal-flow chevron mist eliminator has proven very effective in eliminating droplets with a minimum of maintenance. These eliminators can be sprayed with water, if necessary, to keep them clean.

9.5 MAINTENANCE

Many of the items on the preoperation checklist should be checked during routine maintenance. Access doors should be opened and scrubber internals visually inspected. The maintenance performed generally includes unplugging lines, nozzles, pumps, and so on; replacing worn pump parts, erosion/corrosion prevention liners, and instruments (level indicators, pH indicators, etc.); and repairing damaged components when this is practical from the standpoint of labor and materials.

The following checklist is based on problems encountered in typical scrubber operation. These should be checked routinely and corrected in accordance with the manufacturer's recommended procedures.

Check for wear (abrasion/erosion). Heavy wear usually occurs in areas downstream of gas and liquid acceleration.

Check for corrosion on all scrubber internal surfaces.

Check for excessive buildup, in particular in the wet/dry zone.

Check for excessive scaling. This may be caused by process changes such as changes in temperature, pH, chemical composition of the dust, or chemical composition of the makeup water; reduced liquor recycle rate; increase in the inlet loading; or failure of solids removal system.

Check the nozzles for buildup and/or damage. Repair or replacement may be necessary.

Check for solids buildup in blowdown lines. Cleaning may be effected without system shutdown, and a flush connection may be installed to prevent this condition in the future.

Check for corrosion and leaks in lines and vessel, in particular where protective liners may have deteriorated.

Check operation of the mist eliminator. Formation of droplets can be caused by excessive gas flow rate, plugged drains from the droplet eliminator, or condensation in the outlet duct. Check structural supports for structural integrity and smooth operation.

Check pumps for wear, seal water, packing, and smooth operation.

Check dampers and damper linkages for proper positioning, wear and/or buildup.

Check fan for lubrication, fan-bearing coolant, belt wear and belt tension, and impeller erosion/corrosion.

Inspect all interior surfaces and condition of holding tanks during major outages.

Inspect exterior for leaks in all process and control lines, ductwork, and expansion joints.

Inspect the condition of all instruments (e.g., level probes, pH elements) with regard to solids buildup. It is impractical and usually impossible to remove solids buildup from the probes. The probes must be replaced.

Inspect for weather-related problems. Freezing weather can cause dampers to lock up and can freeze water and slurry lines (process lines). All lines (process and control) should be properly insulated and heat-traced if necessary.

An adequate supply of spare parts should also be a part of the prevention maintenance program. The minimum inventory for a venturi scrubber system is indicated in Table 9-5. Table 9-6 presents general manpower requirements for maintenance involving scaling and plugging for both the wet approach and liquid-injection-type venturi scrubbers. Table 9-7 indicates the types of personnel generally required to perform maintenance on various parts of the venturi scrubber system.

9.6 IMPROVING OPERATION AND PERFORMANCE

To troubleshoot air pollution control equipment performance problems adequately, a multiphase program should be implemented. As a guideline, the following procedures are recommended:

Phase 1: Problem identification. This phase should incorporate a detailed inspection of the system and culminate with a report listing all observations (positive and negative), provide interpretations of these observations (why things were the way they were), and finally, include recommendations of methods and items to

TABLE 9-5 Spare-parts inventory for venturi scrubber.

Section of the System									
			TYPE OF PARTS						
	Motors	Mist-Eliminator Modules	Seals	Bearings	Impeller	Reamers (50% of total)	Packing Material	Adjustable Throat Damper	None
Scrubber						×	×		
Separator								×	
Fan	×		×	×	×				
Pump(s)	×		×	×	×				
Mist eliminator		×							

TABLE 9-6 Manpower requirements for maintenance involving plugging and scaling of venturi scrubbers.

| | TYPE OF PROBLEM | | | |
| | PLUGGING | | SCALING | |
Type of Venturi Scrubber	*Mechanical Cleaners*	*Cylinder Cleaners*	*Chemical Cleaning*	*Hand Cleaning*
Wet approach	1 worker/shift/ month	1 worker/shift/ month	1 worker/shift/ week	3 workers/ shift/week
Liquid injection	1 worker/shift/ month	1 worker/shift/ month	1 worker/shift/ week	3 workers/ shift/week

improve performance. Troubleshooting guidelines for venturi scrubbers, dry scrubbers, and packed-bed scrubbers are provided in Tables 9-8 to 9-10.

Phase 2: Implementation. After thorough analysis and discussion, the recommendations provided in phase 1 should be implemented. These recommendations may be in the form of design modifications, component and accessory part replacement, and fabrication of new equipment. Following this aspect of the program, repair and replacement of the procured and fabricated components must be instituted. Finally, the entire system must be started up and debugged.

Phase 3: Testing and sampling. A performance test must be executed to evaluate the work done. This may be a stack sampling program and/or measurement of the system in continuous operation.

Phase 4: Preventive maintenance. When the modified and overhauled system has been shown to demonstrate satisfactory performance, a preventive maintenance program should be implemented to help maintain continuous and reliable operation.

TABLE 9-7 Type of maintenance required—venturi scrubber systems.

| | TYPE OF MAINTENANCE WORKER | | | | |
Section of the System	*Laborer*	*Electrical*	*Plumber*	*Wastewater Treatment Operator*	*Mechanical*
Scrubber	×				
Separator	×				
Fan		×			×
Pump		×	×		×
Piping, valves	×		×		
Water treatment equipment	×	×	×	×	×

TABLE 9-8 Troubleshooting venturi scrubbers.

Problems	*Possible Solutions*
Particulate buildup in scrubber	1. Clean liquid distribution system 2. Clean drain 3. Check gas flow and composition against design specifications 4. Increase scrubbing liquor flow rates 5. Improve flushing system
Scrubber differential pressure drop too low	1. Repair plugged, broken, or leaking static tap line 2. Check gas flow against design specifications 3. Adjust fan drive belts 4. Adjust fan speed 5. Clean inlet duct 6. Check inlet static pressure against design specifications
Scrubber differential pressure drop too high	1. Eliminate leakage 2. Check gas flow against design specifications 3. Adjust fan drive belts 4. Adjust fan speed 5. Clean scrubber
Entrainment	1. Improve mist elimination 2. Reduce gas flow 3. Reduce liquid flow 4. Clean (or enlarge) drains 5. Clean mist eliminator 6. Reweld any loose vanes 7. Improve flushing

Troubleshooting Problems Associated with Scrubber Ancillary Equipment

A great deal of the operating problems associated with a particulate scrubbing system are due to the scrubber's auxiliary equipment. The auxiliary equipment includes everything it takes to make a scrubber system operate, such as any pumps, fans, clarifiers, vacuum filters, motors, agitators, ducts, dampers, piping, valves, recirculation tank, instruments, and motor starters.

As in any air pollution control system, a scrubber system is subjected to a harsh environment. The abrasiveness of some types of particulate causes equipment such as fan wheels and valves to wear away. Stack gases may also contain a variety of components that, when wetted, produce compounds that corrode equipment. This is particularly true of any gas stream from a boiler that fires sulfur-bearing fuel and

TABLE 9-9   Troubleshooting tray scrubbers.

Problems	*Possible solutions*
Weeping	1. Select open area proper to eliminate weeping 2. Bleed in air if possible 3. Blank off excess area
Plugging	1. Use spray wash header; check operation 2. Reduce solids content in scrubbing liquid 3. Clean periodically 4. Raise flow rate (check factory)
Poor distribution	1. Check, inspect trays; lack of uniformity of condition indicates poor distribution; baffle as required to offset poor gas flow 2. Use end weirs on the downcomers to retain extra water on the tray 3. Clean weir boxes
Weir sizing	1. Size for full flow, maximum conditions; if too small, enlarge or install supplemental external weirs 2. Keep weirs clean
Mechanical	1. Make certain trays are securely in place 2. Stiffen warped trays 3. Tighten all fittings 4. Reweld baffle strips, caps, keeper plates, etc., or replace trays if broken

of gases from many chemical processes. All scrubber components must be designed and maintained with these considerations in mind [4].

Rotating machinery   Rotating machinery consists of all items such as fans, pumps, clarifiers, rotary vacuum filters, motors, and agitators. Because these items are in constant motion, a great deal of wear can take place. These components require some amount of regular attention to assure the proper operation of the system. Key areas of maintenance on rotating machinery are the bearings and any components rotating in the fluid stream.

There are also many other areas of maintenance unique to each type of equipment. Wherever practical, spares should be included. This is especially true for high-maintenance items such as pumps. Wear resulting from any abrasive solids in the liquid stream often dictates frequent inspection and repair. When mounted before the scrubber, fan wheels may also exhibit a great deal of abrasion from the particulate in the gas stream. Unfortunately, however, it is rather impractical to utilize spare fans in a scrubber system due to the difficulties in duct arrangements and the high capital cost of fans.

A stock of the more failure-prone components should be maintained. These include such items as bearings, seals, bushings, pump housing liners (if pumps with replaceable liners are used), pump impellers, fan wheels gaskets, makeup-water control valves, pH control valves, pH probes, recirculation-tank level-control sensors, and vacuum filter cloths.

TABLE 9-10 Troubleshooting packed-bed scrubbers.

Problems	Possible Solutions
Poor gas distribution	1. Use an injection-type support grid
	2. Allow extra vertical height-inlet to grid
	3. Use less than 8 ft/sec vertical velocity
Poor liquid distribution	1. Install redistributors every 4–6 ft of packing
	2. Rearrange headers and liquid entry
	3. Use a reflux distributor grid
Packing sized improperly	1. Check with packing manufacturer, not scrubber designer; many companies ignore recommendations
Too high velocity	1. Put another tower in parallel with existing unit
	2. Cut back in flow rate if possible
Too low velocity	1. Restrict bottom part of grid
	2. Use small packing
	3. Use redistributors
	4. Raise water rate

Fans

As in all rotating machinery, various components of the fan must be periodically lubricated. These components include all bearings on the wheel shaft, couplings, and dampers. Lubrication schedules and types of lubricants should follow the equipment manufacturer's guidelines. Generally, the lubrication schedule for these items ranges from quarterly to annually. The frequency of inspection of wear-prone components of the fan will vary with the severity of the application. On a fan handling relatively clean gas, the wheel and housing may go up to one year without any need of inspection. Fans on dirty applications may require monthly or even weekly inspection. The fan wheel and housing should be inspected for signs of corrosion, abrasion, and particulate buildup. If the fan is located upstream of the scrubber and the gas temperature is above 250°F, corrosion is generally not a problem. However, if the fan is downstream of the scrubber, and the gas contains any corrosive compounds, or if the fan is before the scrubber and the gas temperature is below the dew point of any corrosive components in the gas, corrosion will be a maintenance consideration. If the fan is downstream of the scrubber, moving the fan upstream of the scrubber may eliminate the problem of corrosion if the gas temperature is high enough. Should rearranging the fan not be feasible, the fan will have to incorporate more corrosion-resistant materials of construction. This requires going to a higher alloy on the fan wheel and possibly installing a corrosion-resistant liner on the housing. Attempting to reduce corrosion by collecting the corrosive gas in the scrubber system generally will not work. This is due to the fact that any condensation on the fan wheel will tend to pick up any residual corrosive gas that gets through the scrubber system. This corrosive component, when collected by the condensation, will eventually react with the fan materials no matter how small the concentration.

Abrasion on the fan may be lessened by the installation of replaceable wear liners on the wheel. Replacing the existing fan with a larger fan capable of the same flow rate and static pressure but at a lower speed may help to reduce the abrasion problem. If the fan is located before the scrubber, the installation of a high-efficiency mechanical collector before the fan may help to reduce any abrasion problems. Should a mechanical collector already be installed prior to the fan, the mechanical collector's tubes should be checked for pluggage or excessive wear. Should the existing mechanical collector be found to be working properly, consideration might be given to replacing it with a higher-efficiency mechanical collector or going to a two-stage mechanical collector.

Material buildup on the fan wheel may require a spray wash system to be installed in the fan. If this fails to improve the situation enough, moving the fan upstream of the scrubber may alleviate the problem.

Fan bearings should be inspected once per shift, if possible, for oil level, oil color, oil temperature, and vibration. Bearings found to be lubricated inadequately may require a greater frequency of lubrication or the installation of a forced lubrication system. The bearing housing should be inspected weekly for leaks, cracks, and loose fittings. Excessive bearing wear may be caused by a higher fan operating temperature than was originally specified, or by excessive vibration due to material buildup on the fan wheel. Misalignment of the bearing mountings could also account for this problem. A major annual inspection should be made to check all bearing clearances and to detect any sign of wear, pitting, or scoring on the bearings.

Dampers should be inspected semiannually for wear, corrosion, cracking, or looseness. All stuffing boxes should also be inspected. Annual damper inspection should include the tightening of all fastenings. The travel of actuator and linkage versus damper leaf and linkage travel should also be checked at this time. Shaft seals on the fan should be inspected visually for wear monthly.

Pumps

All bearings on the pumps and pump couplings should be lubricated periodically in compliance with the manufacturer's instructions. Many pumps include oilers, which should be maintained at a specified oil level. Some pumps are equipped with grease-lubricated bearings. These should also be lubricated according to the manufacturer's specifications.

As with fans, the frequency of inspection will vary with the severity of the application. Normally, pump bearings should be inspected at least as frequently as fan bearings with minor inspections once per shift, if possible, for bearing oil levels and for seal leaks and vibration. Semiannual inspection should be made for leaks, cracks, and loose fittings. A major inspection for wear, pitting, scoring, and clearance should be made at least once per year.

The pump housing and impeller, along with the seals, are components that are subject to abrasion and corrosion. A properly operating mechanical collector preceding the scrubber will aid in reducing the amount of particulate to be collected by the scrubber. This, in turn, will reduce the amount of abrasive solids required to be pumped through the scrubber recirculation system and should reduce any abrasion in the pump.

Increasing the bleed stream from the scrubber system will also reduce the concentration of suspended solids handled by the pump. Should neither alternative be available, a pump with a replaceable rubber liner may be of benefit. This should aid in minimizing pump replacement costs. The installation of a water flush in the seals may help to reduce wear on the seals.

Corrosion may be combated by installing a pump constructed of a high grade alloy, or a rubber-lined pump, depending on the chemistry of the recirculation system. Increasing the amount of chemicals added to a scrubber system can help neutralize the recirculation stream and help preserve the equipment. However, this may require a significant expenditure for purchasing chemicals. For example, the addition of caustic will increase the scrubber's efficiency to collect a corrosive gas such as sulfur dioxide. As more sulfur dioxide is collected, the pH of the solution drops and increased amounts of caustic are needed for neutralization until an equilibrium is reached between the scrubbing efficiency and the desired pH level. At a recirculating solution pH of about 7, the scrubber's sulfur dioxide removal efficiency using caustic is about 90% as compared to a 60% efficiency when scrubbing with water at a pH of 4. Increasing the scrubber's bleed rate may help to lessen corrosion problems by decreasing the concentration of the corrosive compounds in the recirculation system.

Clarifiers and Drag Chain Tanks

Clarifiers and drag chain tanks fall into a general category of gravity separation devices. A clarifier or drag chain tank may be used to remove the particulate matter from the scrubber bleed stream when the particulate is sufficiently large and has a short settling time.

Clarifiers rake the settled particulate to the center of the unit, where it is removed from the bottom in a concentrated slurry. The slurry from the clarifier is pumped to a final discharge or to an additional dewatering device if a drier final discharge is required. The pipeline from the clarifier must be periodically inspected to make sure that it is not plugged.

The rake mechanism in a drag chain tank moves the solids up an incline. As the rake breaks the surface of the liquid, water falls off the rake and a relatively dry solid is moved up the incline. The solids continue to be pushed up the incline and over the edge of the tank, where they fall onto a pile along the side of the tank. The pile of solids must be regularly removed from the side of a drag chain tank at a rate dependent on the application. A front-end loader may be used for this chore; however, it is usually better housekeeping to have the solids discharged into a dumpster and then have the dumpster hauled away when full. The rakes of this equipment require little maintenance since the speed of rotation is very slow. However, sludge buildup may put excessive torque on the rake mechanism. If this should occur, the clarifier or drag chain tank must be shut down, drained, and hosed out.

Bearings and gear reducers must be lubricated and inspected according to the vendor's schedule. This is generally done semiannually. In addition to the inspection of the bearings, the drive mechanism must be inspected for temperature rise, sprocket alignment, sprocket wear, chain tension, and oil level. These inspections should generally be done after the first 100 hours of operation and every 500 hours thereafter.

Rotary Vacuum Filters

Rotary vacuum filters are becoming a more common part of wet scrubber particulate control systems, most often used on the bottom discharge from a clarifier to obtain a more concentrated sludge discharge before ultimate disposal. In a particulate control system, the fine particulate matter may pass through an ordinary rotary vacuum filter. A precoat on the rotary vacuum filter may be required for such an application. In a precoat filter a layer of filter aid is formed on the filter drum. Once the filter has been coated with the filter aid, the liquid to be clarified is fed into the unit. The knife-edge then shaves off the particulate from the filtered mass together with a small portion of the precoat. After the precoat has been removed down to a depth where it can no longer effectively filter, the filter must be shut down and recoated. Precoating of the filter may take as long as 1 hour and the filtering cycle may last from 16 hours to 1 week, depending upon how fast the precoat is removed from the filter. The depth of the precoat must be visually checked periodically. A spare filter or multiple filters should be installed to ensure continuous operation of the particulate removal system.

Periodic lubrication of all knife-edge bearings, filter bearings, motor bearings, gear-speed reducers, and vacuum pump bearings must be done in accordance with the vendor's schedule. The filter cloth, knife-edge alignment, and drive belt must be inspected occasionally and adjusted or replaced as required. The frequency of inspection will again vary with the severity of the application, ranging from once per week to once per year.

Motors—Mechanical

Motors are basically quite simple and reliable from a mechanical standpoint. However, as with all rotating machinery, the bearings require some degree of care and attention. Motor bearings should be lubricated in accordance with the manufacturer's specifications. The schedule for this may vary from 750 to 8000 hours of operation depending on the motor speed and the severity of the conditions to which the motor is exposed. Extreme care must be exercised to avoid excessive greasing of the motor bearings. Excessive greasing can cause the bearings to overheat and can contaminate the motor's windings. When greasing motor bearings, the bearing drain or flush holes must be unplugged to help guard against over greasing. Some manufacturers supply bearings that are prelubricated and sealed for life. This is advantageous in that overlubrication is perhaps the greatest cause of motor-bearing failure. Bearings should be inspected in the same manner as described for fan and pump bearings.

Agitators

Agitators may be required on systems where sedimentation of the particulate in the recirculation tank is a problem, or if additional mixing is required to mix any additives into the scrubbing liquor. The bearings and gear reducers should be inspected and lubricated according to the manufacturer's specifications. The agitator's shaft and propeller should be inspected occasionally for wear and corrosion. Should the shaft or propeller exhibit excessive wear or corrosion, it should be replaced with a higher alloy. On side-entering-type agitators, the packing and seals should be inspected for signs of leakage and wear.

Static equipment

Ductwork

Ductwork should be inspected occasionally for leakage and excessive flexing. The frequency of these checks will vary with the severity of the application, but should be done no less than semiannually. Holes in the ductwork can be the result of a number of causes, but are most often the result of corrosion or abrasion. For ductwork located after the scrubber, corrosion will be a primary cause of failure. The deterioration of the duct due to corrosion may be alleviated by lining the duct with a corrosion-resistant liner. This, however, may not be possible if the duct has corroded too greatly or if the gas temperature is too high. Should this be the case, the duct will have to be replaced with an appropriate alloy or with fiber glass if the temperature is low enough. Fiber glass has proven to be an appropriate material of construction when located downstream from the scrubber. The duct can be protected from temperature excursions due to water outages in the scrubber and supplying it with water independent of the normal scrubbing recirculation system.

Abrasion is a problem common to ductwork located upstream from the scrubber. The severity of this problem can be lessened by increasing the duct diameter, thus slowing the velocity through the duct. The addition of a mechanical collector prior to the scrubber system will lessen the particulate loading through the ductwork and therefore aid in reducing ductwork abrasion. Should a mechanical collector already be in the system and abrasion is a problem, the collector should be inspected to ensure that it is in proper operating condition. If it is found to be working properly, consideration should be given to installing a higher-efficiency mechanical collector. If excessive flexing or cracking should develop, an expansion joint should be added to the ductwork at that point.

Dampers

Because they are operated infrequently, dampers present special maintenance problems. Guillotine zero-leakage dampers are perhaps the most effective for air pollution control systems. Louver and butterfly dampers, however, may be appropriate for less severe or demanding service. Since the maintenance on all types of dampers is similar, concentration will be given to guillotine-type dampers.

Lubrication is an important part of any damper maintenance program. The scheduling and types of lubricants should follow the manufacturer's specifications. Lubrication is normally done annually on all components of the drive mechanism and bearings. The blower motor bearings should also be lubricated at that time if one is included with the damper.

A program of regular inspection must be instituted to help ensure that the damper will operate when required. The damper blades should be moved in and out of the duct 6 to 12 in. several times, at a frequency of no less than once per month. More severe services may require a greater frequency of this checking. Checks for leakage around the flanges should be made at least once per month. All portions of the damper's sealing elements should be inspected for wear semiannually. On a yearly basis, all drive components, bearings, damper blades, and blowers should be inspected for wear or leakage. These inspections should be carried out with strict adherence to the manufacturer's instructions.

Guillotine dampers should be equipped with an air purge to prevent buildup of debris in the blade seating surface trough when the blade is in an open position. It is important, to ensure proper damper operation, that the manufacturer's schedule for activating these purges be followed. The frequency of these purges will depend again on the severity of the application; however, it is recommended that it be done at least daily using six air bursts of 5 seconds duration with 1 to 1½ minutes between each burst.

Piping

Piping should be inspected at regular intervals depending on the application. Piping is prone to three common problems: abrasion, corrosion, and pluggage. Abrasion is common to any scrubber system collecting a hard, insoluble particulate, such as coal-fired boiler flyash. Abrasion problems in the recirculation piping are best alleviated by reducing the concentration of the particulate in the recirculation system. This can be done by increasing the bleed rate from the system, by installing a mechanical collector ahead of the scrubber if none already exists, or by ensuring the proper operation of an existing mechanical collector. Reducing the pipe velocity by installing a larger-diameter pipe may also help relieve an abrasion problem. It must be noted, however, that a velocity which is too slow may promote settling and pluggage of the pipes. A pipe velocity of 4 to 7 ft/sec is a reasonable compromise between preventing abrasion and preventing pluggage.

Corrosion will be a problem in recirculation systems where the scrubber is operating on a gas stream containing corrosive gases. The methods of alleviating a corrosion problem in the recirculation system are similar to those discussed for pumps. The addition of caustic, for example, can ultimately raise the pH and neutralize the scrubber solution. Replacing the existing piping with piping constructed of a more corrosion-resistant material may be a more economical alternative in the long run. FRP pipe or plastic-lined cast-iron pipe has been successfully used where corrosion is a problem and the particulate being collected is water-soluble, such as sodium sulfate. Plastic-lined cast iron has an advantage over FRP in that it is more resistant to accidents on the outside, such as with forklift crashes. On a recirculation system where both corrosion and abrasion are problems, stainless steel and rubber-lined cast iron pipes have been shown to be effective.

Pluggage can be a problem on recirculation systems handling a slightly soluble particulate or if the pipe velocities are too low, thus promoting settling. On systems where the former is the problem, procedures such as increasing the bleed rate or raising or lowering the pH may help to alleviate the pluggage. Should the latter be the case, an increase in the pipe velocity by decreasing the pipe diameter may help to eliminate this problem. Piping on a system that is prone to pluggage should be flushed out with clean water on a yearly basis at a minimum. Consideration should also be given to rodding out the piping if this is possible.

Valves

Valves will be subject to most of the same problems that have been discussed for piping: that is, abrasion, corrosion, and pluggage. The solution to these problems in valves will be similar to those solutions for the equivalent problems in piping. Installing larger-diameter valves, increasing the bleed rate, and installation of a mechanical collector before the scrubber will all aid in reducing abrasion. Increasing system pH and installing valves of corrosion-resistant material will help to eliminate corrosion problems. Pluggage can be alleviated by increasing the system bleed rate, adjusting pH, or increasing the valve velocity in cases of settling. On a system that exhibits both corrosion and abrasion, rubber pinch valves have been found to be effective. The rubber provides resistance to corrosion, and the design of the valve provides a smooth passage for the liquid stream containing the particulate, thus minimizing abrasion.

Air-operated control valves may exhibit sticking problems. The problem may stem from one of two sources. The valve may exhibit some pluggage due to the particulate in the recirculation system, or it may have some water buildup at the air connection. The latter problem may be reduced by ensuring a supply of dry air to the valve or by regular inspection and drying of the air connection. The frequency of this cleaning will depend on the amount of moisture in the air supply. Since a control valve can be a rather vital component in the operation of a scrubber system, it is recommended that the piping be arranged so that any system containing a control valve can be temporarily placed on manual operation while the control valve is repaired or replaced. This is particularly important in the caustic addition system, where loss of caustic may severely damage the scrubber system's components. It is also important in the scrubbing liquor makeup system, where a malfunction of the control valve may cause the scrubber to run dry, thus damaging the equipment from excessive temperature. A malfunctioning control valve in the makeup line may also cause the recirculation tank to overflow. A failure of either of these control valves and the inability to resort to manual operation may require shutdown of the entire scrubbing system.

Recirculation Tank

The scrubbing liquor recirculation tank in most systems will require very little maintenance other than an occasional inspection of the general condition of the tank. Corrosion may be exhibited on a tank handling an acidic scrubbing medium. Adding caustic, lining the tank with a corrosion-resistant liner, or replacing the existing tank with one constructed of FRP or an appropriate alloy can alleviate the problem. If the scrubber is handling a large, easily settleable particulate and if the tank has a relatively long retention time, particulate may build up on the bottom of the tank. This buildup may plug the recirculation pump nozzle and foul the instrument probes. The addition of an agitator may help to solve this problem. Lowering the retention time by lowering the liquid level may also help, provided that this can be done without causing cavitation in the pump. Certain contaminants collected in the scrubber may cause a foaming problem. This problem may be reduced through lowering the scrubbing liquor pH by reducing the amount of caustic added to the system or by adding a slight amount of acid if the scrubbing solution is originally basic. Should the scrubber system not be able to tolerate a lower pH, foam-breaking sprays can be installed above the tank or antifoaming agents may be added to the tank. Adding more freeboard to the recirculation tank may also be of some benefit in relieving a foaming problem.

Pump cavitation may be a problem if the liquid level in the tank is not high enough for the pump flow rate and the diameter of the recirculation pump nozzle. A baffle or bellmouth may be used to prevent such cavitation.

Electrical

Instruments

Instruments usually require very little maintenance. The philosophy is generally to replace them as they fail. The accuracy and calibration of analog instruments should be checked annually. All probes should be cleaned at a frequency dependent on the application. Level controllers tanks handling highly soluble particulate may never require cleaning, whereas the pH probes on a lime slaker, for example, should lime be used as a neutralizing agent, may require cleaning daily. It is therefore important to select probes which will be appropriate for the frequency of mainte- nance. A flange-mounted differential pressure transmitter, used for level control and located at the bottom of a recirculation tank, may be appropriate for a scrubber system operating on a highly soluble particulate. However, if the particulate tends to settle easily, this type of level controller might require that the tank be frequently drained in order to clean the probe when a sufficient amount of solids have settled around it. A top-entering float-type level controller may be more appropriate for this type of application. All pneumatic lines should be cleaned and dried of moisture regularly at intervals depending on the quality of the air supplied to them. Orifice plates should be inspected at least annually and may require more frequent inspection should the fluid contain any abrasive solids. Excessive wear on orifice plates may necessitate replacement with a harder material of construction. Ink and paper in the recorders must be checked frequently.

Motor Starters

Motor starter switches should be lubricated occasionally according to manufacturer's specifications. Cleaning of all electrical equipment is important to ensure reliable operation. Dirt in electrical equipment may decrease heat dissipation and cause poor ventilation. Overheating and insulation breakdown may result from this. Insulation breakdown may also occur if the dust is electrically conductive. All contacts, transformer insulation, and cooling fans should be inspected, cleaned, and dried annually. Dirty environments may require more frequent inspection and cleaning.

Motors—Electrical

Motors require little attention from an electrical standpoint except for an occasional cleaning and check of temperature rise. Dust may be removed by using dry low-pressure compressed air and a dry cloth. When mixed with dirt, grease can form a gum that is difficult to remove. Oil can cause a deterioration of the insulating varnish, which in turn can lead to a burnout. Any grease or oil can be removed with a solvent such as naphtha. A fresh coat of insulating varnish should be reapplied after such a cleaning.

Measurement of the temperature rise can provide an indication of problems such as overloading, shorts, or burned insulation. The three methods available to measure the temperature rise are thermometer, resistance thermocouple, and embedded thermocouple. The thermocouple method simply consists of placing a thermocouple as close as possible to the windings and core of the motor. This may be done by removing a head bolt or eye bolt and inserting a common mercury or alcohol thermometer. The resistance method requires that a Wheatsone or Kelvin bridge be used to measure the winding resistance at room temperature and then once again after the motor has operated for 4 to 6 hours at full load.

The embedded thermocouple method involves connecting a bridge-type instrument to a thermocouple that has been embedded in the windings. This will usually give a reading directly in degrees Celsius. The actual temperature rise should then be compared against the manufacturer's recommended temperature rise for that type of motor. A motor can be too hot to touch and still be within the proper operating temperature. Therefore, it is important not to attempt to measure the motor temperature by hand. On motors of 300 hp and above, the thermocouple method should be used with a recorder to monitor the temperature continuously.

9.7 CONCLUSIONS

Wet scrubbers are probably the most widely used and most often abused of all air pollution control devices. The proper operation and maintenance of these systems, in many cases, can enhance particulate removal control efficiency and reduce operating costs. An effective preventive maintenance program will also maximize the time between scheduled shutdown periods, significantly minimize unscheduled breakdowns, extend equipment life, and reduce maintenance and repair costs. Proper system design and adherence to a routine maintenance program will go a long way toward eliminating many of the problems that have plagued the operation of particulate scrubber systems.

9.8 NOMENCLATURE

A	function of mean free path and particle diameter
C	Cunningham correction factor, dimensionless
d_0	droplet diameter, ft (or μm)
d_p	particulate diameter, ft (or μm)
G	gas flow rate, actual ft^3/min (acfm)
k	correlation coefficient in Equation (9-1)
k'	correlation coefficient in Equation (9-3)
L	liquid flow rate, gal/min
N_t	number of transfer units, dimensionless
ΔP	pressure drop through scrubber, in. water
P_g	gas stream contacting power, hp/1000 acfm
p_L	liquid inlet pressure, psi
P_L	liquid stream contacting power, hp/1000 acfm
P_T	total pressure loss, hp/1000 acfm
V_G	gas velocity at venturi throat, ft/sec
α	correlation coefficient in Equation (9-8)
β	correlation coefficient in Equation (9-8)
η	collection efficiency
ρ_p	particle density, lb/ft^3
μ_G	gas viscosity, lb/(ft) (sec)
ψ_I	inertial impaction parameter
λ	mean free path of gas molecules

9.9 REFERENCES

[1] JOHNSTONE, H. F., AND ROBERTS, M. H., *Ind. Eng. Chem., 41,* 2417, 1949.

[2] JOHNSTONE, H. F., FEILD, R. B., AND TASSLER, M. C., *Ind. Eng. Chem., 46,* 1601, 1954.

[3] THEODORE, L., AND BUONICORE, A. J., *Industrial Air Pollution Control Equipment for Particulates,* Chap. 6. CRC Press, Inc., West Palm Beach, Fla., 1976.

[4] CZUCHRA, P. A., "Operation and Maintenance of a Particulate Scrubber System's Ancillary Components," personal communication.

10

Electrostatic Precipitators

Peter Paul Bibbo

Precipitator Product Manager
Research-Cottrell, Inc.
Somerville, N.J.

10.1 DESCRIPTION OF CONTROL DEVICE

Electrostatic precipitation is the most popular method in use for removing fine solids and liquids from gas streams. The first successful commercial-scale precipitators were installed by Frederick G. Cottrell on two California chemical plants in 1907. One of these units treated 5000 cfm of gas, and recovered precious metals and sulfuric acid. In 1910, Cottrell installed a successful precipitator on a 250,000-cfm smelter. The technology grew rapidly, and successful applications were soon found on a variety of industrial processes and in the power generation industry. Precipitators have been used to treat over 5,000,000 cfm of gas and have attained removal efficiencies approaching 99.99%.

Fundamentals

Compared to other methods of gas cleaning, electrostatic precipitators are as elegant as they are efficient. Instead of performing work on the entire gas stream in the cleaning process, the forces in a precipitator are applied directly to the suspended particles themselves. The result is high collection efficiency at a moderate cost in power.

There are three fundamental steps in the electrostatic precipitation process:

1.   Charging the particles suspended in the gas stream
2.   Collecting the charged particles
3.   Removal of the collected particles into an external receptacle

Charging is accomplished by applying a high dc voltage to an electrode system. The presence of a grounded electrode near the charged electrode gives rise to the formation of corona and a unidirectional electric field. Particles traveling between the two electrodes acquire charge from the corona discharge and are driven toward the collecting electrode by the electric field. All commercial precipitators use negative polarity.

Particles arriving at the collecting electrode do not have to be captured immediately for successful operation and may actually be recollected several times before they are removed. Current flows in the collection process are large enough to describe the operation as electrical in nature, the term "electrostatic" being one of convention rather than description. Precipitators may be single stage (or Cottrell type), or two-stage, in which the charging and collecting fields are formed independently. Also, the gas stream may flow through tubular collecting electrodes or between parallel plates. The parallel-plate single-stage precipitator is by far the most common.

Particles remain on the collecting electrode, or plate, until they are dislodged by a mechanical blow, called rapping, and fall by gravity into an external vessel, usually some form of hopper. Rapping is not as simple as it may sound, and indeed must be as effective as the charging and collecting steps for efficient precipitation.

Physical Description

The heart of the electrostatic precipitation process is the discharge electrode system. It must produce a strong, uniform corona while maintaining the correct distance and alignment with respect to the collecting electrodes to prevent imbalances in the electric field and to avoid unnecessary arcing discharges. Discharge electrode sizes and shapes vary mainly by manufacturer, but variations among different applications or in different sections of the same precipitator are feasible. Round, straight wires about 0.1 in. in diameter are the most common discharge electrode in use. They may be hung individually and freely with a suspension weight at their bottom end, or they may be held in a structural framework which is rigidly attached to the precipitator structure. Discharge electrodes may also be mast-type or formed elements, where rigidity, mechanical strength, and corona properties are all embodied into the same member. Figures 10-1 to 10-3 illustrate these three principal electrode designs.

High-voltage rectifiers providing pulsating dc waveforms are in use almost

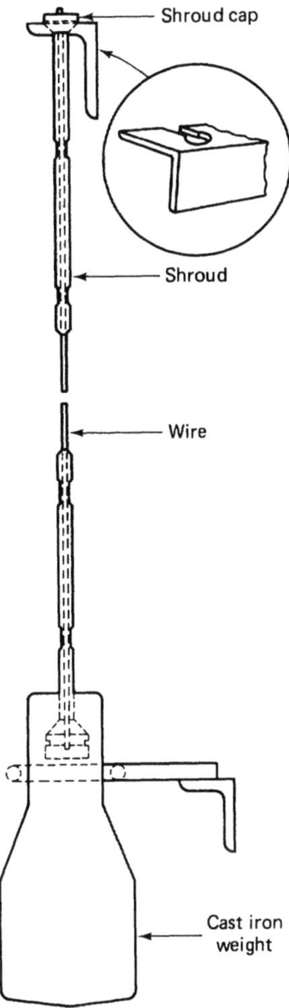

FIGURE 10-1 Weighted wire discharge electrode. (Courtesy of Research-Cottrell, Inc.)

exclusively due to the higher voltage and current attainable under sparking conditions when compared to pure direct current. With few exceptions, the discharge electrode system is subdivided into discrete sections, each being energized by a separate transformer–rectifier (TR) set. This "sectionalization" is important in matching corona currents and voltages to the TR set, and to promote reliability and stability under arcing conditions. TR sets are comprised of a high-voltage transformer and bridge rectifier, with typical secondary winding rms ratings between 53 to 66 kV and 250 to 2000 mA. Most TR sets can be connected to the precipitator discharge electrode system in either full-wave or double half-wave, as shown in Figure 10-4.

An important adjunct to the discharge electrode system is the automatic regu-

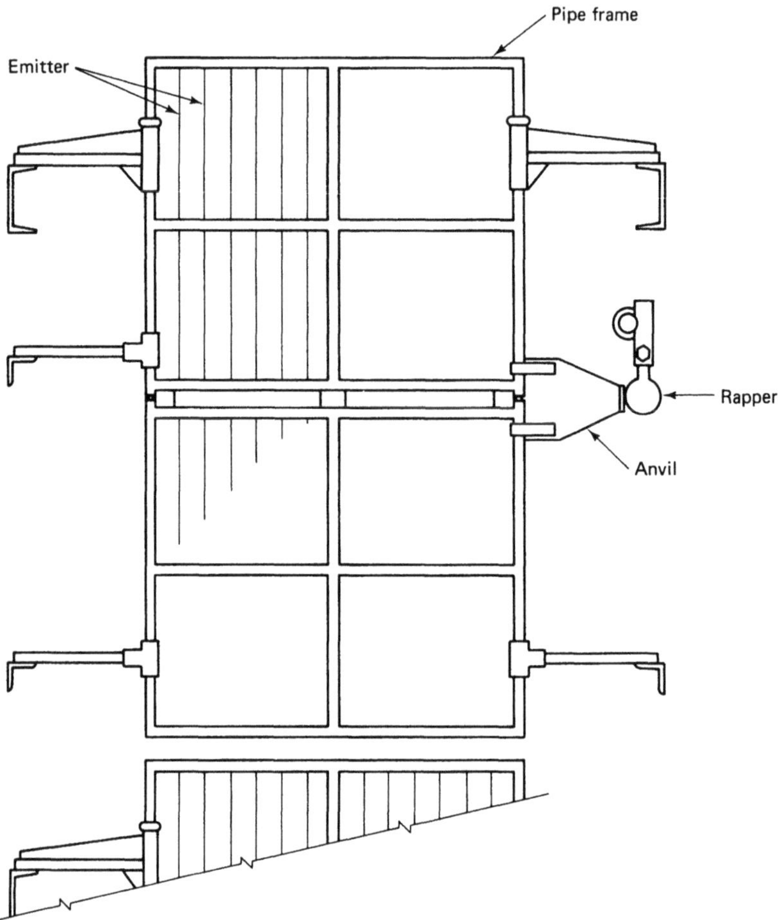

FIGURE 10-2 Rigid frame discharge electrode. (Courtesy of Research-Cottrell, Inc.)

lation of the high-voltage input to the precipitator, because only in rare cases does nature permit ideal operation. The earliest precipitators had no means of voltage regulation, but state-of-the-art advancements such as the silicon-controlled rectifier, solid-state construction, and digital control circuits have enabled precipitators to perform efficiently under the most adverse conditions.

In dry, parallel-plate precipitators, the collecting electrodes (plates) are suspended from the top of the precipitator, parallel to and in proper alignment with the discharge electrodes. These plates and their attachment must be strong enough to support collected particulate weight, yet light enough to react in a "lively" manner to dislodge particulate when rapped, and durable enough to withstand millions of rapping blows without fatigue failure. For these reasons, collecting plates are typically made of light-gauge metal and rigidly fastened to the precipitator structure only at their top ends. Most designs incorporate baffles to provide quiescent zones where the probability of particle collection is enhanced. These baffles are integrated into vertical stiffeners, which are required because collecting plate heights up to 50 ft are in use. Figure 10-5 illustrates one style of collecting plate.

Hoppers are best thought of as temporary holding bins to store collected particulate until permanent disposal can be scheduled. They take a variety of shapes, and are also sectionalized to facilitate handling large quantities of dust. It is a mistake to think the precipitation process stops at the hoppers. Removing collected material from the hopper is just as important as getting the material to fall into the hopper in the first place.

The discharge electrodes, collecting plates, and hoppers are all contained and supported by the casing, or shell. This structure must provide a gas-tight envelope in which the process takes place and must also hold the two electrode systems in proper alignment, sometimes under cyclical load conditions. In nearly all economical designs, essentially all the important auxiliary equipments are attached somewhere directly to the casing. Insulators that support the discharge electrodes system are made almost exclusively from procelain of fused alumina, and are contained in individual or grouped insulator compartments or all the high-tension support insulators may be housed in a top housing or penthouse. In many cases, the man-

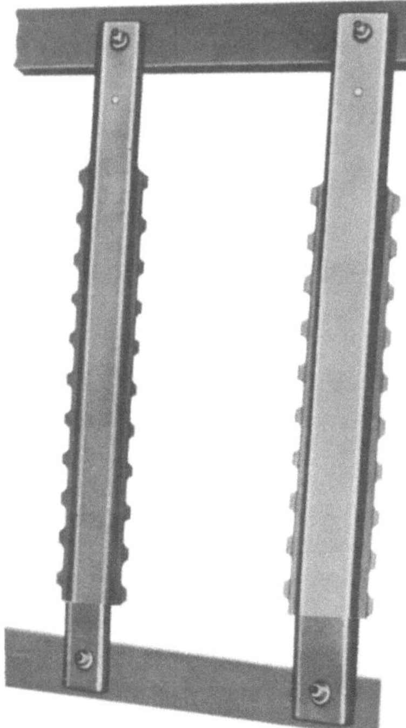

FIGURE 10-3 Rigid discharge electrode. (Courtesy of Research-Cottrell, Inc.)

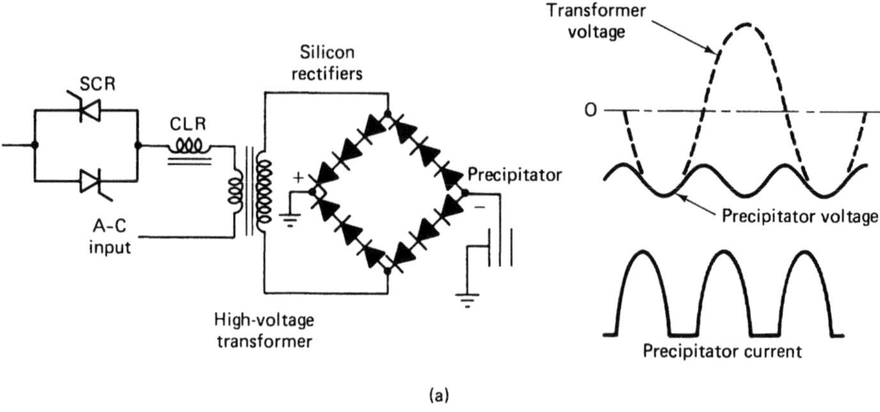

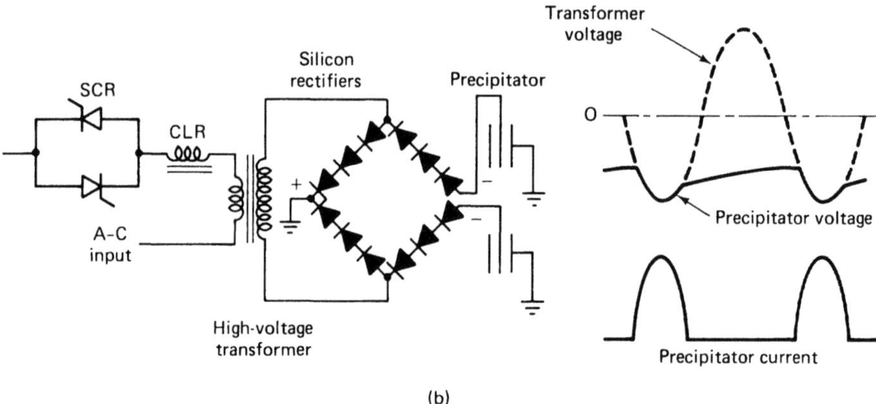

FIGURE 10-4 Precipitator energization equipment: (a) full-wave; (b) half-wave. (Courtesy of Research-Cottrell, Inc.)

ufacturer requires these insulators to be heated and ventilated at certain times during operation of the precipitator.

The widest variation in design among manufacturers comes from rapping. In the simplest sense, rappers are either impulse (single blow) or vibrating (multiple blow) types. One impulse type consists of swing hammers striking the lower portions of the collecting electrodes or an intermediate portion of a rigid frame discharge electrode system. The swing hammers are actuated by a cam shaft driven by an electric motor, and swing from the peak of the cam action by gravity to strike the electrode systems in a horizontal direction. Another type of impulse rapper is the drop hammer, which may be actuated by an electromagnetic solenoid located on top of the precipitators, or a motor-driven mechanical linkage. The rapping blow is gravity actuated or may be spring-assisted, striking the tops of the systems in a vertical direction, or nearly any combination among these practices. Vibrating types are used almost exclusively on the discharge electrode systems, especially on industrial processes where weighted wire geometries are most common. Vibrators may be electric or pneumatic, the latter being extremely powerful and potentially destructive. Figures 10-6 and 10-7 depict two impulse-type rappers.

Other key components in a precipitator include the high-voltage bus and guard that deliver the TR output to the discharge electrodes; gas distribution diffusers at the inlet and outlet face; access systems, and a host of other systems, subsystems, and components necessary to support the operation by providing electrical power distribution, control, insulation, and so on. Most of these features are shown in Figure 10-8.

Applications

Of all the particulate control devices available, the electrostatic precipitator has by far seen the widest usage, with successful applications in all the basic industries and in a few exotic ones. The major reasons are that finely divided particles of all

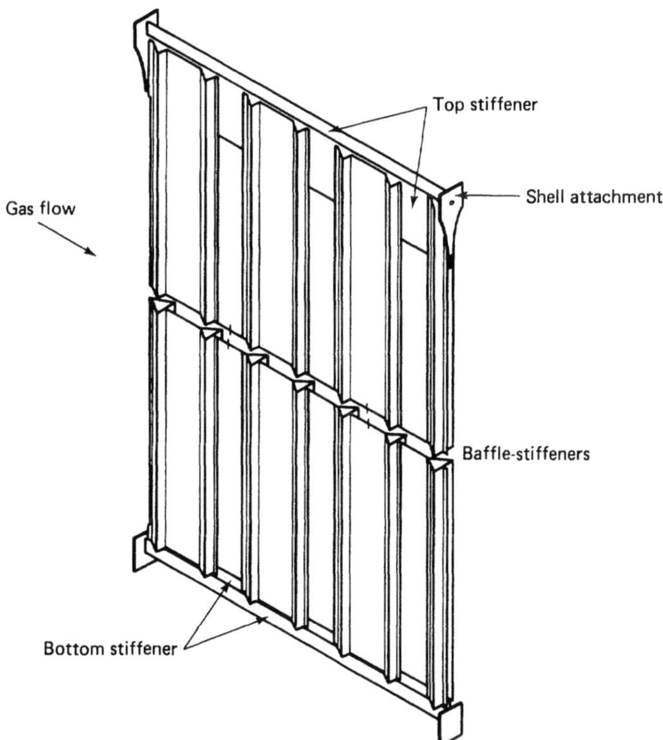

FIGURE 10-5 Collecting plate. (Courtesy of Research-Cottrell, Inc.)

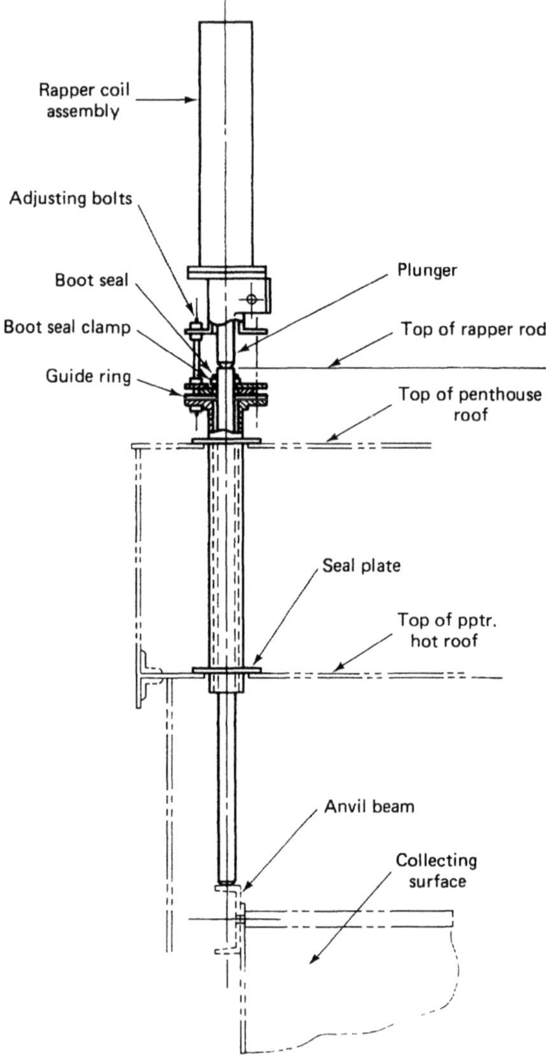

FIGURE 10-6 Magnetic impulse rapper. (Courtesy of Research-Cottrell, Inc.)

types acquire electrical charge quite readily, and that precipitators can handle large volumes of gas over a wide range of operating pressures and temperature. Units have demonstrated collection at 2000°F and several atmospheres.

By far the greatest quantity of gas treated and material collected is in the power-generation industry. It is estimated that the material collected in the power industry in 1970 alone exceeded some 40 million tons.

Precipitators have been used successfully in the ferrous and nonferrous metallurgical industries. Iron-, and later steelmaking, witnessed successful precipitator applications, although variability in the heat cycle on basic oxygen furnaces makes the steel application one of the most difficult. In the nonferrous metal industry, precipitators have been used in copper, lead, zinc, and aluminum smelting, and in a variety of sinter plants.

Two other large application areas are in the cement industry, where precipitators have been used to collect calcium and silicon oxides, and in the pulp and paper industry, where salt cake is recovered in the firing of black liquor in the kraft process, and where a considerable amount of fly ash is collected from coal and hog fuel-fired power boilers.

Other industries that have typically employed electrostatic precipitators in their recovery operations include the petroleum industry, where precipitators are used to recover fines from the fluid catalytic cracking process and to detar fuels; the chemical industry, primarily in the production of sulfuric acid; agriculture and feedstock plants; and gypsum plants.

Table 10-1 is a summary of the applications areas for precipitators in the United States.

Conventional Terminology

The descriptive terminology that will be used in the remainder of this chapter is selected from present convention and defined briefly below. Also, the remainder of the discussion on precipitator design, operation, and maintenance will generally

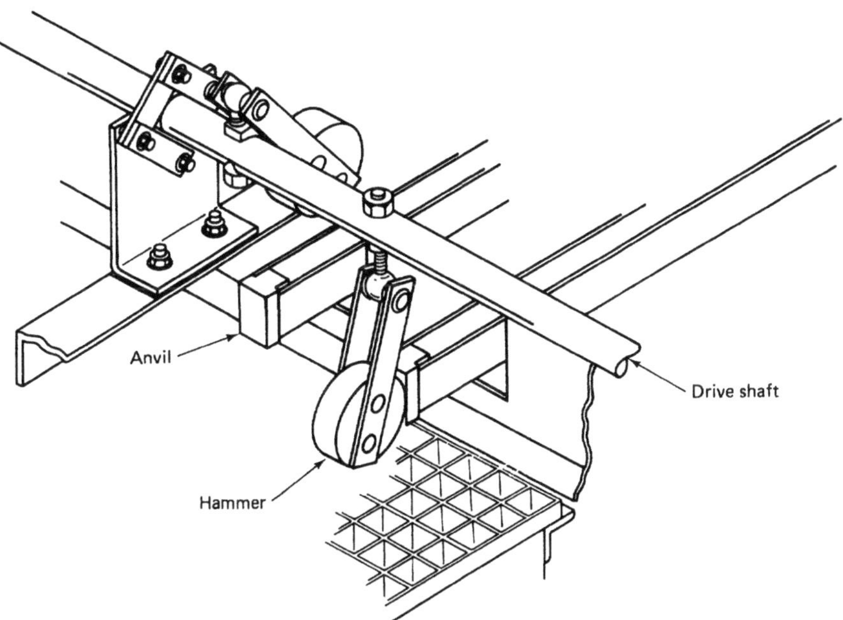

FIGURE 10-7 Internal swing hammer impulse rapper. (Courtesy of Research-Cottrell, Inc.)

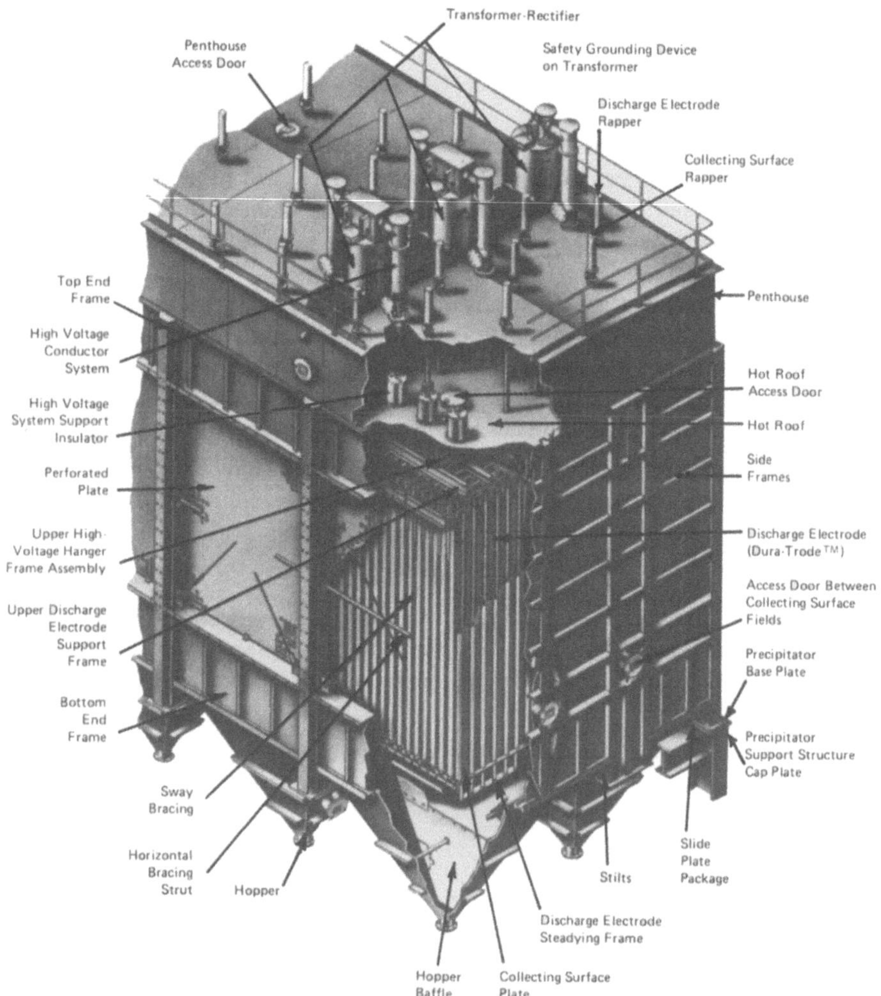

FIGURE 10-8 Electrostatic precipitator. (Courtesy of Research-Cottrell, Inc.)

reflect power-generation-industry experience. This implies no loss of realism to the reader, however, since most aspects of design and use are common among all applications.

Active height: Vertical length of collecting electrodes (plates)

Active length: Horizontal length of energized precipitator, excluding empty spaces between fields

Active surface: Total collecting surface area energized

Aspect ratio: Active length divided by active height

Bus section: Smallest portion of a field that can be deenergized

Chamber: Gas-tight longitudinal subdivision of a precipitator

Collecting surface area: Surface area of energized collecting plates .
Distribution plate: A device installed at either the inlet or outlet of a precipitator to achieve optimum gas flow distribution
Effective cross-sectional area: Active height $\times$ total number of gas passages $\times$ gas passage width
Field: Transverse subdivision of a precipitator formed by parallel bus sections
Gas passage: Volume enclosed by two adjacent collecting plates
Migration velocity: Theoretical speed of particulate normal to the direction of flow in the process of collection; called "effective" migration velocity when calculated from empirical data, but "collecting rate parameter" is less misleading
Power density: Ratio of total power input to collecting surface area or gas flow rate, usually expressed in watts/ft^2 or watts/1000 cfm
Precipitator: Arrangement of electrodes and all other equipment for one independent casing
Specific collecting area: Ratio of collecting surface area to gas flow rate, usually expressed as ft^2/1000 cfm
Treatment time: Active length divided by treatment velocity
Treatment velocity: Gas flow rate divided by effective cross-sectional area, expressed in ft/sec

10.2 DESIGN PROCEDURES

There are many ways to describe the design of any industrial machine, and precipitators are no exception. However, in almost all cases, precipitator design can be effectively divided into the four following aspects, each equally important over the life of the plant:

1. Performance
2. Reliability

TABLE 10-1 Precipitator applications areas.

Industry	Raw Material Consumption (tons)[1]
Power generation	397,000,000
Agriculture	8,400,000
Iron and steel	345,000,000
Cement	74,000,000
Pulp and paper	27,900,000
Primary nonferrous metals	10,948,000
Calciners	5,800,000
Reverbs	1,904,000
Sintering	1,079,000
Roasting	728,000
Secondary nonferrous metals	1,975,000
Asphalt and carbon black	1,300,000
Petroleum catalytic cracking	1,190,000,000 bbl

[1] 1971.

3. Operability
4. Maintainability

Performance Design

Performance design begins with sizing, or selection of the correct amount of collecting surface area, its arrangement, how it is energized, and how it is rapped. A fundamental knowledge of sizing is important to the operator for understanding the basic variables that control precipitator performance. From empirical observations by Anderson in 1919 and theoretical derivation by Deutsch in 1922, the first expression for precipitator performance (Deutsch–Anderson equation) took the form

$$E = 1 - e^{-(A/V)w}$$

where E = collection efficiency

A = total collecting surface area

V = gas flow rate

w = migration velocity

Since E and V are usually known, the amount of collecting surface A can be calculated from a judicious selection of the term w. For decades designers selected w by analogy, that is, from experience with similar types of dusts and the performance that resulted. For each of these cases, w was presumed constant. Sizing by analogy and some inherent inaccuracies in the Deutsch–Anderson equation related to particle size and uniformity led to some disappointing results, especially for very high efficiency requirements. It is virtually impossible to find a truly analogous situation, since industrial dusts are composed of many different-size particles, and w is in fact not constant but dependent upon many process variables.

An attempt to account for the sensitivity of w on process variables, especially particle-size distribution, appeared in 1957 and was later revised by Allander, Matts, and Ohnfeldt, who derived the expression

$$E = 1 - e^{-(Aw_k/V)^n}$$

The second exponent (in this, the so-called modified Deutsch equation) provides a more accurate prediction of performance at high-efficiency levels, but becomes too pessimistic in certain situations. A more useful relationship, developed by Feldman in 1975 [7], accounted for such variables as particle-size distribution at the precipitator inlet, changes in size distribution as particles are collected in the precipitator, and operating voltage levels. Figure 10-9 compares the three models for a selected case. These and many other models have been proposed to predict

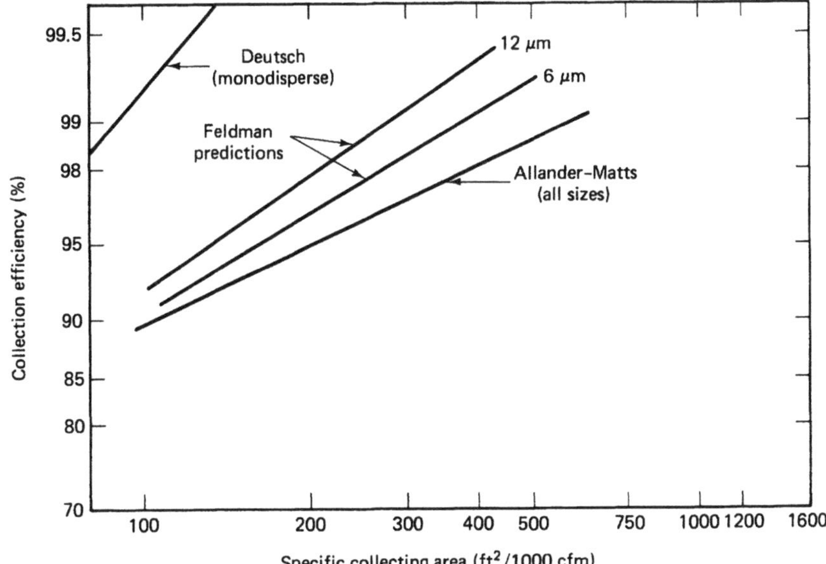

FIGURE 10-9 Comparison of three Deutsch-type prediction models. The Feldman model accounts for differences in particle-size distribution, expressed as mass median size (in µm). Note the optimism of the earlier Deutsch equation.

particulate collection in an electrostatic precipitator, and while special advantages can be argued for each depending upon which critical variables they account for, the fact remains that no model exists that accounts for all the variables that limit migration velocity in all situations. This is attributable mainly to the large number of variables that exist and because many of them are interrelated. For this reason, manufacturers size precipitators partly by theory and partly by analogy, and in all cases the proprietary techniques used are closely guarded. In the final analysis, the goal is still to determine the correct amount of collecting surface area, and that usually depends upon the proper selection of w.

The method used to determine w in precipitator sizing is not of particular importance to the operator, but knowledge of the variables that influence its value can be useful. The following is a brief discussion of the principal factors that influence effective migration velocity and which manufacturers must contend with in the design stage. The operator should be able to identify which factors he has the most operational control over and which ones, although fixed by design, provide information in understanding precipitator behavior under different operating conditions.

Energization   Usually expressed as power density related to the total collecting surface area or unit of gas flow rate, the amount of useful power delivered to the precipitator is one of the chief variables that controls performance. Useful corona power is the power absorbed in collecting particulate, not the power consumed in sparking or simply the theoretical rating of the TR set. One empirical relationship between efficiency and power density is shown in Figure 10-10. Voltage is the important parameter, but reasonable levels of corona current are necessary, since without current flow, collection would not take place.

Sectionalization Energization levels are sensitive to differences in gas temperature, dust concentration, particle-size distribution, and resistivity. Variations in these parameters are common and even expected across the width and along the length of a precipitator. No sound theoretical relationship between sectionalization and migration velocity is in use, but empirical results have been observed which strongly support the need for proper sectionalization. Ramsdell reported a relationship between efficiency and the sectionalization of complete precipitators, shown in Figure 10-11, and Figure 10-12 shows results obtained by Darby relating effective migration velocity to bus section size.

Dust resistivity Determined by ash chemical composition and gas temperature and moisture levels, the resistivity of collected ash limits the rate of current flow and thereby affects the total useful corona power that can be delivered to the precipitator. Figure 10-13 shows resistivity as a function of temperature for some selected ashes. Sulfur content of the coal plays a major role in determining resistivity level. The debilitating effect of high resistivity on energization can be seen in the

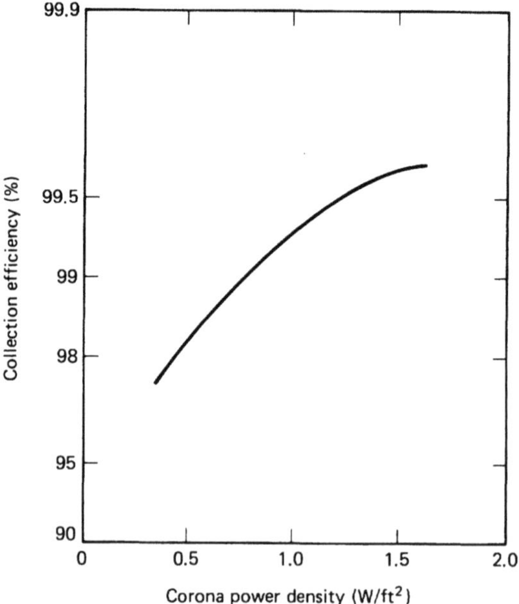

FIGURE 10-10 Power density vs. efficiency.

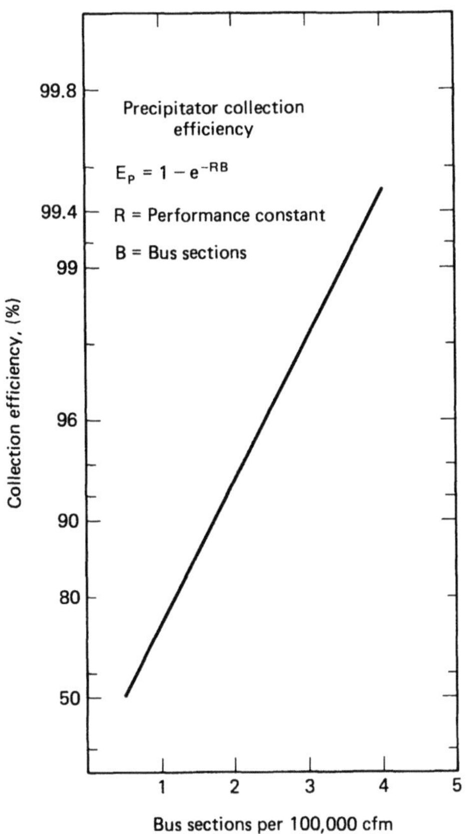

FIGURE 10-11 Precipitator sectionalization. Although capital and operating costs must be considered in a final selection, sectionalization is important in achieving high collection efficiency [13].

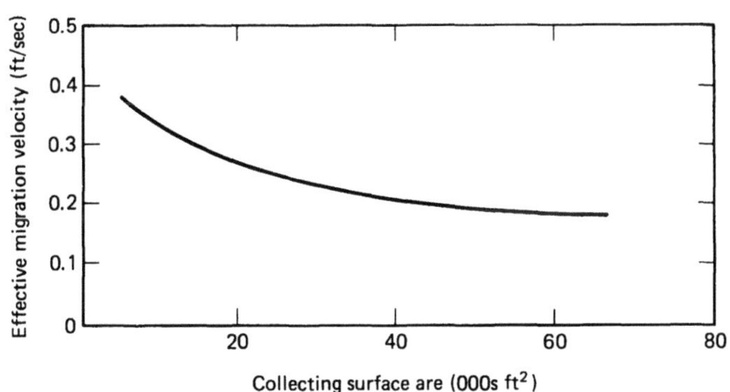

FIGURE 10-12 Sectionalization—bus section size. (Adapted from [5].)

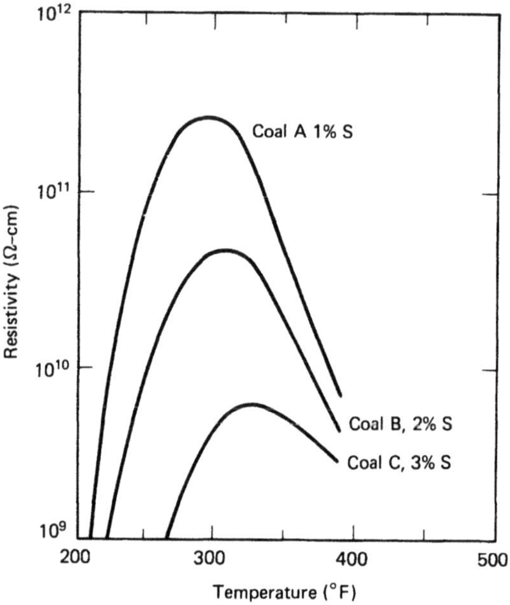

FIGURE 10-13 Resistivity vs. temperature for three different coals.

voltage–current (V–I) curves of Figure 10-14. Here, a moderately resistive ash, around 1×10^{11} Ω-cm, does not impair useful power input, and a healthy current density of 20 mA/1000 ft^2 of collecting area is attained. However, the shape of the curve for the high-resistivity ash breaks sharply around 3 mA/1000 ft^2, after which voltage decreases with increasing current. None of the power above the breakpoint is useful; in the case shown, the power above the breakpoint is indeed detrimental, since it is indicative of a well-established formation of back corona. Back corona is the emission of positive ions from the collected dust layer that occurs when the electrical breakdown strength of the interstitial gases is exceeded by the voltage drop across the collected dust layer. The presence of enough positive ions in the gas passage can destroy the normal precipitation process, but the most common yet least appreciated detriment of high resistivity is that it forces unequal distribution of corona on the discharge electrode, nullifying collection in areas with no corona, and causing premature sparking in areas of intense corona.

Particle size distribution   For a single particle of size d, its migration velocity under the influence of charging field strength E_0 and collecting field strength E_p can be defined as

$$w = \frac{\varepsilon_0 E_0 E_p a C d}{3\mu}$$

where μ = gas viscosity, dependent upon temperature

           a = a term for the charge acquired by the particle

           ε_0 and C = correction constants

From this expression, it is apparent that different size particles will have different migration velocities. Large particles, over 1.0 μm diameter, are charged mainly by direct collision with ions and free electrons moving toward the collecting plate along electric field lines. This field charging mechanism is very efficient for large particles, but the velocity of large charged particles toward the collecting plate is impeded by viscous drag forces. Small particles, less than 0.5 μm, become charged mainly by random thermal motion of ions. This diffusion charging mechanism is less efficient, but charged small particles can proceed toward the collecting plate with relative ease since the drag forces opposing their motion is small. Experimental results examining the effects of particle size are shown in Figure 10-15. The ability of a precipitator to collect the large and small particle sizes is essentially equal, but more difficult for the intermediate-size particles. A precipitator working on a dust comprised mainly of particles in the intermediate-size range will be less efficient

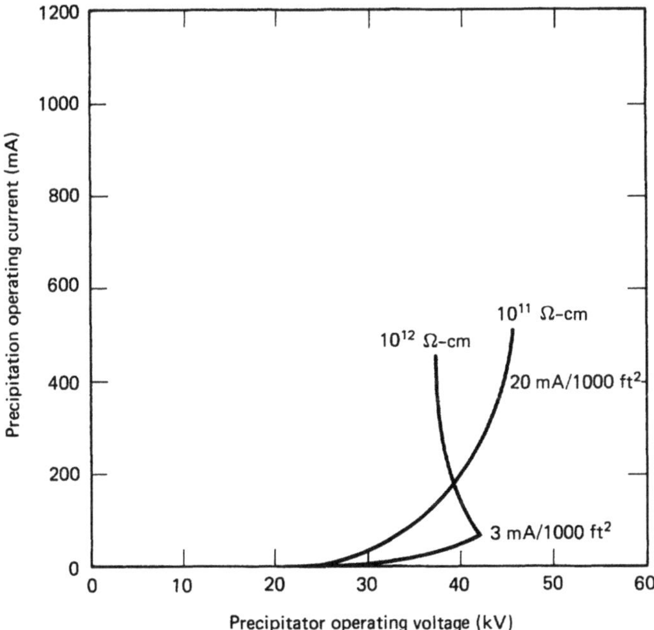

FIGURE 10-14 Voltage–current characteristic, illustrating corona current limitation and onset of back-corona gas passage. Back-corona is well developed above 3 mA/1000 ft² for the higher-resistivity ash.

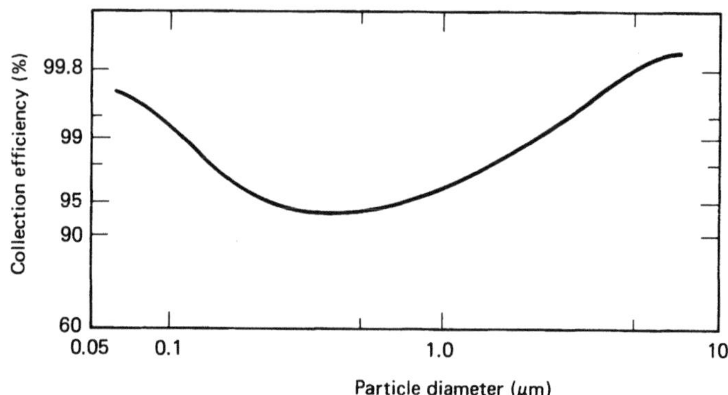

FIGURE 10-15 Effect of particle size on performance.

than an equally sized precipitator working on either predominantly large or small particles.

Rapping   Intensity of the rapping blow, the time interval between raps, and the duration or on-time of vibrators when they are used determine the cleanliness of electrode systems, which in turn can have a profound effect on energization. Overzealous rapping can be a detriment, contributing large dust quantities to reentrainment.

Alignment   Misalignment between discharge and collecting electrodes causes premature sparking and a reduction in the corona power that can be delivered.

Collecting electrode spacing   Direct relationships between spacing and effective migration velocity have been found from empirical observations, Figure 10-16 illustrating one of them. Applied voltage has to be increased with wider spacings.

Discharge electrode geometry   The voltage–current relationship in any precipitator depends in part on the geometries and spacings between electrodes. Discharge electrodes of small radii generally produce less voltage at a given current than do large-diameter electrodes. A comparison of V–I curves for a 0.109-in. wire and a large rigid electrode is shown in Figure 10-17.

Treatment velocity   Variation in migration velocity with gas velocity for a relative case in the precipitator is shown in Figure 10-18. Above some critical velocity, aerodynamic forces on particles contribute to reentrainment.

Gas flow distribution   The best operating condition for a precipitator will occur when the treatment velocity distribution is uniform. When significant maldistribution occurs, the higher velocity in one collecting plate area will decrease efficiency more

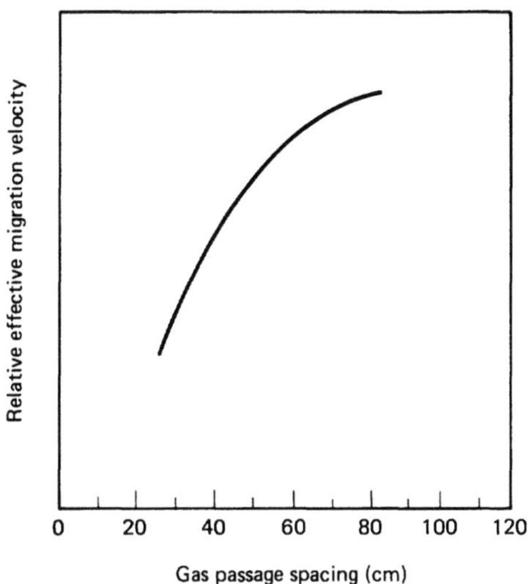

FIGURE 10-16 Effect of gas passage width on effective migration velocity.

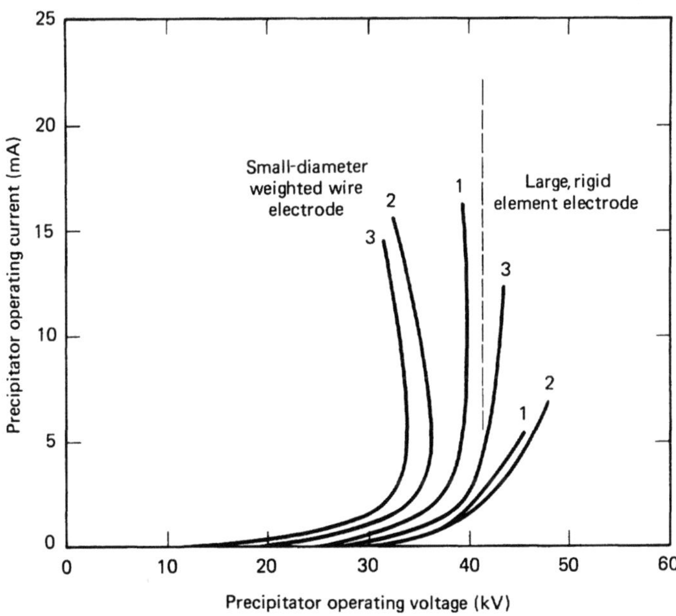

FIGURE 10-17 Comparison of voltage–current characteristics for two different discharge electrodes in a pilot precipitator.

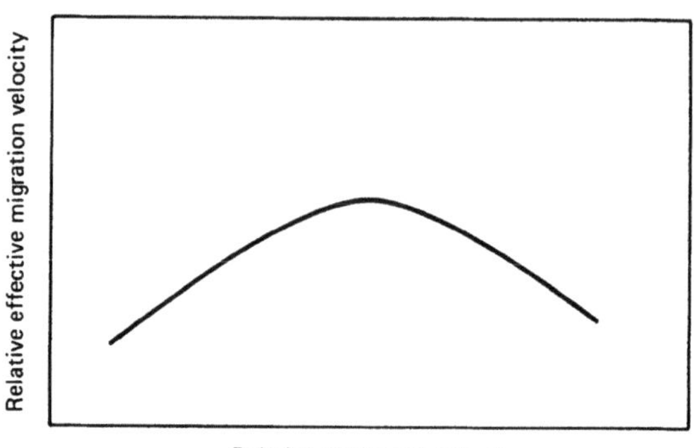

FIGURE 10-18 Effect of treatment velocity on effective migration velocity.

than a lower velocity at another equal area will increase the efficiency of that area. Small particles tend to follow flow streamlines better than large particles, so a maldistribution in particle size distribution in areas of the precipitator will also occur. Gross flow maldistribution also contributes to reentrainment losses and sneakage of raw gases around collecting plates.

Dust concentration Higher dust concentrations in the gas stream inhibit the free flow of current made up of ions and free electrons toward the collecting plate. This suppression in current flow drives the voltage up and can cause sparking. Obviously, the highest dust concentration occurs in the inlet fields, and energization of these fields will be affected most.

Aspect ratio Lacking any theoretical base, the importance of aspect ratio is related to the time necessary for particles agglomerated on the collecting plates to fall into the hoppers when rapped. If the aspect ratio is too low, contributions to reentrainment will occur. This situation is aggravated by excessively high treatment velocity. Aspect ratio is most important for high-efficiency requirements, with values ranging from 1.0 to 1.5 in common use for efficiencies over 98%.

Precipitator length The longer a precipitator is, the finer the particle-size distribution will be in the outlet fields. This aspect of design and selection is most important for very high, on the order of 99.6 to 99.9%, efficiency requirements.

Reentrainment This term represents the reintroduction of previously collected ash into the gas stream by any one or all of the mechanisms described above. Gas sneakage around collecting plates or through hoppers can be considered part of the losses constituting reentrainment. Perhaps the worst type of reentrainment is caused by cross-flow in hoppers, resulting from improper flue geometry. Hopper cross-flow can sweep efficiently collected and rapped dust in large quantities, exiting the hopper as an irretrievable loss. Reentrainment is perhaps responsible for the largest differences between theoretical and effective migration velocity.

A summary of the effects on precipitator design performance of the parameters described above is shown on Table 10-2.

Reliability

A precipitator properly sized and capable of high efficiency cannot be considered a success if it demonstrates this capability only once during its lifetime. A key issue in assuring long-term operational success is designing for reliability. Designs should prevent subsystem deficiencies, which usually reduce full energization of the pre-

TABLE 10-2 Effects of performance design parameters.

Parameter	*Property of:*	*Effect on:*
Energization	Electrode geometry, TR design, gas and ash properties	Collecting efficiency
Sectionalization	Precipitator arrangement TR design	Energization
Resistivity	Ash and gas properties	Energization
Particle-size distribution	Ash	Energization, collection efficiency
Rapping	Precipitator design	Energization reentrainment
Alignment	Precipitator condition	Energization
Spacing	Precipitator design	Migration velocity
Discharge-electrode geometry	Electrode design	Energization
Treatment velocity	Precipitator design, gas flow rate	Migration velocity
Gas flow distribution	Precipitator design, gas velocity	Migration velocity
Dust concentration	Gas distribution, boiler/fuel/flue design	Energization
Aspect ratio	Precipitator design	Collection efficiency
Precipitator length	Precipitator design	Particle-size distribution
Reentrainment	Treatment velocity, rapping, energization (sparking), precipitator/flue geometry	Collection efficiency—theoretical and effective migration velocity

cipitator (performance reliability), and physical integrity of the components themselves (mechanical reliability). Although every component in the precipitator system contributes to overall reliability, certain major components and design elements are usually considered critical. These are the casing, electrode systems, rapping systems, hoppers, energization equipment, sectionalization, and corrosion control.

In addition to its primary functions of electrode support and alignment, the casing must be designed with sufficient strength to withstand temperature or pressure excursions that may occur during process upset conditions. If the casing is distorted by thermally induced stress, the alignment of the attached equipment also becomes distorted. In the case of hot precipitators, thermal stress must be accounted for, usually by derating allowable design stresses in main structural members. All precipitators expand and move on their supporting steel structure, so a liberal use of slide plates permitting free expansion with a minimum of frictional stress is required. Finally, the structural design should prevent large thermal gradients from occurring at high-stress areas in the casing and should also prevent high temperatures in the area of the slide plates or in the supporting structure. These features can readily be accommodated by using a stub-column (or stilt) casing design, as shown in Figure 10-19. The clearance between the casing and the support structure afforded by the stub column permits full insulation and ventilation of highly stressed hopper transitions, provides an isolated area for thermal gradients, and prevents the slide

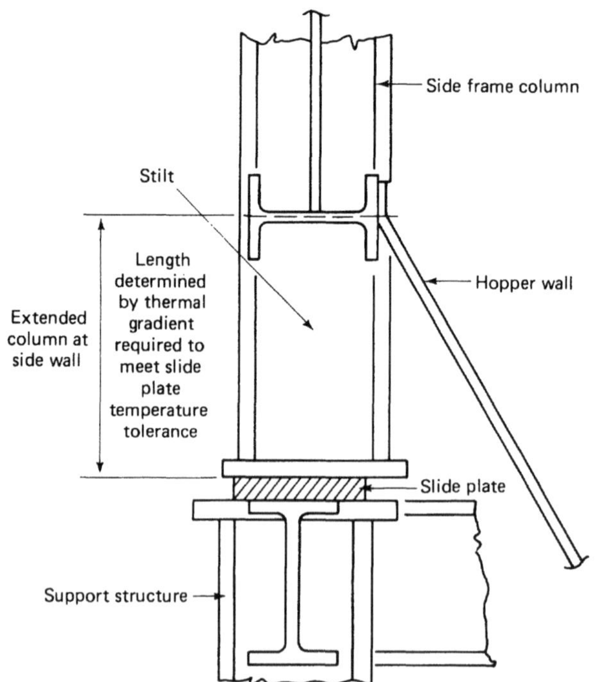

FIGURE 10-19 Typical stub column. (Courtesy of Research-Cottrell, Inc.)

plates and support steel from overheating. A more challenging structural design problem occurs when the precipitator is intended for duty on cycling service. The total number of expected starts during the life of the unit and the excursion values should be included in the design.

Experience has shown that the primary component that controls precipitator reliability is the discharge electrode system. Performance or mechanical reliability deficiencies in this system lead directly and rapidly to diminished precipitator performance. Designs should prevent electrode failure and ineffective rapping.

Four major causes of discharge electrode failure are electrical arc erosion, mechanical fatigue, instability, and corrosion. Arc erosion is caused by misalignment and instability. The mechanism is most detrimental to small-diameter wire electrodes. Experience has shown that all discharge electrode systems are reliable provided that their design limitations are not exceeded. However, when experience indicates that arc erosion is a cause for concern, rigid frame and rigid electrode designs are generally preferred over weighted wire systems. Rigid frame designs are less susceptible to misalignment, and the large mass of rigid electrodes tend to make them impervious to arc erosion. Fatigue failures are primarily caused by cyclic rapping stress, failure usually occurring at or near the shroud crimp in weighted wire systems and at points of high mechanical stress in rigid systems. Prudent design practices, testing, and quality control are sufficient to prevent this problem. Instability leads to arc erosion when long unsupported electrode length permits excessive oscillation under influence of the electric field. Instability is most prevalent in weighted wire systems because the entire span of the small wire is unsupported. The practical limitation for weighted wire electrode length is 36 ft with a 35-lb suspension weight. The design is further improved when stabilizers are located as close as possible to the bottom termination of the electrode. Rigid electrode designs normally integrate their high mass and rigid end connections to gain stability. Rigid frame designs typically resort to four-point suspension and a minimum unsupported length between wire electrode suspension points.

All discharge electrode system designs must be adequately rapped to prevent fouling and premature sparking. Rapping densities usually range from 1500 to 3000 lineal feet of corona emitter per rapper, depending upon experience with different dusts. The magnitude of the rapping force is also a function of experienced dust behavior. Input force ranging from 5 to 20 ft-lb is generally adequate for power generation applications, but some industrial processes, particularly smelters and salt cake recovery boilers, may require considerably higher levels.

Collecting plate mechanical reliability is affected by fatigue and corrosion, and performance reliability is affected by alignment and rapping. Fatigue failures occur when the collecting plate attachment to the casing cannot withstand peak cyclic rapping stresses. The greatest risk of failure occurs when old designs are adopted for more rigorous rapping duty without careful examination of the strength of the design. Design revisions, which may require exotic finite element and dynamic stress analyses, and prototype life testing are necessary to assure a reliable design.

Misalignment of collecting plates with respect to the discharge electrode system causes premature sparking and a reduction in operating voltage and efficiency. A design that prevents in-service warping, presents no special installation difficulty, and receives adequate quality control during construction will usually be successful.

Collecting plate rapping is perhaps the most critical aspect in precipitator performance reliability. The most direct way to optimize rapping is to utilize a collecting plate design that achieves the highest acceleration response with the least input force. The direction of the blow imparted to the collecting plate is not critical as long as a minimum average impulse acceleration normal to the collecting surface is achieved. Selection of the minimum acceleration level should depend upon operating experience, but a range of 60 to 100*g* is typically specified for power-generation applications. Figure 10-20 shows a typical *g*-level topography demonstrating the minimum acceleration value.

Rapping systems can generally be divided into two generic groups: internal mechanical and external mechanical or magnetic. Both designs can be reliable as long as their limitations also are not exceeded. Every component in each system should be designed so the peak operating stress is well below the allowable design stress.

The inability to evacuate collected dust from hoppers is one of the most common causes for forced partial outages in all precipitators. For a large power boiler, the ash removal system may be faced with the formidable task of evacuating up to 30

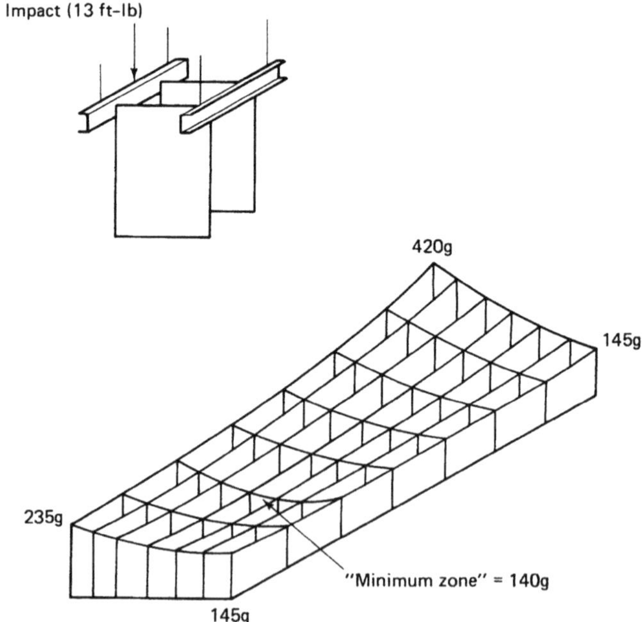

FIGURE 10-20 *g*-level topography map measured zero to peak. (Courtesy of Research-Cottrell, Inc.)

tons of reluctant material through a 12-in. hole in a relatively short time. Some dusts are naturally difficult to move, and for any material, the situation can be aggravated by ambient conditions. In order of increasingly difficult dust, the following is a list of design features that may be built into pyramidal hoppers.

1. 55° minimum valley angle, up to 70° if space permits
2. 12-in. minimum outlet diameter, minimum internal obstructions
3. Poke holes, vibrators, two-zone nucleonic level alarms
4. Strike plates, heaters for lower one-third of surface area
5. Round corner fillets and up to full surface heating with separate zone thermostat control
6. Hopper deck skirt and continuous evacuation

Industrial applications usually make use of trough or flat-bottom hoppers with mechanical bulk transfer equipment such as screw or drag conveyors. These systems are mechanically reliable when adequate attention has been given to operating stress and torque levels. Aeration stones for any application should not be considered unless they will be provided with heated and dried air.

Sectionalization is important to the performance reliability of any precipitator, yet it is one of the least understood design principals. Apart from performance design benefits, a high degree of sectionalization promotes reliability directly by minimizing the amount of collecting area lost from internal grounding, and by reducing the detrimental effects on energization caused by troublesome operating conditions. In practice, the amount of sectionalization is often a compromise between performance reliability and cost.

Operability

Operator knowledge cannot be designed into a precipitator, but good precipitator design accounts for essentially the full range of operating variables that can be expected, and provides a means for the operator to adjust or monitor the precipitator operation over the full range of variables encountered. Aspects of precipitator operation that are instrumental in optimizing performance include:

1. Energization
2. Rapping
3. Hopper control
4. Information

Clearly, energization is one of the most influential aspects of precipitator performance. Proper energization starts with sectionalization. The best automatic voltage controls cannot compensate for excessive or improperly arranged surface area energized by one TR set. The next step is ensuring the TR set rating matches the expected operating electrical load for each bus section. This usually requires that the TR operate at no less than 50% of its current rating. With every flashover, there is at least a momentary loss in the electric field, and excessive sparking can seriously reduce precipitator performance. The TR primary circuit impedance, furnished by a linear reactor, is critical in limiting the magnitude and duration of sparking. Higher impedance will require the use of smaller TR set current ratings but will result in less detriment from sparking. The automatic voltage control should at a minimum provide the following features:

1. Quenching of high-current arc discharges
2. Rapid response and recovery from sparking interruptions in the voltage waveform
3. Sense and retard voltage recovery when a burst of sparks occurs
4. Regulate spark rates below 10 sparks/min without loss of control and with minimum reduction in operating voltage

Finally, provision of full-wave/half-wave switching is a convenient and timesaving feature for the operator.

Rapper operation is a major step in fine-tuning for maximum precipitator performance, but it usually cannot be defined until after the unit has been placed in service. For this reason, rapper systems and controls provide for wide adjustment of rapper intensity and time interval between raps. With most precipitator designs, the operator can do nothing to adjust the quality of gas flow entering the precipitator. However, provision of rappers and controls on flow distribution devices will permit him to prevent deterioration in flow uniformity.

Even the best hopper designs are at times defeated by an especially difficult dust, and it is reasonable to declare that all operators will have some unpleasant moments on the hopper deck. Provision of automatic vibrator and dump valve controls with manual override, strike anvils, and poke holes, although a meager arsenal in the worst situations, provide some extra measure of ensuring precipitator operability.

To make correct decisions, the operator must be furnished with essential and reliable information, and the design should not discount the ability of the operator to make correct decisions. Essential information can be provided with the instrumentation listed on Table 10-3.

Maintainability

Consideration of preventive and corrective maintenance procedures and costs should be considered in the initial design. Access is a key element in all maintenance actions. Convenient access to facilitate preventive maintenance and inspections should be provided to:

1. Upper and lower extremes of electrode suspensions
2. Discharge electrode support insulators
3. Hoppers and their interiors
4. Rapping equipment

TABLE 10-3 Precipitator instrumentation.

Energization
  Transformer primary voltage
  Transformer primary current
  Total average precipitator (secondary) current
  Precipitator (secondary) voltage for each transformer output
  Spark indicator (spark rate optional)
  Trip indicator
Rapping
  Cycle time
  Operating rapper identification
  Rapper intensity (ft-lb)
  Trip or fault-alarm indicator
Hoppers
  Ash high- and high-high-level alarm
  Hopper pulling sequence and cycle time
  Activated hopper indicator
  Empty hopper indicator
  Vibrator operating indicator
Performance
  Transmissometer (opacity)
  Alarm panel
  Remote control console

5. Energization equipment and controls, including control cabinet layout to eliminate shock hazards

Outages are costly for all precipitator applications, so features that minimize downtime are especially valuable after the unit is placed in operation. Electrode systems that minimize the efforts required to repair distorted alignment not only reduce downtime but enhance the probability that maximum performance is maintained. Downtime is minimized with ease of replacement of failed components.

Field reports indicate that considerable maintenance time is spent on removing dust dropout deposits in flues. With the possible exception of some industrial processes, notably kraft mill salt cake recovery, flue dropout rarely impairs precipitator performance unless it seriously distorts gross gas flow distribution. Before dropout reaches such heights, it will likely threaten the structural integrity of the flue. Experience in designing and comprehensive model studies can define the necessary means to prevent excessive dropout and eliminate unnecessary maintenance.

10.3 INSTALLATION PROCEDURES

To a large extent, long-term reliability for a precipitator can be assured by dimensional stability of the casing and all internal components attached to it. This means that all design tolerances, especially those affecting electrode alignment, must be built into the final structure in such a way that thermal expansion, wind loads, and all other stresses the precipitator will experience during its lifetime will not cause dimensional distortion. Most precipitators are large structures assembled in the field using a variety of trade labor crafts and heavy equipment. Construction of them is a project, often requiring 24 months or more to complete, with the natural concerns for cost and time taking priority. Quality of construction and dimensional tolerance under these conditions can be assured best by people not responsible for schedules or cost overruns. This approach is generally the one taken. Of all the categories of interest in a construction project, the ones that influence dimensional stability are procedures, tolerances, and quality assurance.

Procedures

Construction is a stepwise process, beginning after the foundations have been poured, cured, and checked for elevation and perimeter dimensions. A supporting open structure is then completed and the main precipitator structural members are built upon it, followed by the remainder of the shell and finally the internal components. Most of these activities overlap each other in time, with major checkpoints along the way to ensure that necessary prerequisites for tolerances are met before construction of the remainder of the unit gets too far along.

The erection sequence ordinarily requires that the load-bearing columns of the precipitator make full, flat contact with the slide plates on top of the supporting structure before hoppers or internals are attached. When the open shell is essentially complete, struts or sway bracing is generally attached and the entire structure is checked for plumbness and squareness. This check is extremely important, since it will control the ease of gaining alignment of all subsequent parts. Benchmarks are established prior to installation of the electrode systems. Positioning of the electrodes is made relative to these benchmarks. Collecting and discharge electrode systems may be hoisted into the casing at the same time, but the collecting electrodes are normally completed first, and the discharge electrodes positioned relative to them. When the electrodes have been installed, slide plate contact may again be checked, and a major check, for square, plumb, and level electrodes and the alignment between electrode systems, is completed. Rapping equipment and insulators are installed and aligned, the roof is set, and the remainder of the precipitator construction proceeds toward mechanical completion. Mechanical completion usually includes the items listed in Table 10-4.

Tolerance

It comes as a surprise only to the uninitiated that a precipitator structure 200 ft wide, 90 ft long, and 80 ft high can be built within tolerances that require almost perfectly flat and level foundations, elevations on vertical steel members to within ⅛ in., and dimensions on squares and diagonals within ¼ in. throughout the entire structure. These tolerances are necessary, and although care must be taken in the

TABLE 10-4 Mechanical completion items.

☐ All collecting and discharge electrodes complete, within alignment tolerances, and welded out
☐ All rapping equipment complete, aligned, and ready for test
☐ Protection removed from insulators; no tracking or weld spatter
☐ Pressure test completed
☐ Field wiring complete and verified, megger tests done
☐ Instrumentation and rotating equipment checked and ready for test
☐ Key interlocks checked out, spare keys destroyed

construction process, they are usually achieved. Dimensional tolerances specified by one manufacturer are shown in Figures 10-21 and 10-22.

The American Institute of Steel Construction allows the correction of minor misfits by reaming, chipping, cutting, and the use of drift pins. Methods other than these are not recommended, but if they appear necessary, construction may have to be interrupted until the problems are corrected. Construction methods have to be reviewed during the progress of work to prevent distortion of previously obtained tolerances. Backstep welding may be required to avoid excessive heat distortion of components essential to proper alignment. Care must be exercised when handling or erecting discharge or collecting electrode panels and pieces to prevent warping or bowing. Labor forces must also be cautioned against using alignment-sensitive precipitator components as fulcrums or jacking attachments.

Quality Assurance

Assuring structural integrity and dimensional stability of the precipitator takes careful planning and dedication throughout the construction project. Quality assurance interfaces with all aspects of the construction phase, from material delivery to final air-load test. Receipt of materials is an important aspect because in the interest of cost and schedule, materials may be shipped to the field site up to 12 months before they will be erected. The site logistics may require a dedicated laydown area adjacent to the crane-operating side of the precipitator and a preassembly area nearby. Erection is easier and quality better controlled when the largest practical components can be shipped to the site or assembled on the ground. Material storage areas may require up to 20,000 ft^2, including warehouse storage for electrical equipment, and protective coverings for electrode components. Most other materials can be stored in the open. Care must be exercised in unloading material, and defective or damaged shipments should be corrected at once.

Quality assurance for the erection process begins with qualification or certification of welders, since the manufacturers' design specifications call for a variety of welding techniques, involving certain minimum skills that must be demonstrated. The manufacturer will have preferences for the sequence of erection, but quality assurance checkpoints must take priority. Permanent accesses, when installed as soon as the degree of shell completion permits, not only minimize hazardous working conditions, but promote quality by reducing fatigue and fear.

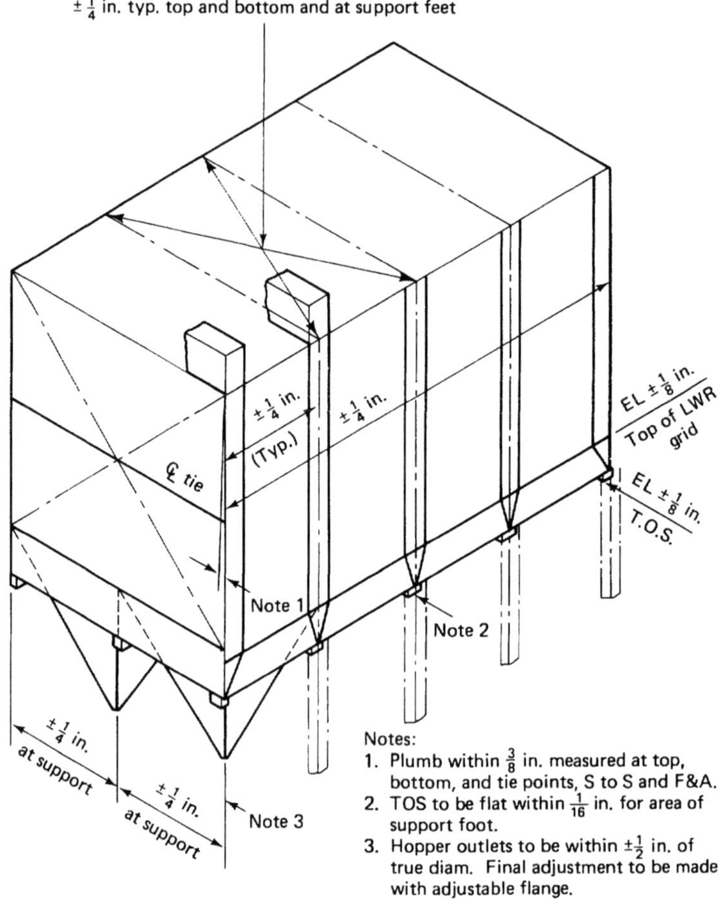

FIGURE 10-21 Precipitator shell tolerances. (Courtesy of Babcock and Wilcox Co.)

Usually, dimensional tolerance checks and tests for mechanical completion are performed by the manufacturer's service department near the end of the construction period. All prior checks are usually made by the erection superintendent. It is a customary and sound practice to avoid major corrections by holding off final welding until the service representative has completed his work.

10.4 OPERATION

In the sense of operations, it is probably better to consider the precipitator as a process system rather than as a piece of hardware. Outwardly simple in design, the electrostatic precipitator takes part in a process that is complex. Many things can influence its behavior, but most of these influences can be overcome if the precipitator is well designed and properly operated. Operation is, in fact, equally important to design. To obtain maximum results, the operator must know how to properly start up, shut down, and operate the precipitator; learn how to distinguish between normal and abnormal behavior; and know what steps to take in case of trouble.

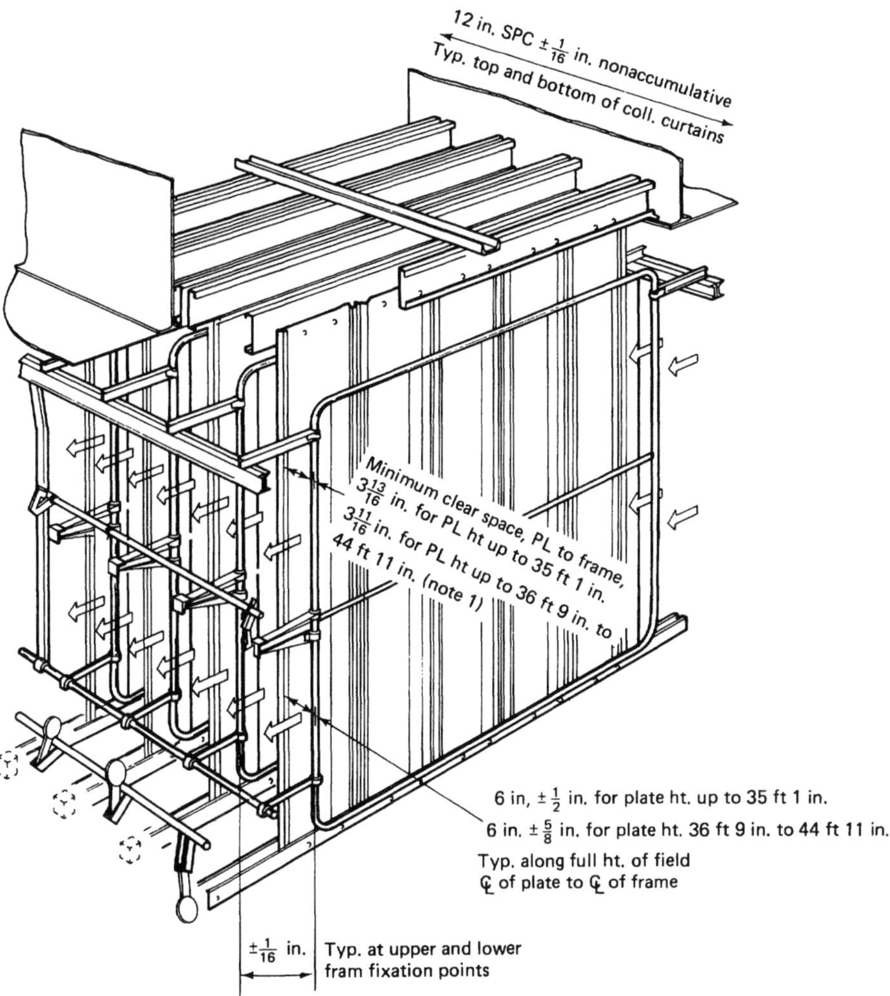

Note 1: The minimum clearance shown from face to frame tubing to face of plate flange must be held for full height at full depth of field.

FIGURE 10-22 Erection tolerances for precipitator internal components: 1, discharge electrodes; 2, collector plates; 3, discharge frame. (Courtesy of Babcock and Wilcox Co.)

Safety

Personnel assigned to operate or maintain precipitators must be well trained in special safety procedures to avoid injury. Like many other industrial machines, precipitators have claimed lives, but in every one of these cases tragedy could have been avoided if logical safety precautions had been observed. Safety is everyone's responsibility, but one person in the plant should be assigned the responsibility for monitoring and upgrading safety standards and equipment, and for training others whose duties include working with or inside the precipitator system. Table 10-5 describes important special safety precautions for precipitators.

Pre-Startup

Prior to energizing the precipitator and its auxiliaries for the first time, a pre-operational inspection should be conducted to verify that all equipment is ready to operate. Basic inspection items are listed in Table 10-6. This list can be adopted into a routine checklist, as many of the inspection items are applicable following any unit shutdown of more than 24 hours, or following a shutdown where work was done inside the precipitator. When the checklist has been completed, an air load should be performed.

Air load is simply the test energization of the discharge electrodes in air. The purpose of this test is to check the proper operation of the rectifier control unit, verify that all grounds, misalignments, and debris have been cleared, and to check safety and alarm conditions. It should be performed with only natural stack draft or minimum forced-air flow, and it also provides a convenient time to test operate all auxiliary equipment and the ash conveying system.

Air-load electrical readings of the TR sets are important, and the data should be recorded and retained for future reference. Readings should be taken of the voltage and current levels in stepwise fashion from corona onset (the voltage at which corona current flow is indicated by the panel secondary current meter or current waveform ripple if tested with an oscilloscope) to maximum power attainable without excessive sparking. Air-load readings are most useful when secondary voltage (kV) can be read, but primary voltage is an adequate substitute. The readings should be more frequent at lower values of corona current, since this is where the voltage changes most rapidly. The results can and should be plotted on a voltage–current curve for each TR set as shown in Figure 10-23. Ideally, curves should be generated on individual bus sections if time permits. Referring to Figure 10-23, the values at which readings were taken are shown as points on each curve. This particular set of data indicates a different starting voltage for each bus section, and a difference in the shape of each curve. Field 1 has an elevated starting voltage compared to fields 2 and 3, indicating discharge electrode buildup or a residual dust layer of high resistivity on the collecting plate. The shape of its curve, with a steep rise in current that occurs before the other two curves, and a tendency to bend backward, is indicative of back-corona, furthering the suspicion of a high-resistivity

TABLE 10-5 Important safety precautions.

Wiring and controls
1. Prior to startup, double-check that field wiring between controls and devices (TR sets, rapper prime motors, etc.) is correct, complete, and properly labeled.
2. Never touch exposed internal parts of control system. Operation of the transformer–rectifier controls involves the use of dangerous high voltage. Although all practical safety control measures have been incorporated into this equipment, always take responsible precautions when operating it.
3. Never use fingers or metal screwdrivers to adjust uninsulated control devices.

Access
1. Use a positive method to ensure that personnel are out of the precipitator, flues, or controls prior to energization. Never violate established plant clearance practices.
2. Never bypass the safety key interlock system. Destroy any extra keys. Always keep lock caps in place. Use powdered graphite only to lubricate lock system parts; never use oil or grease. Never tamper with a key interlock.
3. Use grounding chains whenever entering the precipitator, TR switch enclosure, or bus ducts. The precipitator can hold a high static charge, up to 15 kV, after it is deenergized. The only safe ground is one that can be seen.
4. Never open a hopper door unless the dust level is positively below the door. Do not trust the level alarm. Check from the upper access in the precipitator. Hot dust can flow like water and severely burn or kill a person standing above the door. Wear protective covering.
5. Be on firm footing prior to entering the precipitator. Clear all trip hazards. Use the back of the hand to test for high metal temperatures.
6. Avoid ozone inhalation. Ozone is created any time the discharge electrodes are energized. Wear an air line mask when entering the precipitator, flues, or stack when ozone may be present. Do not use filters, cartridge, or canister respirators.
7. Never rod hoppers with an uninsulated metal bar. Keep safety and danger signs in place. Clean, bright signs are obeyed more than deteriorated signs.

Fire/explosion
1. In case of boiler malfunction that could permit volatile gases and/or heavy carbon carryover to enter the precipitator, immediately shut down all transformer–rectifier sets. Volatile gases and carbon carryover could be ignited by sparks in the precipitator, causing fire or explosion, damaging precipitator internals.
2. If high levels of carbon are known to exist on the collecting surface or in the hoppers, *do not open precipitator access doors until the precipitator has cooled below 125°F (52°C).* Spontaneous combustion of the hot dust may be caused by the inrush of air.
3. If a fire is suspected in the hoppers, empty the affected hopper. If unable to empty the hopper immediately, shut down the transformer–rectifier sets above the hopper until it is empty. Use no other method to empty the hopper. Never use water or steam to control this type of fire. These agents can release hydrogen, increasing the possibility of explosion.

dust buildup. The curve for field 2 appears normal, since full-rated current was obtained without sparking (reaching full-rated voltage is a rare occurrence), and the voltage level increased some amount with every increase in current. The shape of the curve for field 3 is similar to field 2, but full current rating could not be obtained without excessive sparking. This may indicate a close clearance or heavy dust buildup somewhere in the bus section.

TABLE 10-6 Precipitator inspection items.

Preoperational Checklist

Collecting plates
1. Free longitudinal and horizontal bows
2. Free of burrs and sharp edges
3. Support system square and level
4. Spacer bars and corner guides free
5. Free from excessive dust buildup
6. Gas sneakage baffles in place and not binding

Discharge electrodes
1. No breaks or slack wires
2. Wires free in guides and suspension weight free on pin
3. Rigid frames square and level
4. Rigid electrodes plumb and straight
5. Free from excessive dust buildup and grounds
6. Alignment within design specifications

Hoppers
1. Scaffolding removed
2. Discharge throat and poke holes clear
3. Level detector unobstructed
4. Baffle door and access door closed
5. Heaters, vibrators, and alarms operational

Top housing or insulator compartments
1. Insulators and bushing clear and dry with no carbon tracks
2. All grounding chains in storage brackets
3. Heaters intact, seal-air system controls, alarms, dampers, and filters in place and operational
4. Seal-air fan motor rotation correct, or vent pipes free
5. All access doors closed

Rappers
1. All swing hammers or drop rods in place and free
2. Guide sleeves and bearings intact
3. Control and field wiring properly terminated
4. Indicating lights and instrumentation operational
5. All debris removed from precipitator
6. All personnel out of unit and off clearances
7. All interlocks operational and locked out
   a. No broken or missing keys
   b. Covers on all locks

Transformer–rectifier sets
1. Surge arrestor not cracked/chipped and gap set
2. Liquid level satisfactory
3. High-voltage connections properly made
4. Grounds on: precipitator, output bushings, bus ducts, conduits

Rectifier control units
1. Controls grounded
2. Power supply and alarm wiring properly completed
3. Interlock key in transfer block

316

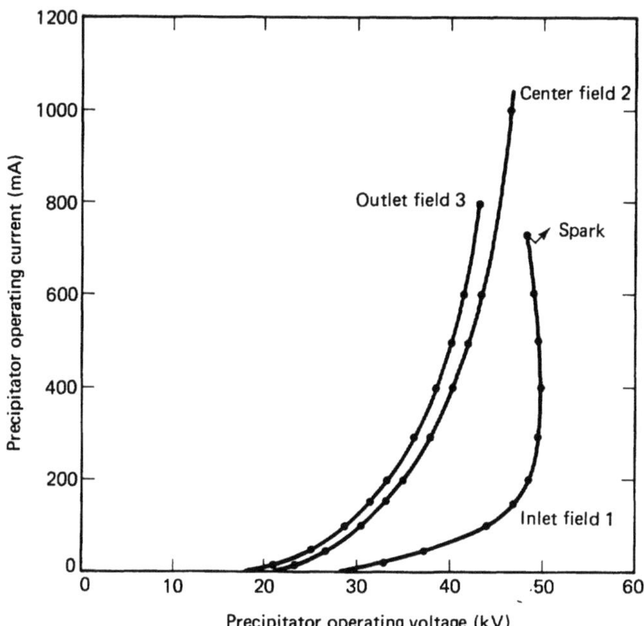

FIGURE 10-23 Air load voltage–current characteristic.

The air load voltage–current *(V–I)* curve is characteristic for a certain bus section under local ambient conditions of temperature and humidity in air. On subsequent tests under similar conditions, similar curves should result, and if the discharge and collecting electrode geometries and spacings are the same for each bus section, similar curves should be generated. A difference between curves indicates a difference inside the precipitator. On some days it is impossible to obtain an air load without sparking, and any similarity between air load and actual gas load curves is usually coincidental. When the air load tests are complete, the precipitator is ready to start up.

Startup

Startup should be a relatively simple operation, requiring the turning of a few switches to the ON position, and letting the automatic controls take over. Under ideal conditions, startup would not be initiated until the process had reached stable, normal full-load conditions for several hours. However, environmental regulations do not permit discharge of pollutants at any time when the gas process is on-stream, including startup and shutdown. Startup on oil when oil is not the primary fuel and cycling operations compound startup problems.

If the precipitator must be energized prior to reaching minimum operating temperature, high gas humidity and unburned carbonaceous material prevailing under transient startup conditions can lead to fouled electrodes and tracked insulators. Clinker formation on discharge electrodes is common. To avoid these hazards, the startup should be programmed sequentially.

Bushing and hopper heaters, seal-air systems, and rapping equipment can be started up to 12 hours prior to light-off and left in operation. Rapper intensity should be higher than normal at light-off. At light-off, the ash removal system should be started and operated continuously. The TR sets can also be energized (after purging), but only to the extent necessary to remain in emissions compliance. Startup mass emissions are much lower than the precipitator was designed for, and low corona power levels are sufficient to control them. This can generally be achieved by setting controls to operate at the spark threshold and energizing as few fields as possible. Energization starting with the inlet field is frequently favored, since subsequent operation with coarser ash will help scour off remaining startup deposits. However, this procedure will also foul downstream fields by deposition of penetrating charged particulate, and coarser ash, if it is eventually generated, will reach the outlet field as well. The main concern should be to minimize the potential for fouling buildups and corrosion, and this can only be done by starting with the outlet field.

Startup on nondistillate fuels for prolonged periods of time presents a more difficult problem due to the higher fouling tendency of the emission, especially in a system not designed to burn oil under normal conditions. The potential for fire, or explosion if a flame-out occurs, is increased, and the operation has to be monitored closely, keeping spark rate at or near zero. Energization beginning with the outlet field helps confine the potential for fire in a known area. Hot precipitators present the worst case. If process startup is difficult, requiring prolonged use of heavy oil, shutdown, inspection, and water washing may be required. Under cyclic operations, the situation is similar, but generally less hazardous with regard to fire and explosion.

Normal Operation

When normal full-load operation is achieved, precipitator energization, rapping, and ash removal can be operated normally. For energization, this still means using only as many bus sections as required, mainly for economy, and keeping spark rates very low (less than 10 per minute), and optimizing voltage and current on the TR sets in use. Energization can be either full- or half-wave, depending upon which mode yields the best results. Contrasting to air load characteristics, gas load V–I curves will likely be different by field, as particulate physical and chemical properties will largely control the power levels that can be attained. For example, in Figure 10-24 are plotted three curves for a relatively conductive and coarse dust. Highest dust quantity naturally occurs in the inlet field. This results in a high concentration of charged particulate in the space between the discharge electrode and collecting plate that is migrating toward, but not necessarily reaching the collecting surface. The resultant "space charge" suppresses the flow of corona current, driving the voltage up to a level that exceeds the normal sparking voltage,

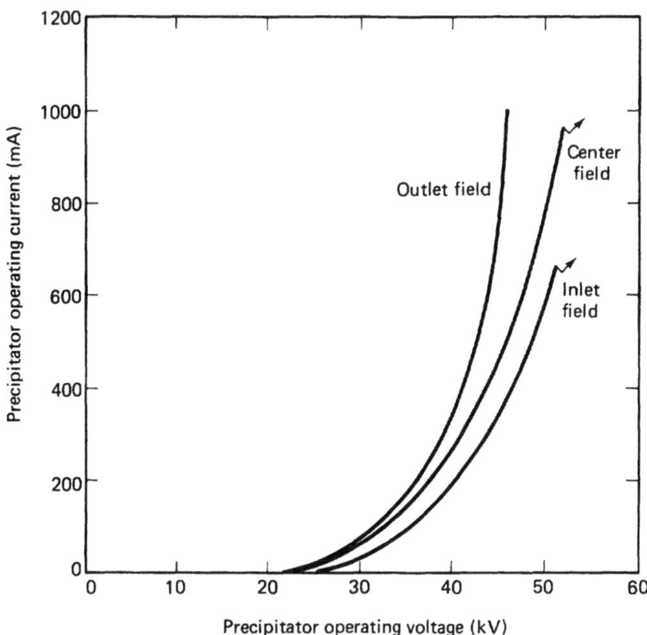

FIGURE 10-24 Voltage–current characteristic: 9-in. gas passage, low-resistivity ash at 280°F.

and sparkover occurs prematurely, at low current levels. The situation grows less severe through the precipitator, outlet field V–I characteristic shapes eventually approximating those from air load. Figure 10-25 represents operation with a relatively high resistivity and fine particle size distribution around 300°F. In the inlet field, severe corona current suppression is also due to the resistance to the flow of current caused by the collected dust. In the rear field, back corona is clearly present. The dashed lines in this figure illustrate typical plots for a hot precipitator, with characteristically low starting voltage and fast rising current. A precipitator operating with any of the foregoing energization characteristics could achieve emissions compliance. The important aspect to recognize is that changes in the V–I characteristic are caused by a combination of related parameters, and experience with these changes will be useful in recognizing cause and evaluating solutions.

Rapper operation is second only to energization in controlling precipitator performance. Often, energization and rapping must be optimized together. There are four fundamental variables in rapper operation: pattern, intensity, duration, and time interval between raps. Since precipitator performance is least sensitive to pattern, this variable can be fixed by design. The duration of impulse rappers is instantaneous. This leaves intensity and interval, or cycle time. Most of the collection in a normally operating precipitator occurs in the inlet field. Compared to downstream fields, the inlet rapping intensity should be higher and the cycle time

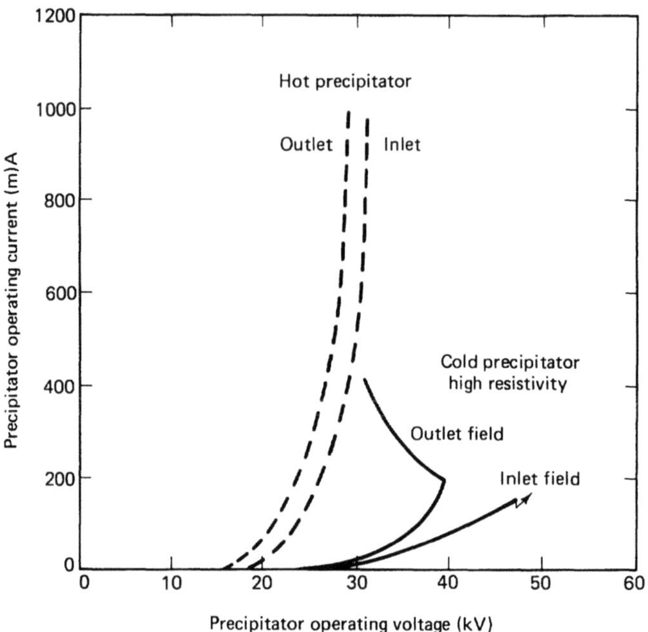

FIGURE 10-25 Voltage–current characteristic: 9-in. gas passage, for high-resistivity ash at 300°F and typical hot-side precipitator.

should be shorter. In addition, higher-resistivity dusts, or dusts that display a relatively small particle-size distribution, will generally require higher rapping intensities and longer cycle times than will coarse, conductive particulates. Vibrating-type rappers are most often used on weighted wire discharge electrode systems. In addition to cycle time and intensity considerations, the duration of each vibration is important, since it must be long enough to establish the small oscillations and random nodal points critical to the cleaning action. A minimum necessary duration of 8 seconds was observed on one 30-ft weighted wire system.

Evacuating ash from hoppers is an important operational variable that, unlike other parameters, has little available control. The interval between pulling cycles is the major variable. Compromise between operating costs for conveying equipment and hopper storage capacity is a logical approach to plant practice, but it is difficult to recommend storage times greater than 12 hours. The reason for this concern becomes obvious when one considers the effort that would be required to remove, for example, up to 20 hours of collected ash from a plugged hopper. Part of the operation should include frequent walking tours of the hopper deck. If a hopper level alarm is not energized and the hopper throat is cool, the hopper is almost certainly plugged.

Performance

Precipitator performance is dependent upon many variables, but fortunately, process by-product or fly ash and operating conditions are reasonably consistent within the plant, lessening the problem for the operator. However, changes in particulate resistivity, size, and quantity can change even in the same process or from firing coal from the same mine. Precipitators of different manufacture may be installed at the same plant, and if their electrode geometries are appreciably different, their operating characteristics will be different. The chart contained in Table 10-7 can be helpful in recognizing normal behavior under changing operating conditions. When operating conditions deviate from the norm, the effects of the abnormality will almost always be detectable from energization symptoms. Table 10-8 contains common operating problems and their probable causes that can be inferred from energization characteristics.

Shutdown

When the process gas source equipment will be temporarily shut down, it is always desirable to deenergize the precipitator and its auxiliaries, mainly to prevent waste of the large amounts of power that precipitators can consume.

The shutdown process under routine operating circumstances is the reverse of startup. Fields can be deenergized beginning with the inlet field as the process gas source is being turned down, leaving only as many bus sections in service as necessary to maintain emissions compliance or by-product recovery rates. Usually, shutdown progresses more rapidly than startup, but the deenergizing sequence should be followed, especially if the precipitator is upstream of induced-draft fans. When a TR set is deenergized, the electric field that was forcing collected particulate to adhere to the collecting electrode suddenly disappears, but repulsive charges on adjacent particles will remain. In many cases, this residual charge can force large amounts of material off the collecting electrodes and back into the gas stream. If the outlet field is deenergized first, all the reentrained material will reach the ID fan, and can cause overload or fan blade erosion over a period of time. Shutdown usually does not require the firing of an auxiliary fuel, so fouling is generally not a concern during shutdowns.

The last bus section can be deenergized as soon as particulate is no longer being emitted from the gas source equipment. Seal-air fans can be shut down as soon as the main process air or gas stream is terminated. Ash removal systems and rappers can be left on for several hours to enhance electrode cleanliness and assure empty hoppers in preparation for the next startup. Figure 10-26 is an example of a startup and shutdown procedure used by the operating company of a utility generating plant.

On cyclic processes, shutdown is rapid and time may not be available to follow the sequence recommended above. The cost savings under these rapid conditions

TABLE 10-7 Interrelationships of operating parameters.

EFFECTS—CHANGED PARAMETERS (columns *Gas Volume* through *Performance—Opacity*)

Parameter Change—Cause	Source Documents	Changes—Notes	Gas Volume	Gas Temperature	Gas Humidity	Type of Fuel	Dust Concentration	Dust Resistivity	Dust Particle Size	Carbon Carryover	Energ.—Voltage	Energ.—Current	Plate Cleanliness	D.E. Cleanliness	Flow Distribution	Air In-Leakage	Hopper Ash	Spark Rate	Performance—Opacity
Inlet Conditions																			
Gas volume	Boiler data	+ Increase—higher velocity	N	N	N	N	N	N	N	N	N	N	−	−	N	+	+	+	+
	Coal analysis	− Decrease—less total ash	N	N	N	N	N	N	N	N	N	N	+	+		−	−	−	−
Gas temperature	Boiler data	+ Increase	+		−	N	−	+	N	N	−	+	N	N	N	+	N	+	N
		− Decrease	−		+	N	+	−	N	N	+	−	N	N	N	−	N	−	
Gas humidity	Boiler data	+ Increase	+	−		N	I	−	N	N	+	−	−	−	N	I	+	+	E
	Coal analysis	− Decrease				N	I	+	N	N	−	+	+	+	N	I	−	−	
	Test report																		
Type of fuel	Coal analysis	+ Better rank	−	E	E		−	−	+	N	N	N	+	+	N	N	N	−	−
	Ash analysis	− Poorer rank	+	E	E		+	+	−	N	+	−	−	−	N	N	N	+	+
Dust concentration	Coal analysis	+ Increase	N	N	N	N		N	N	N	+	−	−	−	N	N	+	+	+
	Boiler data	− Decrease	N	N	N	N		N	N	N	−	+	+	+	N	N	−	−	−
Dust resistivity	Ash analysis	+ Increase	N	N	N	N	N		N	N	+	−	−	−	N	N	N	+	+
	Test report	− Decrease	N	N	N	N	N		N	N	−	+	+	+	N	N	N	−	−

Parameter	Source of data	Change																		
Dust particle size	Coal analysis	+ Larger size	N	N	N	N	N	N		N	−	+	+	+	N	N	I	−	−
	Test report	− Smaller size	N	N	N	N	N	N		N	+	−	−	−	N	N	I	+	+
Carbon carryover	Ash analysis	+ Increase	N	N	N	N	+	−	+		−	+	+	+	N	N	+	I	+
		− Decrease	N	N	N	N	−	+	−		+	−	−	−	N	N	−	I	−
Internal Conditions																			
Energization—voltage	Electrical data	+ Increase	N	N	N	N	N	N	N		E	−	N	N	N	E	+	−	
	V–I curves	− Decrease	N	N	N	N	N	N	N		E	+	N	N	N	E	−	+	
Energization—current	Electrical data	+ Increase	N	N	N	N	N	N	N	N	E		−	N	N	N	+	−	−
	V–I curves	− Decrease	N	N	N	N	N	N	N	N	E		+	N	N	N	−	+	+
Collecting plate cleanliness	Rapper log	+ Cleaner	N	N	N	N	N	N	N	N	−	+		N	N	N	N	−	−
	V–I curves	− Dirtier	N	N	N	N	N	N	N	N	+	−		N	N	N	N	+	+
Discharge electrode cleanliness	Electrical data	+ Cleaner	N	N	N	N	N	N	N	N	−	+	N		N	N	N	−	−
	Rapper log	− Dirtier	N	N	N	N	N	N	N	N	+	−	N		N	N	N	+	+
Flow distribution	Test report	+ More uniform	N	N	N	N	N	N	N	N	N	N	N	N		N	N	+	−
		− Less uniform	N	N	N	N	N	N	N	N	N	N	N	N		N	N	−	+
Air leakage	Boiler data	+ More leakage	+	−	I	N	N	I	N	N	+	−	N	N	−	N		+	+
		− Less leakage	−	+	I	N	N	I	N	N	−	+	N	N	+	N		−	−
Hopper ash level	Electrical readings / Hopper log	+ High level potential	N	N	N	N	N	N	N	N	−	+	N	N	N	N		+	+
		− Low level	N	N	N	N	N	N	N	N	+	−	N	N	N	N		−	−
Spark rate	Electrical data	+ Increase	N	N	N	N	N	N	N	N	−	−	+	+	N	N	N		+
		− Decrease	N	N	N	N	N	N	N	N	+	+	−	−	N	N	N		−

Notes: The following are assumed: all at typical full load; cold-side ESP; and changes are from "normal." Opacity: +, increase; −, decrease; N, no direct effect; I, insignificant effect; E, either may occur.

TABLE 10-8 Troubleshooting guide—precipitator energization symptoms.

Symptom	*Probable Cause*
No primary voltage No primary current No secondary current	TR control set failure, such as reversed wiring to or shorted SCRs, blown fuse, failed A.V.C. component, misadjustment of control or internal ground, causing trip-out.
No primary voltage Normal primary current Normal secondary current	Short circuit in system after TR. Check for ground in bus duct, insulator housing or precipitator; full hopper; broken discharge electrode; broken support insulator.
Low primary voltage Normal primary current Normal precip. current	High-resistance short to ground in precipitator. Check for insulator tracking; clinker or foreign material on discharge electrode; full hopper.
Low primary voltage Low primary current Low secondary current No apparent sparking	Control set misadjustment or maloperation. Spark rate indication, if installed, may be high. Check A.V.C. and remainder of TR control.
Low primary voltage High primary current Low secondary current	Fault in TR secondary, such as rectifier diode failure, secondary winding failure, shorted surge arrestor, or ground in control circuit.
High primary voltage No primary current No secondary current	Open TR secondary. Replace TR.
Unstable voltage and current	Control circuit misadjustment or maloperation.
Normal primary voltage Normal primary current Half normal secondary current	Shorted surge protector.
Rhythmic fluctuation in voltage and current	Broken or slack discharge electrode, unstable discharge electrode system, or dust buildup near discharge electrode.
Spark bursts at regular intervals	Dust buildup near discharge electrodes or on collecting plates.
Higher-than-normal voltage, decreasing current	High-resistivity dust buildup on discharge electrodes or collecting plates.

TABLE 10-8 Continued

Symptom	Probable Cause
Gradual voltage drop with steady current and sparking	Discharge electrode erosion.
Gradual voltage drop and corresponding current rise	Tracking insulator, hopper fouling.
Abnormally low or no voltage, high current	If hopper throat cool— plugged hopper. If hopper throat warm—shorted insulator or broken discharge electrode.
Higher spark rate in one chamber only	Ambient air in leakage.

would be small anyway. Some of these units, especially in the power-generation industry, "bottle up" the boiler and precipitator by closing dampers in the main gas stream, holding heat in the system and permitting a rapid return to full-load operation. If the temperature can be maintained at or near the acid dew-point of the gas, fouling hazards are greatly minimized, obviating the need for a carefully programmed precipitator startup at the beginning of the next cycle.

Record Keeping

Maintaining efficient and reliable precipitator operation is dependent on the operator's knowledge and the amount, type, and use made of the data he keeps. Analysis of properly recorded data can heighten the operator's knowledge of precipitator behavior. Fortunately for the operator, the parameters that he can measure and adjust are not complex, and to the extent that the manufacturer has followed good design practices and provided the means for adjustment, these parameters can be effectively controlled to maintain precipitator performance at optimum levels.

Referring to Table 10-7, the column headed "source documents" indicates where the operator can find information useful in determining cause-and-effect relationships. The table is divided into the categories of inlet conditions and internal conditions. Adequate source documents for inlet conditions include boiler data sheets, coal analyses, ash mineral analyses, precipitator performance test reports, and reports of the manufacturer's field tests of gas flow distribution. Coal analyses should include ultimate as well as proximate results on an as-received basis to be most meaningful, and should also indicate the name of the mine, seam, and geographical location together with the date the sample was obtained. Ash mineral analyses should be accompanied by the same reference data.

Although performance test data are obtained only once per year and the manufacturer's gas flow distribution test once in the life of the precipitator, these are the only sources that contain the most comprehensive set of related data useful in assessing precipitator performance. Performance test data should include precipitator

---

Startup and Shutdown Procedures

Normal operation

Startup

Preoperational checks—at least 2 hours prior to gas load:

1. All maintenance/inspection items complete.
2. All debris removed from ESP.
3. Safety interlocks operational and all keys accounted for.
4. No personnel in ESP.
5. Lock out ESP and insert keys in transfer blocks.

Prestart (at least 1 hour prior to gas load):

6. Check hoppers
   a. Level indicating system operational.
   b. Ash-handling system operating and sequence check—leave in operational mode.
   c. Hopper heaters on.
7. Check top housing seal air system
   a. Check operation of seal-air fan—leave running.
   b. Bushing heaters on.
8. Check rappers
   a. Energize control, run rapid sequence, ensure that all rappers operational.
   b. Set cycle time and intensity adjustments, using installed instrumentation—leave rappers operating.
9. Check T/R sets
   a. Check half-wave/full-wave operation (half-wave operation is recommended for lignites on cold-side ESP).
   b. Keys in all breakers.
   c. Test-energize all T/R sets and check local control alarm functions.
   d. Set power levels and deenergize all T/R controls.
   e. Lamp and function-test all local and remote alarms.

Gas load:

10. After gas at temperature at 200°F has entered ESP for 2 hours:
    a. Energize T/R sets.
    b. Check for normal operation of T/R control.
    c. Check all alarm functions in local and remote.
    d. Within 2 hours, check proper operation of ash removal system.
    e. Deenergize bushing heaters after 2 hours (hopper heaters optional).

Shutdown

11. When broiler load drops and total ash quantity diminishes:
    a. Deenergize ESP by field, starting with inlet field to maintain opacity limit.
    b. Deenergize outlet field when all fuel flow ceases and combustion air flow below 30% of rated flow.
    c. Leave rappers, ash removal system, seal air system, and hopper heaters operational.
    d. Four hours after boiler shutdown, deenergize seal-air system and hopper heaters. Secure ash removal system.
    e. Eight hours after boiler shutdown, deenergize rappers.
    Note: Normal shutdown is a convenient time to check operation of alarms.

Cold start and emergency shutdown

Cold start

When it is not possible to admit flue gas at 200°F for 2 hours prior to energizing controls, proceed as follows:

12. Perform steps 1–9 above. Increase rapping intensity 50%.

Figure 10-26 Continued

Startup and Shutdown Procedures

---

13. Energize T/R sets, starting with inlet field, setting Powertrac voltage to a point just below sparking.
14. Successively energize successive field as load picks up to maintain opacity, keeping voltage below normal sparking (less than 10 flashes/min or spark indicator).
15. Perform step 10d above.
16. After flue gas at 200°F has entered ESP for 2 hours, perform steps 10b, c, d, and e above. Set normal rapping.

Emergency shutdown

17. Deenergize all TR sets.
18. Follow steps 11c, d, and e above.

---

inlet dust measurements as often as possible. The operator can estimate gas volume on his own without taking actual samples. A rough estimate of the gas flow rate (volume) V_w, including moisture, can be calculated from

$$V_w = \frac{9.1(\text{MW}) \times [K + 102(X_a)] \times (460 + T)}{60 \times 520}$$

where   MW = gross generator output

X_a = operating excess air level

T = economizer or air heater cold end gas temperature

K = fuel constant, defined as 11,300 for bituminous, 11,800 for lignite, and 11,000 for anthracite

The terms in the denominator are conversions for time and temperature to give a solution in actual cubic feet per minute.

Source documents for precipitator internal conditions include electrical energization readings, voltage–current curves, rapper operating data, ash-removal-system operation log, internal inspection reports, and a replacement-parts log. Rapper data should include the cycle time and intensity by field, and a chart should be kept of the current rapping pattern. Recording descriptions, failure dates, and replacement dates in a parts-replacement log will show whether components fail on a regular or random basis, and can be used to ascertain the cause of failures.

There is an understandable and natural reluctance for an operator to generate what may seem to be a mountain of useless paper. If the data, once recorded, get filed in a drawer never to be seen again, a useless pile of paper is what it will be. What it should be is a written memory, separately filed and easily retrieved, that can be used to identify and correct real occurrences of precipitator malperformance.

10.5 MAINTENANCE

While pollution control regulations provide for some upsets in process operation, they have little tolerance for malfunctions in air pollution control equipment. Forced outages to repair defective equipment are costly. For this reason, specifications demand high component reliability and design redundancy, with the expectation that sufficient performance reliability will be attained so that major maintenance can be performed during regularly scheduled outages. However, assurance of adequate performance reliability transcends design. Maintenance proficiency at the plant, reluctance or inability of manufacturers to provide experienced service advice or personnel when needed, and similar inertia on the part of operating and maintenance personnel at the plant to learn essential methods and procedures on their own all contribute to the actual reliability of the equipment. Precipitator designs vary according to manufacturer and they have all demonstrated a relatively low maintenance requirement when properly applied, designed, and maintained. In those areas where reliability has been disappointing, it is reasonable to conclude that additional operator knowledge and well-planned and executed preventive and corrective maintenance programs would result in measurable improvement.

Preventive Maintenance

Any action on a piece of equipment that is planned and executed to prolong availability by detecting or preventing impending failures is preventive maintenance. In addition to potential reduction in downtime cost and improved reliability, a major advantage of preventive maintenance is that since it is planned, it can be scheduled. Key elements in a successful preventive maintenance program are promoting the concept of recognizing trouble before it happens, following the preventive maintenance schedule without exception, and recording essential data at appropriate times.

Table 10-9 illustrates the rank order of the most common precipitator problem

TABLE 10-9 Rank order of precipitator component failures from 1975 APCA survey and failure consequences.

	Frequency (%)	Impending Failure	Actual Failure
Discharge electrodes	35.2	Diminished capacity	Outage
Ash removal system	31.8	None	Outage
Collecting plates	13.6	None	None
Rappers	5.7	Diminished capacity	Diminished capacity
Insulators	1.1	Diminished capacity	Outage

components, taken from a 1975 survey conducted by the Air Pollution Control Association [1], and the consequences resulting from impending and actual failure. It is apparent from the table that impending failure can be detected in three of these five areas, and in many cases, it is possible to detect impending failure of ash removal systems. A preventive maintenance program can readily be developed for these and more precipitator components. Table 10-10 is an example of a precipitator

TABLE 10-10 Preventive maintenance checklist for a typical flyash precipitator.

Daily
1. Take and record electrical readings and transmissometer data.
2. Check operation of hoppers and ash removal system.
3. Examine control room ventilation system.
4. Investigate cause of abnormal arcing in TR enclosures and bus duct.

Weekly
1. Check rapper operation.
2. Check and clean air filter.
3. Inspect control set interiors.

Monthly
1. Check operation of standby top-housing pressurizing fan and thermostat.
2. Check operation of hopper heaters.
3. Check hopper level alarm operation.

Quarterly
1. Check and clean rapper and vibrator switch contacts.
2. Check transmissometer calibration.

Semiannual
1. Clean and lubricate access-door dog bolt and hinges.
2. Clean and lubricate interlock covers.
3. Clean and lubricate test connections.
4. Check exterior for visual signs of deterioration, and abnormal vibration, noise, leaks.
5. Check TR liquid and surge-arrestor spark gap.

Annual
1. Conduct internal inspection.
2. Clean top housing or insulator compartment and all electrical insulating surfaces.
3. Check and correct defective alignment.
4. Examine and clean all contactors and inspect tightness of all electrical connections.
5. Clean and inspect all gasketed connections.
6. Check and adjust operation of switchgear.
7. Check and tighten rapper insulator connections.
8. Observe and record areas of corrosion.

Situational
1. Record air load and gas load readings during and after each outage.
2. Clean and check interior of control sets during each outage of more than 72 hours.
3. Clean all internal bushings during outages of more than 5 days.
4. Inspect condition of all grounding devices during each outage over 72 hours.
5. Clean all shorts and hopper buildups during each outage.
6. Inspect and record amount and location of residual dust deposits on electrodes during each outage of 72 hours or longer.
7. Check all alarms, interlocks, and all other safety devices during each outage.

specific personnel and dates to each task. In addition to the potential for improved reliability and lower operating cost, a well-executed preventive maintenance program can be an excellent vehicle for training, for enhancing operator knowledge, streamlining work flow and communication, and it is relatively simple to implement.

Corrective Maintenance

As indicated by the APCA survey, maintenance on precipitators will almost certainly include discharge electrodes and their suspensions, hoppers, collecting plates, rappers, and insulators. Other areas that require occasional attention or modification work are casing access doors, control cubicles, alignment, and corrosion.

Discharge electrodes   As indicated earlier, the four major causes of discharge electrode failure are mechanical fatigue, electrical arc erosion, wire instability, and corrosion. Mechanical failures usually occur near the points of attachment to the high-voltage support structures, where the wire is twisted about itself in some designs, or near the terminus of a shroud if one is used. Most fatigue failures occur as a result of cyclic movement of the electrode, causing excessive cold working of the electrode material. Cyclic movement can be caused by rapping, energization conditions, high gas velocity jetting, or a combination of these. In the majority of cases, the problem can be eliminated by obtaining a tighter fit at the points of attachment or on wire electrodes themselves. The tight fit must be accomplished without creating stress concentrations. Failures in shrouds usually occur where the wire emerges from the shroud because design or quality flaws leave a sharp edge to work against the wire. Failures where the wire pulls out at the shroud occur occasionally. In the majority of these cases, adequate design and quality assurance testing should eliminate the problems, and users should contact the manufacturer if the problem is severe.

The majority of discharge electrode failures occur as a result of arc erosion. Arcing can occur at localized high-field regions due to sharp edges or protrusions on collecting plates or between the wire and its support if the connection is not tight and a small air gap exists. This type of failure can be visually recognized by observing a flat and rough surface of the wire directly exposed to the arc. The erosion mechanism is most detrimental to small-diameter electrodes. When localized high-current arcing persists at one point on the discharge electrode, some electrode material is carried away in the discharge. The intense heat cannot be dissipated by the small wire mass, and the electrode flexes and creeps, aggravating more arcing. Eventually, enough material is eroded and sufficient cycling occurs to cause breakage and grounding outage of the affected bus section. This type of failure is commonly found near the bottom of the wire in the vicinity of the edge of the collecting plate, caused by misalignment or wire instability. However, a major factor contributing to arc erosion is the nature of the energization equipment.

The magnitude of arcing current is a function of the TR set size and the stored energy in the bus section. The energy in a system comprised of a 2000-mA set serving 50,000 ft^2 of surface is enormous. If spark rates must be kept high to provide sufficient corona voltage for emissions compliance, or if sectionalization and arc-quenching capability of the control are inadequate, electrode failures will be higher. The high inrush currents associated with sparking can be especially detrimental to a design that contains many small connections between electrodes and their supports. Remedial actions for these types of failures include indentifying the nature of the failure and isolating the cause, which will usually be realignment or modifications to gain better stability. Use of a properly designed shroud and improving the energization design are appropriate measures also.

Instability problems arise when the total unrestricted length of the discharge electrode and its support and steadying frames is excessive. Accumulated tolerances, thermal expansion, and structural deflections that occur during normal operation make it difficult for systems using long wires and many long couplings to maintain alignment. In a noteworthy work by Casiglia and Fletcher [4], it was found that the fundamental mechanism of instability in weighted wire systems involves the resonant interaction of the wire and its tensioning weight. When the fundamental wire frequency and the tensioning weight pendulum-mode frequency approach the same value, large-amplitude oscillations readily occur. Separating the natural frequencies of these two components was successful in reducing instability-induced arc erosion failures in one fly-ash precipitator from several per week to less than two per month. Better stability was gained by increasing the mass and lowering the center of mass of the tensioning weight, and decreasing the linear mass density (weight per unit length) of the wire itself. It was also found that any weighted wire system could be made stable if the suspension weight were restrained from movement, but this is not a practical solution. Some manufacturers use suspension designs where stabilization of the weighted wire structure is close-coupled, such as placing a stabilizer bar between the bottom of the collecting plates and the lower steadying frame.

In a properly erected precipitator, the maintenance and modification actions described above should minimize discharge electrode failures, even in weighted wire designs. Electrode failures of one or more per week have been reported for designs that are identical to those reporting as few as one or two per year. A well-executed maintenance program should keep the electrode failure frequency low.

Hopper   Inadequate ash removal systems can permit buildups in hoppers to the point where they overflow. Of all the causes for misalignment and subsequent discharge electrode arc erosion, overfull hoppers is by far the biggest offender. Weighted wire systems are least sensitive to misalignment by this mechanism, while rigid frame designs appear to be most sensitive. In either case, grounded bus sections and diminished precipitator capacity result whether there is any electrode damage or not.

Major causes of ash removal problems include low gas temperatures, promoting condensation of tacky condensible matter or free water vapor, insufficient insulation, inadequate hopper heat, internal obstructions, air in-leakage, inoperative vibrators, shallow valley angles, faulty level detectors, and evacuation system faults. Evacuation system capacity problems are more prevalent on industrial precipitators employing mechanical conveyors than on dusting equipment using pneumatic conveyors.

The most direct attack against formation of condensibles uses external heat and better insulation. In the worst cases, it may be necessary to apply heat to almost all of the hopper surface instead of the nominal 15% of surface found on many installations. If hang-up at the very bottom is the common fault, heating and insulating the hopper throat and transition pieces can be helpful. Additional heat should be applied in vertical zones and thermostatically controlled. Location of thermostats is important, because if housed in a normally high temperature zone, they may never activate. Care must be taken when selecting heaters to ensure that they have sufficient design margin for temperature excursion if the heaters are energized and the hopper does not evacuate.

Internal obstructions can cause relatively minor residual accumulations of ash to remain after a pulling cycle. These deposits flatten the effective valley angle, promoting additional hang-up on successive cycles and eventual blockage. Anti-circulation and sneak baffles should be terminated as far above the throat as possible. In hopper designs where the ratio of the vertical height to the horizontal span is high (say greater than 2.5), the distance between the baffle and throat can be increased to several feet instead of the 18 in. many standard designs call for. Ladder rungs are an excellent access feature, but must be confined to vertical baffles or the steepest wall, and in no event should they be near corners. Hinged, flexible gas baffles have been used overseas with some success, and rounded corner fillets are recommended in the most severe problem cases.

Air in-leakage is most common at access doors and dump valves. Door gaskets, damaged most frequently by overtightening, should be inspected regularly and changed if worn. Dump valve seats and gates should be checked for erosion and replaced if air leakage can be heard in this area during normal operation.

Two vibrators operating on opposite walls at a relatively low frequency are better than one vibrator operating at a higher frequency. Level detectors should be placed on either side of a gas baffle. If only one detector is installed, a hole can be cut in the gas baffle to allow free communication of dust from one side to the other, but care has to be taken to prevent gas sneakage. This may require installation of additional baffle plates. Poke holes are rarely useful but should be provided for rodding clinkers. Aeration stones should only be used if the air is heated and dried.

Collecting plates   Misalignment and fatigue failures are the most common problems associated with collecting plates. Fatigue failures generally result from the use of rapping forces that the plate attachments were not designed to survive. Manufacturers should be consulted for repairs if fatigue failures of this nature occur.

Rappers   Problems associated with internal swing-hammer rappers include bearing failure due to wear and abrasion by dust in the gas stream, misalignment of the hammers and anvils, and wear or breakage of the anvils themselves. Swing hammer failures on collecting plates can be detected only by changes in precipitator operation, principally energization. On discharge electrode frames, failure of the shaft insulator will stop shaft rotation, which can be detected fairly easily. Failures of electromagnetic drop hammers are associated primarily with the packing gland or boot seal. Gas leaking by the gland can soften the rod tube and corrode mounting bolts. Better packing and seal materials and the use of clamps on the seals can eliminate the problem. For any rapping equipment, frequent inspection is a good maintainence practice. On vertical drop hammers, there is a popular notion that binding between the rod and guide tube is detrimental to delivery of the rapping blow. Actually, measurements have shown that the attenuation caused by this binding is less than 5%, and the condition, if it is noted, can probably be left unmodified. These studies show that a dust layer not much thicker than this page, allowed to accumulate between the hammer and the anvil, can attenuate the blow as much as 50%. If this condition is noted, immediate correction is required. In some installations using magnetic drop hammers, there is a tendency for the hammer to magnetize and stick in the dropped position at low energy levels. Insertion of a flat aluminum plate or a shrink-fit stainless steel cap between the hammer and rapper rod will eliminate this problem. In no event should rubber pads be used.

For the rapper input force to perform the most work, it is essential that the collecting plate attachment to the casing be rigid and tight, and that no energy absorbers are built or allowed to creep into the system. Manufacturers of highly active and "cleanable" plates work hard at evolving designs that ring like a bell when rapped. If loose attachments are found, the manufacturer should be contacted for methods to correct the problem.

Insulators   These components can be made of tile, porcelained ceramic or alumina, and in some cases Teflon, nylon, or Micarta. To be successful, they all must be kept relatively dry, and kept from being exposed to conditions that can cause tracking, a condition where part of the material is carbonized by high-voltage stress.

Drying is accomplished mainly by heated air. In designs that use local resistance heaters, these devices must not be too close to the insulator, or local high temperatures will reduce the stress-carrying capability of the insulator and failure will occur. Thermostat controls must be located near the heaters to sense the heat near the insulator and prevent overheating. A 5-in. clearance between insulator and heater and a thermostat setting around 250°F, but certainly not more than 300°F, are satisfactory. Some support insulators are ventilated. If the original design calls for 100 cfm per insulator and corrosion is evident on the precipitator internal parts below the insulator, this flow rate can probably be reduced to 25 cfm with a corresponding drop in heater temperature to maintain dry insulators and minimize corrosion simultaneously.

Some insulators are contained in small compartments while others are grouped in large penthouses or top housings. Top housings provide better maintenance access, but maintenance activities should pay attention to insulation, ventilation, and ambient dust. If insulation is installed on the floor of a top housing, metal walkways should always be used, and kept in good repair and in their proper location. Ventilation pipes should not impinge on insulator surfaces. If insulator problems occur near ventilation openings, some ducting or diffusers should be installed. In-leakage of dust from the precipitator indicates a deficiency in the seal-air system. Fan filters, vent diffusers, and operating pressures should be checked to verify that the seal-air system is operating properly and providing positive ventilation.

Annuli between support insulators and hanger pipes should be checked for excessive leakage or mechanical binding. Most insulators can tolerate incredibly high compression forces even in dirty flue gas atmospheres, but they generally cannot sustain much shear force. All support and transmission insulators should be checked periodically for misalignment and pinching.

Doors Access doors are usually gasketed to prevent air in-leakage to the precipitator. Overtightening can cut and eventually destroy the gasket material. If left uncorrected, local thermal gradients on the door can cause it to warp and the whole door will require replacement for lack of a properly maintained and inexpensive gasket. Doors with interlocks are the primary precipitator safety feature. Frequent inspection and powdered graphite lubrication are mandatory to maintain key interlocks in proper working order.

Control cubicles The biggest enemies of control enclosures are dust and heat. For this reason, open cubicles are usually equipped with fans and filters, but these may be insufficient, especially in controlling dust. If problems arise, consideration should be given to changing or eliminating the vent fan. If the fan intake is near the floor, reversal of the air flow with an outdoor intake and additional filters can be considered. In the worst case, the cubicle can be sealed, but in TR controls, the linear reactor may have to be removed.

Alignment One of the major maintenance requirements for a precipitator is the inspection for and correction of misalignment between electrode systems. Misalignment can be caused by mishandling of materials during construction, by improper construction techniques or quality checks, by overfull hoppers, or thermal excursions. Misalignment may result from mislocated electrode suspension systems or bending and bowing of collecting electrodes, discharge frames, or the discharge electrodes themselves. Manufacturers have design standards for alignment which are developed from the clean-air sparking-voltage characteristics of their design geometries. Maintenance alignment efforts should strive to obtain these tolerances.

Points of alignment measurement to ascertain compliance with design tolerances for various electrode systems are shown in Figure 10-27. Examples of collecting electrode bowing are shown in Figure 10-28. It should be noted, for example, if the discharge electrode system is properly installed but the elevation of the collecting electrode system is out of design tolerance, the upper end of the two systems could

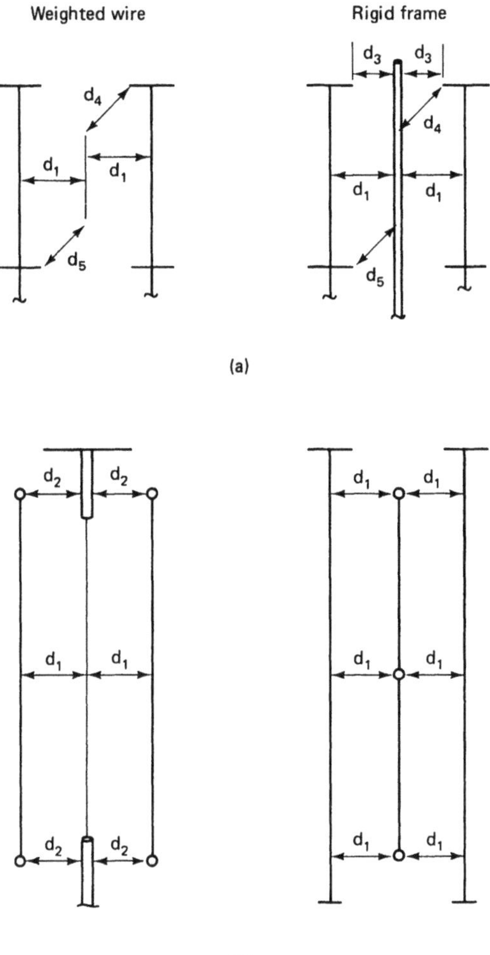

FIGURE 10-27 Alignment measurement: (a) plan views; (b) elevations. (Adapted from [10].)

be within design tolerances, but the bottom ends could be significantly misaligned. A precipitator composed of fields 15 ft long and 45 ft high containing only 10 gas passages will be misaligned by several inches at the bottom when the top alignment is within less than ½ in. For this reason, repairs to alignment should always include dimension checks on diagonals, square, level, and plumb prior to beginning repairs, to define the base cause of the misalignment. If the cause is severe enough, the next step may be to cut the bottom ends of the electrodes loose and let them hang free so that vertical alignment can be reestablished. Bows can usually be worked out by heating the side of the member that is under tension at a few points in the area of maximum deflection. When these cool, the member will often spring back more than to its initial position. Several repeats of this step may be required. In any event, it is always a good step to perform a dimensional check of the misaligned area before and after repairs, unless it is obvious that the misalignment is being caused by bowing only.

It is also a good practice to make a map of a typical field, similar to Figure 10-27, that can be used during internal inspections. Gas passages and discharge electrodes can be numbered or identified in some convenient fashion, measurements recorded, and the record kept to facilitate speedy repairs during a subsequent maintenance availability.

Corrosion   Low-temperature fouling and corrosion of metal surfaces is a problem that affects nearly all industrial process machinery. The process of corrosion itself is a natural one. The makeup of the earth's environment permits the forces of corrosion to be continuous and powerful, and short of attempts to prevent it completely, maintenance efforts are often directed toward identifying and minimizing the rate at which corrosion or fouling take place.

Corrosion and fouling agents vary by process. Although much field work and research has been performed in the general areas of corrosion and fouling, there is no concise formula that enables operators and maintenance people to predict the amount of metal loss or particulate buildup that will occur under changing process conditions. In many processes, particularly the power industry, corrosion depends upon the presence of free water vapor in the gas stream. For example, in power boilers firing sulfur-containing fuels, practically all of the SO_3 generated in the combustion process will react with available water vapor to form gas-phase sulfuric acid. In order for this acid to attack metal surfaces, the gas temperature ordinarily must be below the dew point of the specific concentration of acid formed. However, estimation of dew points in particulate-laden gas streams is highly uncertain, and the presence of particulate complicates the corrosion process. Acids can absorb first

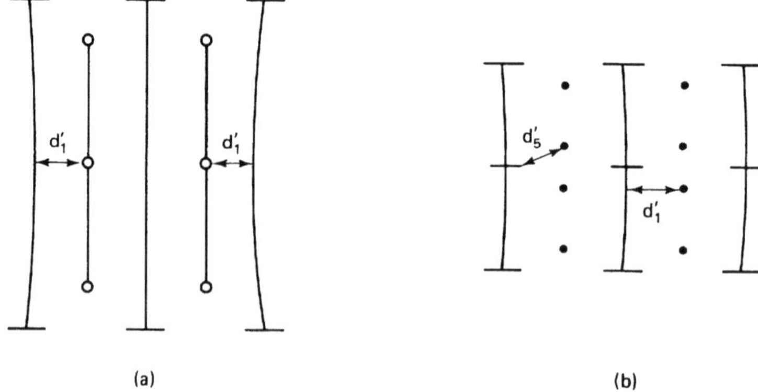

(a)   (b)

FIGURE 10-28 Examples of collecting plate bowing: (a) vertical bowing—rigid frame precipitator (elevation); (b) horizontal bowing—weighted wire precipitator (plan). (Adapted from [10].)

onto particle surfaces before the dust is deposited onto a metal surface. If the particle surface contains neutralizing chemicals, the concentration of acid available will be reduced, and although the ash will be the vehicle by which the corrosive agent contacts the metal surface, its presence can impede the rate at which acid reacts with the metal. The rate of corrosion, being uncertain, can only be roughly estimated in most precipitators. One way of looking at the situation for a coal-burning power plant is shown in Figure 10-29. This plot is based on operations at *or near* the predicted acid dew point.

Preventing the rate of corrosive agent attack on metal surfaces depends upon good design and thoughtful operations and maintenance. The greatest amount of corrosion in precipitator systems occurs when cold pockets are allowed to exist. Figure 10-30 illustrates structural methods that both promote and prevent corrosion pockets. The maintenance program should inspect the percipitator and fluework for potential problem areas and modify them to the best degree possible.

Other physical aspects of the precipitator design that promote corrosion are air in-leakage and inadequate insulation. Older units and designs that contain many expansion joints are especially sensitive to areas of inadequate insulation. Insulation thickness in these areas should be measured and compared to modern industry standards for the normal precipitator operating gas temperature. Air in-leakage sources are structural cracks, access doors, hopper valves, and seal-air systems. Air in-leakage through cracks can be detected audibly during routine walking tours around the precipitator, and even closer audible or visual detection is possible for leakage through warped doors or damaged gaskets. Hopper valve leakage is more difficult to detect this way, but during shutdown, inspection from an empty hopper with the ash removal system operating can reveal the presence of excessively worn valves. Seal-air flow through support insulators can cool gases and metal temperatures below the acid dew point in localized areas where the corrosion rate will be

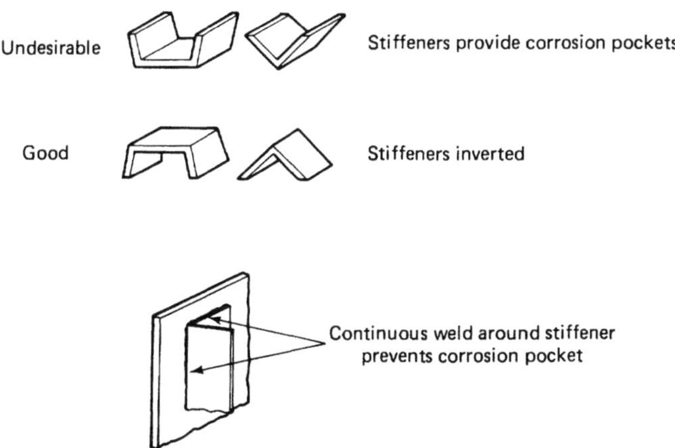

FIGURE 10-29 Corrosion, fouling, and plugging potential for coal applications.

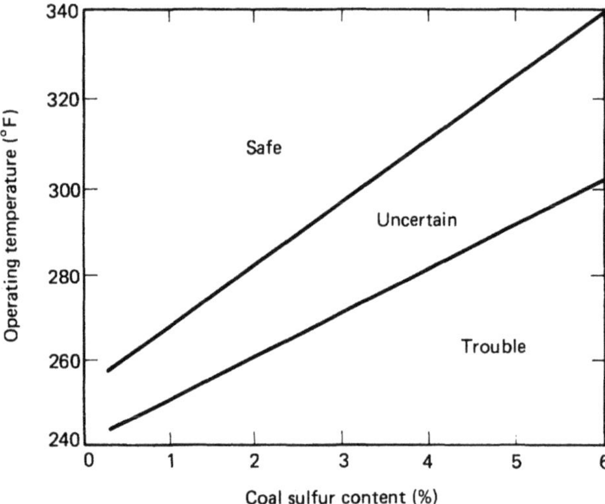

FIGURE 10-30 Some construction aspects in corrosion control. (Adapted with permission from [9].)

much more rapid than in any other part of the system. Areas affected include the upper ends of collecting and discharge electrodes, support systems for these electrodes, local structural components, and the underside of the shell roof. Air in-leakage through packing glands or boot seals will operate the same corrosion mechanism.

Obviously, maintaining gas temperatures well above the acid dew point will accomplish more of the desired results than any other means, but this is usually not a practical solution, especially in cyclic operations, where corrosion can be greatly aggravated. In the most severe cases, special coating or metal preparations may be required.

10.6 IMPROVING OPERATION AND PERFORMANCE

Knowledge that precipitator performance is controlled by many and often interrelated variables may discourage the operator into thinking that there is little he can do to improve the performance of a precipitator already in commercial operation. To the contrary, the sensitivity of precipitators to the quality of energization and rapping provide the operator with controls over performance that can even overcome some design deficiencies. At the very least, examination of these and a few other operating parameters permit rapid and accurate diagnosis of precipitator malperformance.

Efforts at improving operation and performance are as dependent upon approach as they are upon technique. The first steps should always be the complete correction of mechanical/component deficiencies and excessive dust buildups, since their existence will almost certainly inhibit later efforts, utilizing subtle techniques that can identify problems associated with fuels, boiler, or process operation, or these problems complicated by the ravages of age, corrosion, or neglect. In most cases, several different problem areas will exist simultaneously, and proper interpretation of available data, possible only through a systematic approach, is essential for reliable performance. Once mechanical and component deficiencies are corrected, optimization of energization, rapping, and gas flow distribution can proceed toward the goal of optimum precipitator performance within the inherent limitations of precipitator size, or toward definition of alternative methods and contingent technologies necessary to achieve a minimum level of performance.

Energization

Since a large number of process variables directly affect the quality of energization, it is difficult to prescribe a best sequence for its optimization. However, a systematic approach would suggest that examination of the design and application of energization equipment is an appropriate first step since this will define the limitations of the equipment and assist further optimization efforts. A review of Figure 10-4 reveals that the basic energization package consists of a high-voltage transformer, a group of silicon rectifiers, a current-limiting reactor (CLR), and a voltage controller, in this case a silicon-controlled rectifier (SCR or thyristor), although a saturable core reactor could also be used as a control. Of these components, the function of the CLR seems to be the least understood and most often abused. Linear reactors are used to control the precipitator current waveshape, add stability to the power supply under sparking conditions by limiting the peak spark current, and reduce heat losses in the TR set. Larger reactors generally permit greater stability, but since the CLR is a primary circuit inductive impedance, larger reactors also reduce power input to the precipitator for a given TR voltage rating by limiting the TR rated current. The net result is that to obtain optimum stability and TR performance, one normally has to design for a high impedance and low current capacity, and this usually limits the bus section collecting area that can be used. The methods of matching precipitator power supplies to actual operating loads is complex, but if a TR set is energizing over 25,000 ft of collecting surface and TR set reaction to sparking or arcing is extremely sluggish, or if the TR is operating at less than 50% rated secondary current and displays poor recovery after a spark event, the operator should investigate the benefits of better sectionalization. This may simply require a larger CLR if the current capacity of the modified energization equipment is still sufficient. In some cases, this may mean additional TR sets or modifying internals to obtain smaller bus sections. Of course, the reverse of this situation can occur, where the current capacities of the TRs for their respective bus sections are too small. In any event, if trouble is verified, correction will probably involve the original precipitator manufacturer, who is also in the best position to recommend modifications.

Sluggish reaction to sparking may also be caused by a design inability of the automatic voltage control. The automatic control must minimize interruptions in the applied voltage caused by sparking or arcing. Complete bridging of the gap between discharge and collecting electrode by a spark or arc represents a momentary short to ground, during which time the precipitator applied voltage will tend toward zero, with a resulting loss in charging and collecting electric fields. A control should be able to respond to a spark and restore full voltage within 10 to 80 msec. Sparks of low current are usually of natural short duration and extinguish themselves. Extremely fast response should be possible with this sort of spark. High-current sparking discharges, or arcs, generally do not extinguish readily. Or a burst of sparks may occur that can last several seconds if uncontrolled. Under these sorts of discharges, a slower response, up to 80 msec may be required to quench the arc and prevent another immediate occurrence.

The primary debilitating effect of poor sectionalization or control is that applied voltage to the precipitator is limited or interrupted too often to sustain high performance. The effect can be observed on any energization system by connecting an oscilloscope to a convenient secondary ground return current signal and setting the trace on a slow speed. Frequent reductions in the waveform to zero or long recovery times after a spark event are indications that the energization equipment itself can be improved. An illustration of good and poor recovery characteristics is shown in Figure 10-31. Methods and equipment are available to correct the deficiencies in energization design, and the necessary effort should be made prior to attempting additional optimization.

Problems with energization also frequently arise as a result of changing process, fuel, or by-product properties. In industrial processes, each process generally produces the same sort of energization characteristics from one plant to another. For power industry applications, the situation is more complex, because fly ash is truly

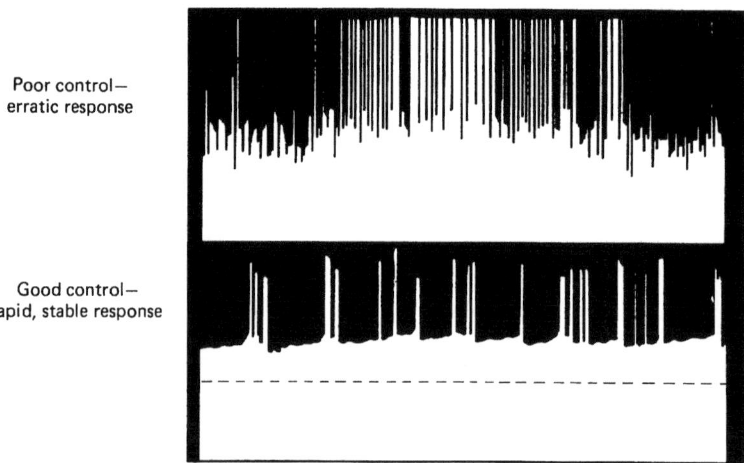

FIGURE 10-31 Current waveforms observed on oscilloscope.

not a general process. All coals and the ashes they produce are different to some extent, and even coal from the same mine can produce vast differences in ash properties. Finally, an identical coal fired in different boilers can produce different energization results. The most frequent problems that arise are associated with ash resistivity, space charge, and corona quenching. Fortunately for the operator, there are techniques useful in diagnosing these problems and fine-tuning the power supplies for optimum precipitator performance.

The single most valuable diagnostic tool available to the operator is the voltage–current relationship. A strong indication of a high-resistivity situation can be gotten by comparison of a clean plate V–I curve to the operating V–I curve. Figure 10-32 illustrates this concept. The clean plate curve can be obtained by leaving a bus section deenergized during startup and recording the V–I relationship at normal operating temperature immediately upon energizating the TR set. Operating V–I curves can be obtained at any time after the clean plate curve, the one shown illustrating a resistivity around 5×10^{11} ohm-cm. There would be an infinite number of curves for intermediate and greater resistivities, and although they may be altered somewhat by changes in particle-size distribution, moisture, and gas temperature, all would show the trend toward lower current at sparkover, and lower corresponding operating voltages. In other words, higher resistivities would shift the curve farther to the right without a large change in starting voltage. The detriment to energization is that the useful corona current density and maximum sparkover

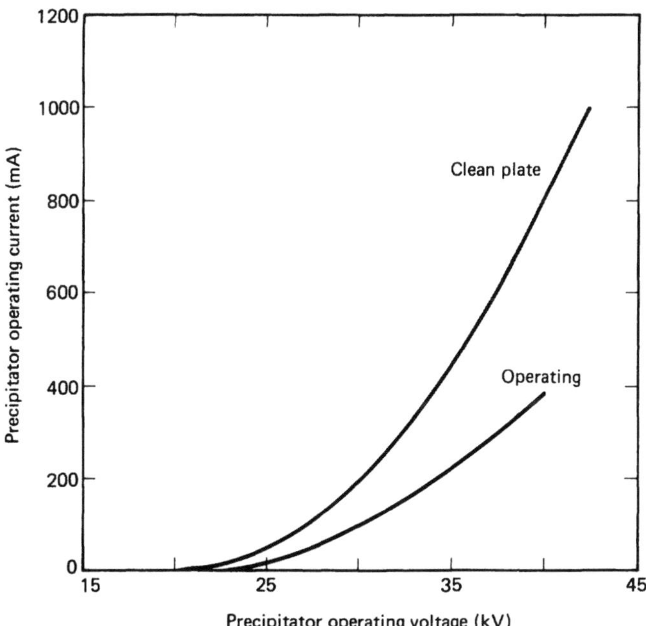

FIGURE 10-32 Voltage–current characteristics: 9-in. gas passage on fly ash at 300°F.

voltage are decreased with increased resistivity, as shown for a given plate spacing in Figure 10-33. The important thing to understand is that forcing the voltage or current above the critical point for a given application can be detrimental because excessive back corona may be generated. In fact, the operator may be surprised to find that performance improves when input power is reduced! Lower precipitator performance at higher power settings is caused by operating above the critical point where back corona is generated, as illustrated in Figure 10-14. The energization level that corresponds to the lowest discharge opacity is the best verification that maximum useful power has been attained. The amount of trial and error needed to reach this point can be reduced by observing the minimum and peak value of the secondary voltage waveform with an oscilloscope connected to TR sets equipped with voltage dividers. The secondary voltage waveform will appear similar to that shown in Figure 10-4. Plotting both the minimum and peak value of the secondary voltage against corona current will generate a set of curves similar to the ones shown in Figure 10-34. The onset of back corona in many cases is related to the point of the kV_{min} curve where the voltage ceases to increase. The optimum value of voltage on the kV_{min} curve does not necessarily fall right on its highest value, and in some cases a point of maximum kV_{min} may not be well defined. However, for many installations, the relationship between the optimum voltage setting and the maximum value of kV_{min} will be relatively consistent. It is not necessary to perform this sort of diagnosis on all TR sets; finding the optimum setting for key bus sections in the same chamber and setting the remaining control points by

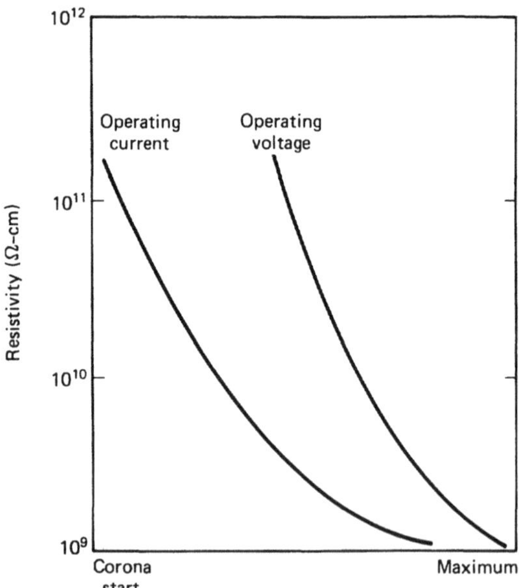

FIGURE 10-33 Effect of resistivity on attainable operating voltage and current. (Adapted from [6].)

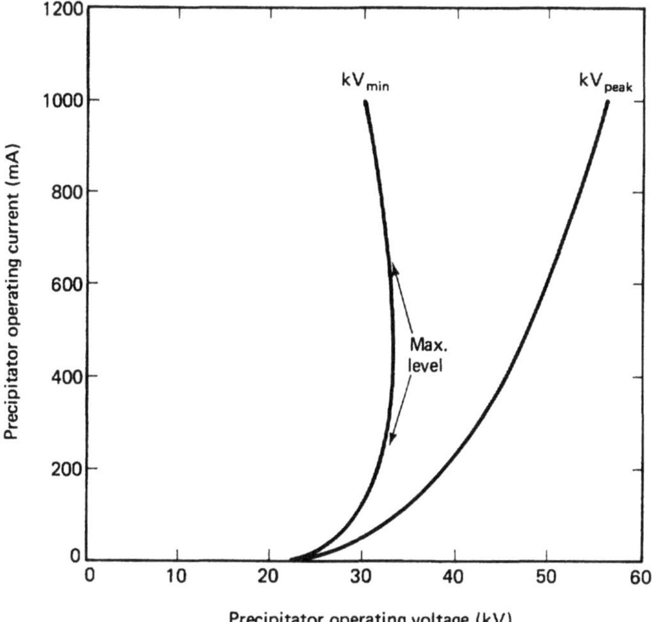

FIGURE 10-34 Voltage–current characteristic observing minimum and peak values of operating voltage in a 9-in. gas passage on moderately high resistivity fly ash.

inference should be sufficient. It should be evident that on applications where resistivity is not high enough to limit power, the kV_{min} curve will continually increase, and the normal *V–I* curve will be similar to the clean plate curve of Figure 10-32.

Space-charge effects and corona quenching can produce energization effects somewhat similar to high resistivity, the major differences being the voltage level that can be achieved prior to sparkover. Figure 10-35 illustrates the different types of *V–I* relationships characteristic of different conditions. Space-charge current suppression and corona quenching effects normally result in sparking at higher voltages than high-resistivity cases, typically near the maximum voltage attainable for a certain gas passage spacing, even though the current may be low. Space-charge suppression is caused by high dust concentrations, and corona quenching may be caused by a high concentration of fine particulate matter. These conditions, although not necessarily good for precipitator performance, are usually not as detrimental as operations with high resistivity. Maximum performance is usually obtained by maximizing voltage while minimizing spark rates.

Full-wave rectification often results in better performance under conditions of corona quenching and operations not limited by high resistivity. For severe space-charge current suppression and high-resistivity situations, half-wave rectification is more beneficial. The reason for this is generally attributed to the longer decay

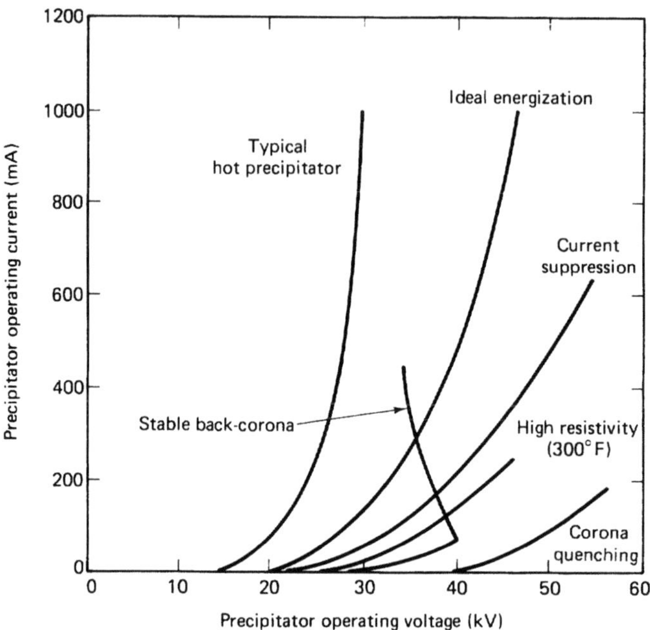

FIGURE 10-35 Fly ash voltage–current characteristics for various conditions. (Adapted from [8].)

period between half-wave voltage pulses, as can be seen in Figure 10-4. The operator who experiments for several hours with half- and full-wave energization may not detect any useful differences in performance, since as many as 4 or 5 days may be required for the effect to be evident.

Rapping

Energization is intrinsically related to the quality of rapping, and the most critical aspect is collecting electrode rapping. The inability to remove certain difficult particulates, commonly associated with dusts that exhibit high-resistivity characteristics, leads to reduced voltage and current input. Even low-resistivity dusts, if allowed to build up to a level where alignment is distorted, can cause premature sparking. For these reasons, rapping and energization often have to be optimized simultaneously.

The first step in rapping optimization is establishing the relative frequency of rapping by field. Since the inlet field in a reasonably well behaved precipitator will collect the most ash, the inlet field should be rapped more frequently than downstream fields. The outlet field should require the least frequent rapping. The best pattern depends upon the number of precipitator fields in operation. A frequency ratio of 4 to 9 raps in the inlet field for every rap in the outlet field is a reasonable starting point, and intermediate values should be used for intermediate fields. The optimum ratio can be refined by trial and error, but precision is not critical for this aspect.

The next step is determining the interval between raps for each field. Experience has shown that longer cycle times are better than short ones, the rationale being that the long time interval permits natural particle agglomeration tendencies to take place, reducing the likelihood of reentrainment. A plot of particulate penetration versus cycle time is shown in Figure 10-36. This represents highly generalized cases, for cycle time is related to the reentrainment potential of the particulate. High-resistivity ashes, usually tenacious when collected and not prone to reentrainment, generally respond best to longer cycle times, on the order of 1 hour or so on the inlet fields to 4 or more hours between raps in the outlet fields. Low-resistivity ashes typically reentrain readily. They respond best to shorter cycle times, as low as 10 to 15 minutes on inlet fields to not more than several hours in outlet fields.

The final step in rapping optimization is finding the optimum intensity by field. High-resistivity and other tenacious ashes require higher-intensity rapping than do reentrainment-prone particulate. Although the actual rapping intensity required will depend upon the design of the precipitator, the best starting point for tenacious dusts is the maximum intensity for which the unit was designed, and variation between fields is not important. Final optimization is reached at the lowest intensity required to maintain power input while prolonging component life. For dusts sensitive to rapping reentrainment, intensity is critical, but again, there may be little required variation between fields. The detriment of overzealous rapping for reentrainment situations is shown in Figure 10-37a and b. Figure 10-37a is the opacity monitor trace output at high rapping intensity for a low-resistivity, reentrainment-prone fly ash. Rapping spikes are frequent, and the average opacity is nearly 50%.

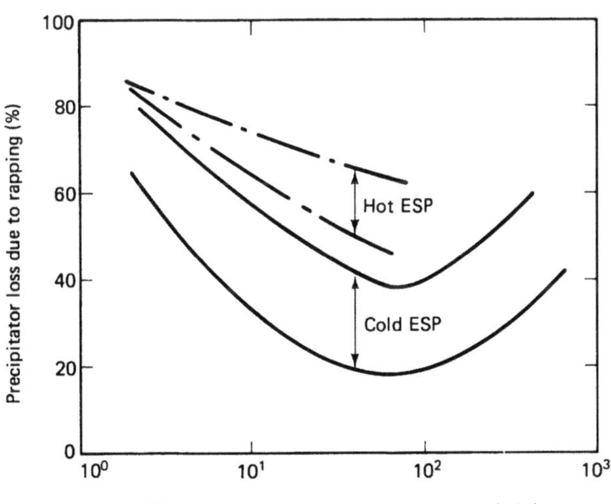

FIGURE 10-36 Percent penetration through precipitator versus rapper cycle time.

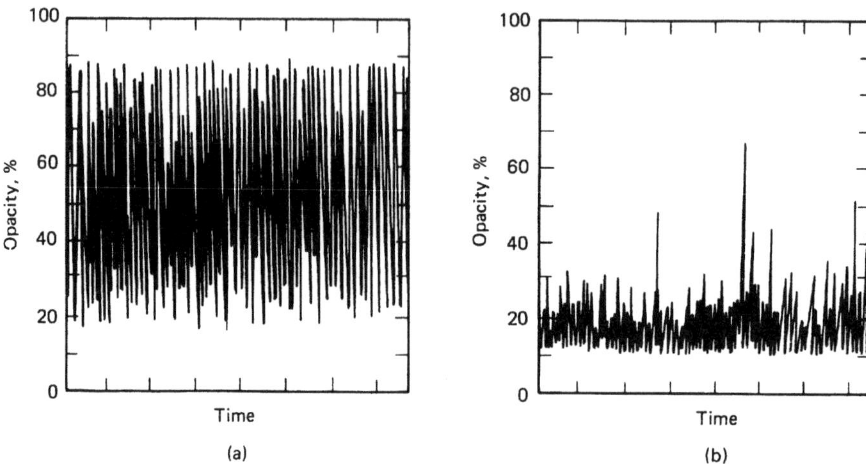

FIGURE 10-37 Stach opacity strip-chart recordings—reentrainment situation: (a) before optimization; (b) after optimization.

Rapping cycle time was increased, and then intensity was modulated to lower values, resulting in the trace shown in Figure 10-37b. The improvement shown in opacity alone represents an approximate 300% decrease in emissions.

If attempts to improve electrode cleanliness and energization by following the guidelines above are not fruitful, it may be necessary to improve the rapping hardware itself. This is especially true where the dust is known to be ordinarily difficult to remove and the rapping density is greater than 3000 ft^2 of collecting electrode per rapper. Correction may entail additional rappers, modifying support systems, or cutting high-tension bus sections, but if the expected end result can justify the means, these options are considered valid optimization alternatives.

Gas Flow Distribution

Uniformity of the bulk flow entering the precipitator is important to maximum precipitator performance, and a customary practice in precipitator supply is the conduct of a complete model study prior to installation to define the flow control devices necessary to ensure compliance with stipulated uniformity criteria. Occasionally, it is found that the uniformity criteria is not attained in the actual field unit, and if subnormal collection efficiency is experienced, there is a natural inclination to suspect that maldistribution of the bulk flow is the cause. However, experience has shown that correction of the gas flow maldistribution rarely achieves the expected improvement in precipitator performance. This experience may reflect the empirical aspects of precipitator performance design, where reduction in theoretical performance is unconsciously accounted for in the sizing. Also, improvements in gas flow uniformity will not result in improved collection efficiency if the fundamental problem is unrelated, such as a limitation of useful energization caused by high dust resistivity. If improvement in flow uniformity in view of these precautions is still considered mandatory, the manufacturer or a qualified consultant should be involved before any decisions or modifications are made.

Elimination of flow-induced reentrainment losses is often more rewarding than small refinements in gross flow uniformity, and in most cases, the operator can detect the source of these losses on his own. The most sensitive area for reentrainment losses in a precipitator is the outlet field. Large-scale reentrainment can be detected visually by using a high-intensity lamp in the outlet plenum. Observations should be made with the gas flow backlighted, accomplished by shining the light through a special port on one side of the plenum and making the observations through a similar port on the opposite side. Precipitators performing as low as 80% efficiency provide a remarkably clear view of the exiting gas, and areas of dense reentrainment are readily detected, pinpointing the source of the reentrainment or indicating an area for closer observation. The course of this sort of investigation may lead to visual observation of flow in the hoppers. This can be done in a manner similar to making plenum observations, but caution must be used to avoid accidental electric shock. No lighting or other device should be inserted through a port and into the hopper. Three categories of flow reentrainment can be detected by visual observation. Dense dust discharge at the junction of the hopper opening and the plenum floor indicates large-scale hopper turbulence. A similar dense discharge congregating at partition walls is a strong indicator of hopper cross flow, illustrated in Figure 10-38. Localized dense discharges well above the floor, and directly adjacent to side or partition walls, indicates raw gas sneakage around collecting plates, caused by loose, inadequate, or completely missing antisneak baffles at the leading edges of the outboard collecting plates. Dense discharges not near walls or the plenum floor are indicative of high-velocity jetting, which may be caused by severely maldistributed flow at the precipitator inlet or by internal obstructions.

Elimination of flow-induced reentrainment is critical to maximum precipitator performance, because the associated dust losses may account for up to 90% of the

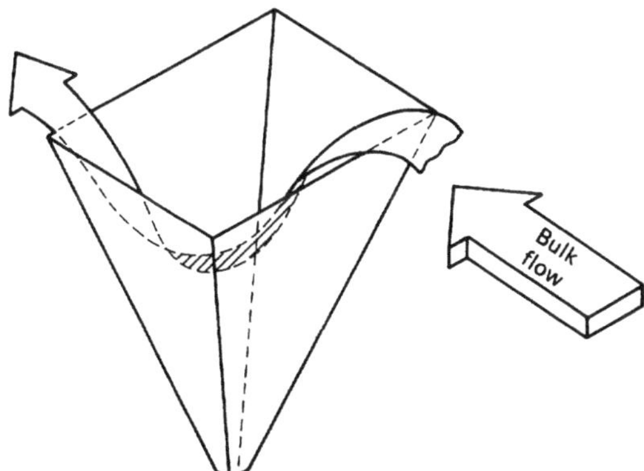

FIGURE 10-38 Hopper cross flow.

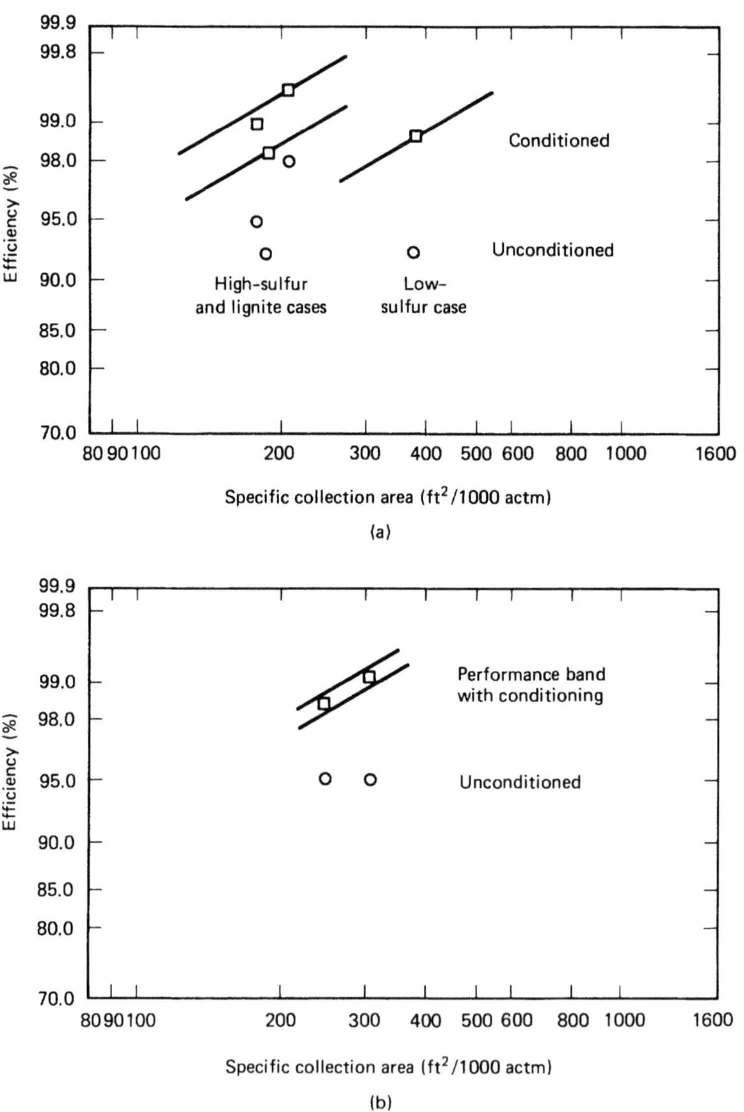

FIGURE 10-39 Performance improvement results with three types of conditioning: (a) NH_3; (b) SO_3; (c) Na_2CO_3.

total emission. Raw gas sneakage is eliminated by properly designed and installed antisneak baffles around the perimeters of the collecting zones. Hopper flows and jetting can be reduced by additional hopper baffling or gas distribution plates at the precipitator outlet.

In summary, the best approach to improving operation and performance involves the following sequence of events:

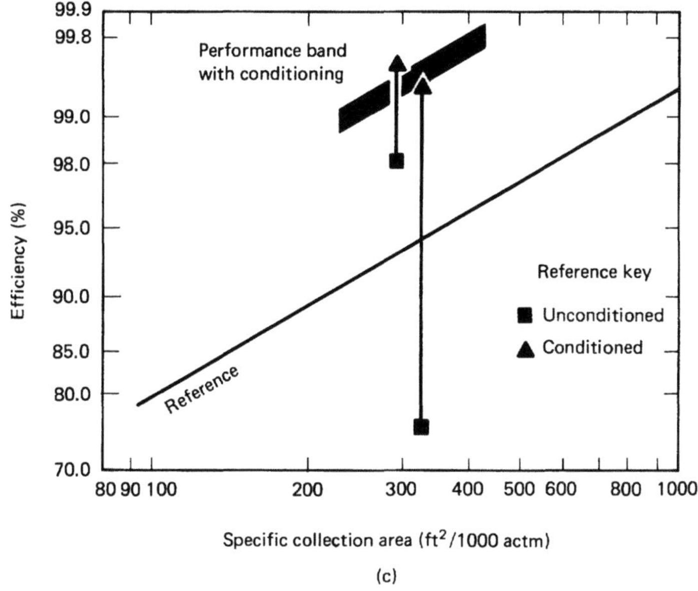

FIGURE 10-39 (con't.)

1. Eliminate known mechanical/component deficiencies.
2. Take care in the proper operation of the precipitator, get to know its normal behavior, and maintain it regularly, preventing misalignment, large dust buildups on electrodes, and full hoppers.
3. Look for and eliminate sources of gas sneakage or large flow-induced reentrainment losses.
4. Get the energization equipment right, matching TR set capacity to actual load and obtaining fast-acting automatic voltage control.
5. Optimize power and rapping.

If the factors that influence precipitator performance are understood and accounted for in the operation of the precipitator, and the sequence above is followed carefully but does not provide the desired results, performance may be limited by precipitator size or inherent design limitations, requiring the use of more precipitator collecting surface or a contingent technology to meet emissions compliance.

Contingent Technologies

There are two areas of contingent technology that have received the greatest attention in upgrading precipitator performance short of backfitting: chemical conditioning and pulse energization.

Dust chemical and physical properties can combine to severely limit the performance capacity of a precipitator. Rather than increase the size of a precipitator

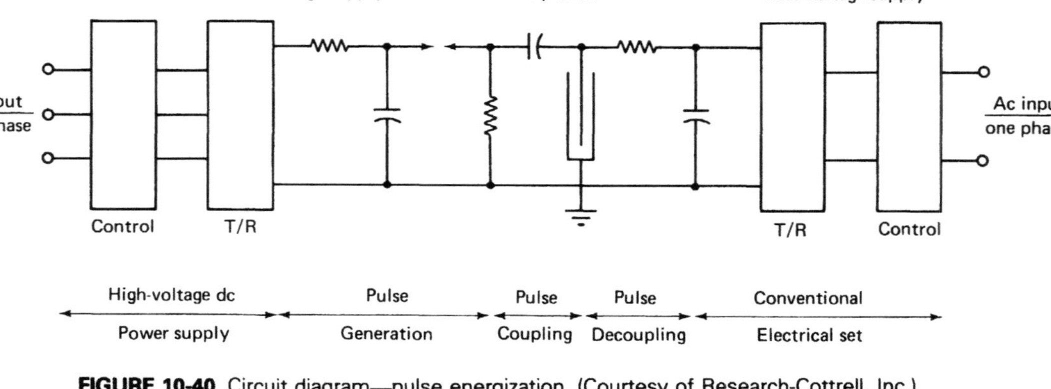

FIGURE 10-40 Circuit diagram—pulse energization. (Courtesy of Research-Cottrell, Inc.)

to treat a difficult ash, chemical conditioning offers a means to modify the ash to suit a given precipitator. Ash conditioning has been used to provide an acceptable solution to difficult problems. Conditioning systems that make use of raw chemical materials should be more cost-effective than methods that use commercially prepared additives, and such systems are available to treat practically every situation that could arise. It is important to know which conditioning agent is best suited for which type of precipitator application. Ammonia appears to have the widest range of use, having been successfully applied to cold-side ESPs with both low- and high-sulfur coals, and on some hot-side units. The conditioning mechanism is not well known, but appears to enhance particle agglomeration and increase space charge. Sulfur trioxide has been most widely used to increase the surface conductivity of low-sulfur coal ashes. Since the mechanism depends on sorbtion, it cannot be expected to operate well above 190°C. Sodium addition has seen the most limited use of these three. The charge-carrying capability of the sodium ions appears to predominate in this method, the carbonate playing no known role. Some results obtained with ash conditioning are illustrated in Figure 10-39.

Pulse energization can enhance the collection efficiency of precipitators where performance limitations are caused by poor energization, either from high resistivity or fine particulate corona quenching. Pulse energization consists of electrical components that may superimpose high-voltage pulses of very short duration on an underlying, relatively constant electric potential, or the pulses may be solely applied and varied automatically with the pulsed parameters, such as voltage, frequency, and width, to tune the precipitator operation and minimize the debilitating effects of resistivity-induced sparking or back corona. The precipitator can then operate with relatively high charging and collecting fields needed for effective particulate removal in the precipitator. Pulse energization produces a profuse corona along the discharge electrodes, resulting in uniform corona distribution, which plays an important role in enhancing particulate collection. Thus, performance enhancement is accomplished by altering the physics of the corona, not by raising the average level of the applied electric field. Figure 10-40 is a schematic circuit of a pulse-energization system.

10.7 CONCLUSIONS

Electrostatic precipitators have demonstrated wide applicability in the control of air pollutants and recovery of valuable by-products from industrial plants, ranging from large units in power-industry generating stations to less familiar processes in the chemical and refining industries. Their almost universal acceptance is attributed to several of their inherent advantages. First, the corona charging mechanism is operable on essentially all particulate matter, regardless of size, consistency, or chemical composition. Second, since the filtration forces work only on the particulate, the cost of energy consumption by mechanical filtration is avoided. Finally, precipitators have for many years demonstrated their performance capability on a wide range of particulate characteristics. They have become an established part of the pollution control inventory worldwide.

The record of precipitator performance and reliability has not been without its problems. A simple listing of the major activities involved in the selection and use of precipitators reveals ample opportunities for something to go wrong:

1. Feasibility study and permitting
2. Specification
3. Proposal
4. Evaluation and selection
5. Engineering
6. Construction
7. Operation
8. Maintenance

Four of these activities are completed before the unit is even designed, and much of the preengineering activity depends upon sizing. Theories that have evolved to advance the state of the art are usually worked out in clean laboratories under controlled conditions, and the invaluable verification data that are obtained on field units is often incomplete and sometimes useless simply because the operators or technicians were unfamiliar with the theory. Despite these inaccuracies and the naturally complex nature of sizing, these theories have produced the means that have made larger and more efficient units possible, with collection efficiencies exceeding 99.9% actually achieved in the field on some difficult dusts.

Much of the success of any precipitator also depends on operation and maintenance, but the importance of these activities is often overlooked in favor of specifying and designing units that are expected to operate automatically and trouble-free. However, a well-trained operator, given the responsibility and authority for optimizing performance and maintenance, has all the capability required to correct many of the field problems that occur. Most of these problems are no fault of nature, rather a shortcoming of people who, driven by values not related to the nature of the process, are either unwilling or unable to exercise the knowledge in design and operation that the state of the art allows.

10.8 NOMENCLATURE

a	charge acquired by a particle, dimensionless
A	collecting surface area
C	Cunningham correction factor
d	particle diameter
E	collecting efficiency
e	base of the natural logarithm

ε_0 — permittivity of free space
E_0 — charging field strength
E_p — collecting field strength
K — fuel constant, dimensionless
MW — gross generator output, megawatts
T — gas temperature
μ — gas dynamic viscosity
V (or V_w) — gas volumetric flow rate
w (or w_k) — migration velocity
$\dot{X}_a$ — excess combustion air, fractional

10.9 REFERENCES

[1] AIR POLLUTION CONTROL ASSOCIATION, ED., *Proceedings of the Specialty Conference on Operation and Maintenance of Electrostatic Precipitators, Dearborn, Mich., April 10–12, 1978.* Air Pollution Control Association, Pittsburgh, Pa.

[2] BIBBO, P. P., "A Programmed Approach to Precipitator Preventive Maintenance." *Proceedings of the 1976 American Power Conference,* Vol. 38. Illinois Institute of Technology, Chicago, 1976.

[3] BIBBO, P. P., AND AA, P. J., "Increasing Precipitator Reliability by Proper Logging and Interpretation of Operational Parameters—An Operator's Guide," paper presented at the 2nd Annual Symposium of the Transfer and Utilization of Particulate Control Technology, sponsored by the U.S. EPA, Denver, Colo., 1979.

[4] CASIGLIA, J. H., AND FLETCHER, H. R., "Unstable Vibration and Power Arc Induced Failures in Electrostatic Precipitator Discharge Electrode Wires," unpublished paper, Detroit Edison Company, © 1971.

[5] DARBY, K., "Criteria for Designing Electrostatic Precipitators," paper presented at the 2nd Symposium on the Transfer and Utilization of Particulate Control Technology, sponsored by the U.S. EPA, Denver, Colo., 1979.

[6] ENGLUND, H. M., AND BEERY, W. T., EDS., "Electrostatic Precipitation of Fly Ash," by Harry J. White, Carmel, Calif. In *APCA Reprint Series.* Air Pollution Control Association, Pittsburgh, Pa., July 1977.

[7] FELDMAN, P. L., "Effects of Particle Size Distribution on the Performance of Electrostatic Precipitators," 68th Annual Meeting of Air Pollution Control Association, Boston, June 15–20, 1975. Air Pollution Control Association, Pittsburgh, Pa.

[8] HALL, H. J., "Solving Problems in the Electrical Energization of Electrostatic Precipitators," *APCA J.,* 28(9) 871–878, 1978.

[9] JAVETSKI, J., "Solving Corrosion Problems in Air Pollution Control Equipment, Part II," *Power,* pp. 80–87, June 1978.

[10] KATZ, J., *The Art of Electrostatic Precipitation*. Precipitator Technology, Inc., Munhall, Pa., 1979, pp. 80–90.

[11] KESTER, B. E., ED., *Proceedings of the Specialty Conference on Design, Operation, and Maintenance of High Efficiency Particulate Control Equipment, St. Louis, Mo., March 29–30, 1973*. Air Pollution Control Association, Pittsburgh, Pa.

[12] OGLESBY, S., JR., AND NICHOLS, G. B., in *Electrostatic Precipitation. Pollution Engineering and Technology Series*, Vol. 8, Richard A. Young and Paul N. Cheremisinoff, eds. Marcel Dekker, Inc., New York, 1978.

[13] RAMSDELL, R. G., "The Prediction of Fly Ash Precipitator Efficiencies," Proceedings of the 12th Annual American Power Conference, 1968, Vol. 30, Illinois Institute of Technology, Chicago.

[14] SZABO, M. F., AND GERSTLE, R. W., *Operation and Maintenance of Particulate Control Devices on Coal-Fired Utility Boilers*. EPA-600/2-77-129, July 1977, pp. 2-1 to 3-51.

[15] WHITE, H. J., *Industrial Electrostatic Precipitation*. Addison-Wesley Publishing Company, Inc., Reading, Mass., 1963.

11

Flue Gas Desulfurization Systems

Timothy W. Devitt P.E.

Vice-President
PEDCo Environmental, Inc.
Cincinnati, Ohio

11.1 DESCRIPTION OF CONTROL DEVICE

Flue gas desulfurization (FGD) is the process of removing sulfur oxides, primarily SO_2, from combustion gases. In wet process FGD systems, flue gases are contacted with an absorbent in a vessel called either an absorber or a scrubber. The SO_2 reacts with the absorbent or dissolves into the solution to produce a slurry or liquid that contains dissolved or solidified sulfur compounds. In all operating systems, water is used to dissolve or suspend the reacting chemicals, although a number of dry absorption systems are currently being commercialized. The advantages of dry scrubbing include potential cost and energy savings and relative ease of operation compared with wet FGD systems. Dry scrubbing, however, requires higher chemical consumption and may be restricted to low-sulfur coals. Because most of the commercial dry scrubbing installations are under construction or in the planning phase, there is a lack of data on full-scale systems. Therefore, this review is limited to wet FGD systems.

Wet FGD processes are grouped into two general categories, regenerable and nonregenerable, depending on whether the sulfur is separated from the absorbent as waste. Nonregenerable processes produce a sludge that requires disposal in an environmentally sound manner. Regenerable processes include additional steps to convert the sulfur into by-products such as liquid sulfur dioxide, sulfuric acid, or elemental sulfur.

Six basic wet FGD systems are used in this country, primarily to treat flue gas from large, coal-fired boilers. Of these, the lime scrubbing, limestone scrubbing, sodium carbonate scrubbing, and dual alkali systems are classified as nonregenerable. The magnesium oxide scrubbing and Wellman–Lord systems are classified as regenerable processes.

Table 11-1 lists the major regenerable and nonregenerable systems and indicates the number of systems that are operational, under construction, or planned for application on both utility and nonutility boilers.

Because most operating FGD systems are nonregenerable (a trend that is likely to continue because of their generally lower capital and operating costs), only the regenerable systems are reviewed here. Furthermore, only the wet lime- and limestone-based systems are discussed, although many of the design, operation, and maintenance considerations are similar for other systems.

The Lime Scrubbing Process

The first step of the process is preparation of a lime slurry. Lime (CaO) is reacted with water in a slaker to produce a slurry of calcium hydroxide $[Ca(OH)_2]$ and water, according to the following reaction: $CaO + H_2O \rightarrow Ca(OH)_2 + heat$. This material is diluted with recycled water to produce a scrubber slurry containing about 15 to 20% solids.

The lime slurry is added at a regulated rate to a recycle tank that contains the scrubbing slurry. Large pumps circulate the scrubbing slurry through the scrubber or absorber vessel. In the absorber, the slurry droplets are brought into contact with the flue gas containing SO_2, and the SO_2 passes from the gas into the droplets driven by a concentration gradient. The dissolved SO_2 reacts with calcium from the lime according to the following reactions:

$$CA(OH)_2(s) \leftrightarrow Ca(OH)_2(aq) \leftrightarrow CA^{2+} + 2OH^-$$

$$2OH^- + CO_2 \leftrightarrow CO_3^{2-} + H_2O$$

$$CO_3^{2-} + CO_2 + H_2O \leftrightarrow 2HCO_3^-$$

$$Ca^{2+} + CO_3^{2-} \leftrightarrow CaCO_3$$

$$CaCO_3(s) \leftrightarrow CaCO_3(aq) \leftrightarrow Ca^{2+} + CO_3^{2-}$$

$$SO_2(g) \leftrightarrow SO_2(aq) + H_2O \leftrightarrow H^+HSO_3^-$$

TABLE 11-1 Distribution of FGD systems by process type.

| | Utility Boiler FGD Capacity (MW)[1,2] | | | | | Industrial Boiler FGD Capacity (scfm)[3,4] | | |
| | Under | | | | | Under Construction/ | | |
Process	Operational	Construction	Planned	Total	Process	Operational	Planned	Total
Aqueous carbonate/spray drying		100		100	Ammonia	164,000	154,000	318,000
Citrate	60[5]			60[5]	Caustic	211,000		211,000
Dual alkali	1,201		842	2,043	Caustic waste stream	665,000		665,000
Lime	8,801	2,060	6,841	17,702	Citrate		142,000[5]	142,000[5]
Limestone	11,437	7,637	17,254	36,328	Dual alkali		1,480,000	1,480,000
Limestone/alkaline fly ash	1,480			1,480	Dual alkali (concentrated)	489,070	512,000	1,001,070
Lime/alkaline fly ash	2,613	1,400		4,013	Dual alkali (dilute)	233,400		233,400
Lime/limestone	20		475	495	Lime	40,000	30,000	70,000
Lime/spray drying	110	1,060	1,813	2,983	Lime/spray drying		72,700	72,700
Magnesium oxide		724		724	Limestone	55,000		55,000
Sodium carbonate	925	330	1,900	3,155	Sodium carbonate	2,467,000	504,500	2,971,500
Sodium carbonate/spray drying		440		440	Sodium hydroxide	1,140,300	12,000	1,152,300.
Wellman Lord	1,540	534		2,074	Sulf-x score		10,000	10,000
Process not selected			6,650	6,650	Process not selected		124,000	124,000
Total	28,187	14,285	35,775	78,247	Total	5,464,770	3,041,200	8,505,970

[1] M. Smith, et al., *EPA Utility FGD Survey: October - December 1980*. Vol. 1. EPA-600/7-81-012a, PEDCo Environmental, Cincinnati, Ohio, October 1979.

[2] Capacities are reported as equivalent scrubbed capacity (ESC); the summation of effective scrubbed flue gas in equivalent MW based on the percent of the flue gas scrubbed by the FGD system(s).

[3] J. Tuttle, et al., *EPA Industrial Boiler FGD Survey: First Quarter 1979*. EPA-600/7-79-067b, PEDCo Environmental, Cincinnati, Ohio, April 1979.

[4] For approximate comparison purposes 1 MW may be considered ≈ 3000 scfm.

[5] The citrate process FGD system on St. Joe Zinc's G. F. Weaton unit was included in both the utility and industrial report since it supplies power to the utility grid as well as steam for the industrial plant. This unit should be deleted from one or the other sections when estimating total process capacities for the industrial-utility marketplace as a whole.

$$HSO_3^- \leftrightarrow H^+ + SO_3^{2-}$$

$$Ca^{2+} + SO_3^{2-} \leftrightarrow CaSO_3(aq)$$

$$Ca^{2+} + SO_3^{2-} + \tfrac{1}{2}H_2O \leftrightarrow CaSO_3 \cdot \tfrac{1}{2}H_2O(s)$$

$$CaSO_3 + H^+ \leftrightarrow Ca^{2+} + HSO_3^-$$

$$H^+ + HCO_3^- \leftrightarrow H_2CO_3 \leftrightarrow CO_2(g) + H_2O$$

$$CaSO_3 \cdot \tfrac{1}{2}H_2O + \tfrac{3}{2}H_2O + \tfrac{1}{2}O_2 \leftrightarrow CaSO_4 \cdot 2H_2O$$

These reactions take place to some extent in every lime scrubbing system; similar reactions occur in limestone systems. Such factors as ionic concentrations, pH, temperature, and retention time in the reaction tank influence the speed and completeness of the various reactions. The net effect, however, is removal of SO_2 from the flue gas, depletion of calcium hydroxide in the slurry, and generation of insoluble calcium sulfite and calcium sulfate. The calcium ions used in the reactions are replenished by dissociation of calcium hydroxide. The use of an insoluble alkaline reagent is the primary difference between the lime and limestone slurry scrubbing systems and the so-called clear solution scrubbers (e.g., double alkali scrubbers), where the predominant reactive alkali (sodium) is more soluble.

Forcing most of the sulfite/sulfate precipitation to occur in the reaction tank requires a certain amount of retention time to permit the slurry to reach chemical equilibrium before it is recirculated to the scrubber. For this purpose, a reaction tank, sometimes an integral part of the scrubber vessel, is usually sized with a retention time of between 5 and 10 minutes.

The calcium sulfite and calcium sulfate that are formed in the reaction tank must be removed from the system. This is accomplished by bleeding a stream from the recirculating slurry to a thickener or clarifier. The calcium sulfite and sulfate settle out and are removed from the underflow of the clarifier in a slurry of about 30% solids. The overflow, a clear liquid of less than 1% solids, is brought back to the system for reuse in the process. The sludge from the underflow is disposed of by one of several means.

One method of disposal is to pump the sludge to a pond. In this case the sludge is pumped from the thickener underflow (or occasionally directly from the reaction tank) to a large pond, where the solids settle. Clear water is recycled from the pond, as necessary, to keep the pond from overflowing. Return of all this clear liquor from the pond to the process is called a "closed-loop" operation.

Another method of disposal is to dewater the sludge in a vacuum filter, chemically stabilize it by the use of generic or proprietary processes, and dispose of it in a landfill. When vacuum filtration is used, the filtrate is returned to the system to create closed-loop operation.

The Limestone Scrubbing Process

Limestone slurry preparation is the only step that is significantly different from the lime scrubbing process. Lumpsize limestone is pulverized to -200 to -300 mesh in a ball mill and slurried with water. The fine size is needed because dissolution of $CaCO_3$ is a slow reaction, the rate being directly related to the particle surface area and therefore to the particle size. For further limestone dissolution, the scrubbing slurry is maintained at an acidic pH, in contrast to lime-based systems, in which scrubbing slurry is slightly alkaline. The overall process reactions are the same as for the lime slurry scrubbing process, except the reactant species is $CaCO_3$, not $Ca(OH)_2$.

Figure 11-1 illustrates a typical limestone process. The lime process is similar, except a lime slaker replaces the ball mill of the limestone process.

Other wet process FGD systems operate similarly. The SO_2 is scrubbed from the gases in a contactor, and reaction products are removed in a separate vessel. Several excellent texts and reports are available that describe the various systems, their chemistry, and status of commercialization; these are listed in Section 11.8.

11.2 DESIGN PROCEDURES

Proper system design is crucial to successful operation of an FGD system. Although this is true for all air pollution control systems, it is especially important for FGD systems. In contrast to most other air pollution control systems, there are two design considerations; in addition to adequate mechanical design, an FGD system must receive careful chemical process design. Many operating problems have developed in these systems as a result of inadequate chemical design.

Table 11-2 presents a partial list of the factors outside the scrubbing system battery limits that influence system design, and Table 11-3 lists the mechanically and chemically related design parameters that must be considered within the scrubber battery limits. A significant body of knowledge has been established concerning proper mechanical design, but the factors bearing on chemical process design and their interrelationships with the mechanical design have only begun to receive adequate consideration in the past few years.

Chemically related problems, primarily scaling and corrosion, are due to improper consideration and/or inadequate control in the operational phase of process chemistry. Mechanically related problems refer to the malfunction or failure of mechanical components in the scrubber and auxiliary systems, such as pumps, fans, and agitators. Failure of field instruments, expansion joints, and piping may sometimes be caused by interrelated mechanical and chemical conditions. Indeed, the most persistent and significant problems with FGD systems have both chemically and mechanically related contributing factors. The design practices being used to minimize these problems are discussed below.

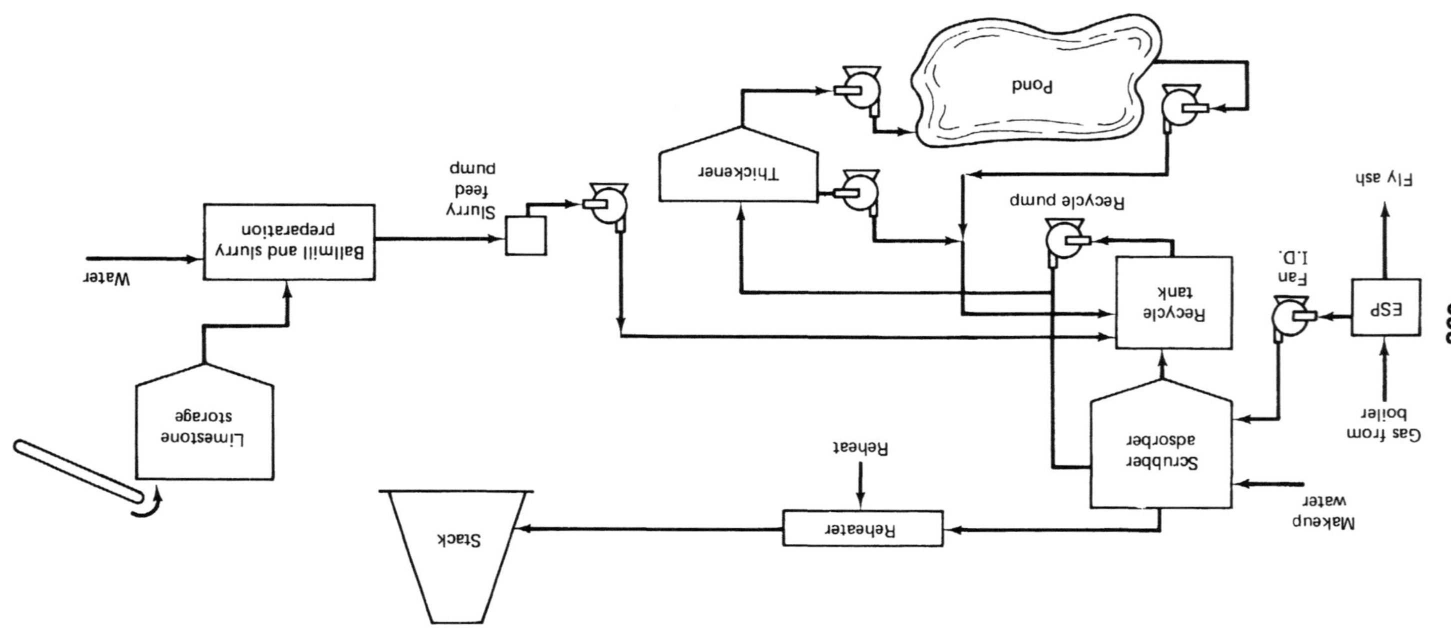

FIGURE 11-1 Flow diagram for a limestone scrubbing system.

TABLE 11-2 Factors influencing scrubber system design outside the battery limits.

Coal properties	Site conditions
Sulfur content	Land availability
Ash content	Soil permeability
Fly ash composition, particle size	Rainfall
Chloride content	Climate
Heating value	Regulations
Moisture content	SO_2 emission/ambient standards
Composition variability	Particulate standards
Boiler design	Plume visibility standards
Type of boiler	PSD/nonattainment considerations
Size of boiler	Water quality standards
Age of boiler	Makeup water
Flue gas flow	Source
Loading characteristics	Chemical composition
Reagent properties	Flue gas
Percent inert	Temperature
Ca, Mg contents	Flow
Reactivity	Dew point
Size	Particulate loading
	Particulate alkalinity
	SO_2 content
	O_2 level

Control of Scale Formation

Scale-free operation can be accomplished by operating in either of two basic modes: coprecipitation or control of supersaturation. Coprecipitation involves the removal of calcium sulfate from the system as part of a calcium sulfite/sulfate solid solution. If a system is operated so that the maximum oxidation in the slurry circuit is about 16% (i.e., 16% of the SO_2, on a molar basis, is oxidized completely to the SO_4^{2-}), the scrubbing liquor can remain subsaturated with respect to calcium sulfate (gypsum). Above this level, more calcium sulfate is formed in the slurry

TABLE 11-3 Factors influencing scrubber system design inside the battery limits.

Absorber type	Makeup-water distribution
Waste slurry disposal scheme	Process chemistry
Degree of redundancy	pH gradient throughout the system
Reheat type and amount	Sulfite to sulfate oxidation
Degree of instrumentation	Chloride balance
Mist-eliminator configuration	Liquid/gas ratio (L/G)
Raw-material reactivity	Point of fresh water addition
Process layout	Reagent utilization
Materials of construction	

circuit than can leave the system in a coprecipitated form. This causes the sulfate concentration to increase and the system to operate supersaturated with respect to gypsum. When the relative saturation levels approach a critical value of 140%, hard scale can form within the system.

Supersaturation can be controlled by providing nucleation sites (seed crystals) for the precipitation of gypsum and calcium sulfite/sulfate solid solution. This maintains the system below the critical (140%) level of saturation with respect to gypsum.

Alternatively, the slurry in a forced-oxidation system contains predominantly gypsum crystals, and saturation levels are maintained at 100 to 110% with respect to gypsum. The use of forced-oxidation systems has been especially beneficial for limestone systems treating flue gas with low sulfur dioxide concentrations. For these applications, the low sulfur dioxide levels, coupled with the long liquid-retention times, result in high "natural" oxidation levels. This makes forced oxidation a particularly attractive method for improving sludge quality, minimizing scaling, increasing sulfur dioxide removal, and improving reagent utilization.

Another method for controlling scaling involves increasing the magnesium-ion concentration. This increases the liquid-phase alkalinity of the scrubbing slurry and the amount of sulfite and sulfate the scrubbing slurry can hold without exceeding solubility limits. The overall effect is a subsaturated operation that produces higher sulfur dioxide removal efficiencies and higher utilization. Organic acids, such as citric, benzoic, and (more recently) adipic, are also being evaluated for use in increasing the total alkalinity of the scrubbing liquor.

Mechanical System Design

Balanced-draft or booster fans   In addition to placing these fans upstream of the FGD system, another development is the use of variable-pitch, axial-flow fans. The main advantage of this design is its consistently higher efficiency (versus centrifugal fans) over the entire boiler operating range, which produces a substantial energy savings. Other advantages are better flow control, easy access and maintenance, less severe construction requirements, and increased reliability.

Dampers   Bypass and isolation dampers are used to regulate the flow of flue gas into and around the FGD system. The primary purpose of isolation is continuance of unit operation while the scrubber modules are under maintenance. Common damper designs include slide-gate (guillotine), single-blade butterfly, and multiblade parallel (louver). Corrosion and erosion of the various types of dampers and damper seals have been common. In some cases, dampers have failed entirely or have been so inefficient that the modules could not be maintained during bypass situations. The current trend is toward two-stage louver dampers having a pressurized seal-air system that maintains a positive pressure between them. Pressurized seal air increases the energy demand of the system, but it contributes significantly to successful damper operation.

Scrubbers The trend in the design of lime/limestone slurry FGD systems is toward the use of simple, relatively open units, such as spray towers, or combinations of venturis and spray towers. Spray towers have few internal components in the gas–liquid contact zone and thereby offer fewer sites for deposition of solids, collected fly ash and unused reagent, and for formation of scale. Spray tower operation has been very successful, but SO_2 removal efficiency may be limited when high-sulfur coal is fired. In such circumstances the use of more reactive absorbents such as lime or magnesium-promoted lime or limestone, or the use of combinations of scrubbers such as a venturi in series with a spray tower, may be required.

Reaction tanks Coinciding with gas-side staging via a venturi/spray tower or similar combination is liquid-side staging, in which the hold tanks are arranged in series to approach plug flow reactor design.

Mist eliminators Chevron and baffle-type mist eliminators continue to be the predominant designs used in U.S. utility FGD systems. The popularity of these separators is due to design simplicity and flexibility, adequate collection efficiency for medium- to large-size drops, relatively low pressure drop, open construction, easy access for maintenance, and relatively low cost.

Reheaters A pronounced preference for stack gas reheat is still evident for systems in service and committed for future operation. In several systems the use of wet stacks has been abandoned in favor of reheat because of problems with stack corrosion, plume dispersion, and localized "rain" caused by plume condensation. Stack corrosion is more pronounced where high-sulfur coal is burned because sulfur dioxide concentrations are higher.

Solids separation (sludge dewatering) The major development in this area is the increased emphasis on dewatering and stabilization of FGD wastes for managed landfill disposal and the corresponding decreased emphasis on ponding. Formerly, a pond was expected to fulfill three functions: clarification, dewatering, and temporary or final sludge storage. The increased emphasis on disposal in landfill and structural fills and on attaining closed water-loop operations further stimulated the use of clarifiers, centrifuges, and vacuum filters. In addition to using these techniques, several installations are using forced-oxidation strategies to enhance solids settling and filtration properties, to improve process chemistry and waste sludge quality, and to decrease land requirements for sludge disposal.

Process control and instrumentation Virtually all full-scale operating systems regulate reagent feed rate by controlling slurry pH. Dip-type pH sensors have been more successful than in-line sensors because they are easier to clean and calibrate; the in-line, flow-through sensors are subject to more wear and abrasion and generally require much more frequent maintenance. Other systems for control of reagent feed are also being evaluated on full-scale systems. One type involves feedforward reagent control based on the inlet flue gas flow rate and sulfur dioxide concentration, with trim provided by slurry pH. Another type involves control of reagent feed rate by use of the outlet sulfur dioxide as the control variable. Limited success has been reported for both systems, primarily because of the difficulty in obtaining accurate and consistent readings from sulfur dioxide gas analyzers; this has been especially true on high-sulfur coal applications. Nonetheless, this type of process control should enable much better control of process chemistry, particularly with limestone systems and variable boiler loading, and should continue to be developed.

Other critical instruments are those that measure and control the density of recycle slurry and the level of reaction tank liquid. In general, the more sophisticated instruments based on nuclear absorption or ultrasonic principles have proved more satisfactory than mechanical bubble-tube devices.

Construction materials   The construction materials generally preferred for use in critical areas of the FGD system are 316 and 316L stainless steels. However, some designers prefer carbon steel with a surface lining or coating that physically shields the bare metal from the corrosive environment. These linings are usually resins applied in liquid or semiliquid form, by spray or trowel, and allowed to cure. The use of lined or coated carbon steel offers the potential benefits of being able to withstand low pH/high chloride environments much better than the 316 steels and being considerably less expensive. Operating experience, however, has shown that these materials are susceptible to damage by high-temperature excursions. In addition, improper application of linings has been the suspected cause of lining breakdown and subsequent base metal corrosion, especially in reapplications, where proper preparation of the metal surface is more difficult.

The use of natural rubber, neoprene, polyvinyl chloride (PVC), fiber-reinforced plastic (FRP), and flaked-glass polyester generally predominates in the liquid and thickener circuits of the FGD system (tanks, pumps, agitators, piping, and thickeners) where the metal parts are 316 stainless steel or Alloy 20. Neoprene, polypropylene, FRP, and Alloy 20 are generally used in filtration systems.

Many systems are also being constructed of more-resistant alloys in trouble spots (e.g., a wet/dry, high-temperature, high-chloride environment such as a prescrubber or presaturator). Alloys such as Hastelloy C-276, Hastelloy G, Alloy 20, Inconel 625, Incoloy 825, 317 low-carbon stainless steel, 904 low-carbon stainless steel, Jessop JS-700, and E-Brite 26-1 are being selected for special problem areas.

The use of stack lining or coating materials has been troublesome, especially where high-sulfur eastern coal is burned. Lining failures have also been reported in systems that included an apparently adequate degree of reheat. Recent developments in this area indicate that elastomer (Coldbran's CXL-2000) has recently been applied with apparent success on the flues at the Bruce Mansfield station, which burns high-sulfur eastern coal without reheat. The use of acid brick also appears to be successful in a similar application. At installations burning low-sulfur coal, adequate linings have not been as great a problem; many installations can get by with unprotected carbon steel flues.

11.3 INSTALLATION PROCEDURES

The installation of an FGD system is a major construction project, and the details that affect operation and maintenance are primarily a part of the engineering design. Unless considerations of operation and maintenance are included in the design, the system will be expensive to operate and maintain.

The most serious design errors result when an item of equipment that should be expected to require eventual replacement or major repair is buried deeply within the assembly. In particular, the rotating machinery, such as pumps, fans, reagent conveyors, and vessel agitators, should be installed with sufficient free space to permit their removal without an extensive disassembly of other parts of the plant. Large dampers also may require eventual replacement. Some compromise of this provision is usually necessary because open space in a process unit greatly increases the original construction cost. Design arrangements can usually be devised, however, in which those items most likely to require major service are located near the outside of the assembly.

Instruments present a special installation problem because the cost of maintaining field instrumentation has been especially high in some FGD systems. Since the instruments are often the most delicate items of equipment in the plant, they will usually require much more frequent maintenance than will the larger pieces of equipment. These instruments can usually be serviced easily and quickly, however, and the time required to service an instrument is often much less than that required to gain access to it and to remove it from the process line or vessel. It is especially important that installation of such higher-maintenance instruments as SO_2 analyzers and pH electrodes permit simplified access and removal.

Installation procedures that simplify FGD system maintenance usually also simplify operation. The system design must be reviewed to ensure that the test connections provided for routine operational use are conveniently accessible to the operators. A partial list of these test connections is provided in Chapter 12.

The continuing maintenance cost of an FGD system can be adversely affected by errors that occur during checkout or startup of a new system. Failure to flush piping thoroughly before the unit is placed in operation can cause damage to pumps or vessel linings that cannot be field-repaired to original condition and therefore create continued maintenance problems. Improper startup can begin the process of corrosion or scale formation, which will show up later as a premature equipment failure. The first few days in the operation of a new system may form the basis for excessive maintenance that may require years to correct.

11.4 OPERATION

Two areas of FGD system operation are addressed here, in which the problems of operation are quite different. Startup of an FGD system involves not only a thorough checklist of the unit; but also a rapid approach to the conditions of normal operation. On completion of the startup phase, the slower-paced period of normal operation begins, in which a fixed routine of activities must be maintained to keep the system performing properly. The following paragraphs first describe the considerations that apply during normal operation and then discuss the activities of startup.

Normal Operation

The routine program conducted during normal operation of an FGD system has two objectives. First, it is necessary to determine the conditions under which the system performs best and to hold the system to those conditions. Second, an organized procedure must be devised and executed to assess continuously the mechanical condition of the unit and each unit component. Although problems and minor emergencies are always changing during normal operation, the routine of operation is neverending.

With high-sulfur coal especially, operation of the reagent feed equipment necessitates close coordination with the operation of the power-generation equipment to permit the FGD system operator to anticipate a change in boiler loading. In such systems, it may be necessary to manually override the automatic instrumentation during excessive load changes to prevent an out-of-limits surge in outlet SO_2 concentration.

Another parameter that must be optimized is the concentration of solids in the scrubbing slurry. Recycle of previously precipitated calcium sulfite and sulfate is necessary to supply nucleation sites and thereby to minimize accumulation of crystals on scrubber surfaces as scale. A high concentration of solids in scrubbing slurry, however, not only minimizes scale formation, but also increases erosive wear in equipment and increases carryover of solids into the mist eliminators. The fresh or clarified water used to wash solids from the mist eliminators becomes mixed with the scrubbing slurry, and thereby reduces solids concentration. In each FGD system, therefore, an optimum balance must be made between the tendency to create scale and the tendency to plug the mist eliminators and exacerbate equipment erosion. This balance is expressed in concentration of solids in the slurry and is measured as density of slurry. Optimum solids concentration is around 8% solids in lime-based FGD systems and 12 to 15% in limestone-based systems. Equivalent densities range between 1.06 and 1.12 g/cm^3.

Within limits, some FGD system operators also consider the cost of reagent in defining optimum conditions of operation. Especially in limestone-based systems, lower PH and higher solids content provides a greater percentage of utilization of the reagent added. However, the quality of the reagent in regard to thoroughness of lime slaking or fineness of limestone grind probably has a great effect on reagent utilization.

The optimum pH and slurry density for a specific FGD system must be established, at least in part, by trial and error. Chemical analyses of scrubbing slurry can establish such characteristics as scaling potential and degree of reagent utilization to speed up the stepwise procedure of defining optimum parameters. It is not yet possible to obtain a complete prediction of these values from theoretical calculations because data that define the precise effect of combinations of various operational and design options have not been fully developed.

System health   The other facet of normal operation is the routine program of checks and examinations that are conducted to spot potential problems before they increase in complexity and result in costly correction. Following is a list of diagnostic checks that constitute a typical program of this type.

1. *Scrubber/absorber (slurry recycle)*. Check the recirculation slurry header pressure at the branch take-offs to the scrubber. If the header pressure has increased at one or more of the branches, it is indicative of nozzle pluggage. If it has decreased, it indicates nozzle wear. The pressure should be checked and recorded once per shift by using hand held gauges (static pressure sensor taps frequently become plugged).

2. *Scrubber/absorber*. Check differential pressure across the absorber section. Slow buildup can indicate scale formation, although scale formation can be quite rapid if control of the process chemistry is lost. Suggested frequency is two to four times per shift. If a low pressure drop is detected, further checks should be made to determine whether there is low scrubbing liquor flow or low flue gas flow due to a partially closed damper, inoperable or damaged variable venturi throat, or some similar constriction. A sudden high pressure drop probably indicates a sudden increase in gas flow rate, possibly due to an improper fan damper setting. If the fan amperage is too high, the gas flow should be restricted immediately to avoid fan-motor overheating and serious damage.

3. *Scrubber/absorber*. Check the SO_2 removal efficiency by using data from the continuous SO_2 monitors. Decreased removal efficiency could indicate a chemical imbalance or poor gas flow distribution. The gas velocity could be exceeding design limits in one or more of the sections. The SO_2 removal efficiency information coupled with pressure drop and slurry chemical composition data can be used to determine whether poor gas flow distribution may be the problem. Suggested frequency of checking SO_2 removal efficiency is twice per shift.

4. *Scrubber/absorber*. Another indication of poor gas flow distribution is entrained solids or mist carryover. If inspection ports are available, the amount of carryover from the scrubber to the mist eliminator as well as downstream of the mist eliminator can be assessed. In severe cases, entrained solids and mist carryover can be observed. Another indication of carryover would be excessive fan vibration due to slurry deposits on the fan blades. Suggested frequency of inspection and notation in the operating log is once per day.

5. *Mist eliminator*. Check the differential pressure across the mist-eliminator section and visually inspect (through portholes) the appearance of blade surfaces. A pressure increase indicates scale buildup or pluggage, which can be verified by visual inspection. Excessive moisture in the scrubber outlet duct is another indication of mist-eliminator malfunction. Suggested frequency is once per shift.

6. *Mist eliminator*. Check the composition of the spray liquor, especially for suspended solids, alkalinity, and pH. Often, the spray liquor source is the overflow

from the clarifier, although many systems use the mist-eliminator wash as the source of freshwater makeup; thus checking the composition of the spray liquor often ties in with assessing the performance of the clarifier. High solids content in the spray liquor can lead to erosion; more significantly, however, high pH or dissolved alkalinity can lead to removal of additional SO_2 in the mist eliminators and scale formation. If the potential for scaling exists, more freshwater makeup should be used, consistent with overall water-balance limitations. The composition of the wash water should be checked and noted several times per week.

7. *Mist eliminator.* Check timing sequence and valve operation. Flow in deluge-type systems can be checked for proper timing by observing valve actuators and by feeling the headers for flow (e.g., temperature). A significant number of deluge-type systems have been found to have leaks in the automatic shut-off valves, which lead to upsets in the overall water balance. Leaks can be detected by observing the pressure in the supply line with flow to all points from that line shut-off and comparing that pressure with the pressure in the line with the supply open to the mist-eliminator header and flow from the mist eliminator curtailed only by the automatic shut-off valves. The ability to perform this check and the frequency of such checks depend upon the overall piping configuration and the degree of concern about the water balance.

8. *Slurry recirculation pumps.* A slow loss of discharge pressure indicates probable impeller erosion. High pressure can indicate restricted flow. Intake restrictions to the pumps can be caused by plugging of the intake lines or screens, which can lead to cavitation. Slurry flow can be checked by touching the pipe; a warm pipe (about 120°F) indicates flow. The discharge pressure should be noted at least once per shift.

9. *Slurry recirculation pumps.* Seal-water consumption should be individually monitored for each pump. Excessive water consumption indicates seal wear, which is significant because it may necessitate restricting the freshwater supply to the mist eliminators in order to maintain the overall water balance. Seal-water consumption should be noted once per shift.

10. *Scrubber slurry.* The density meter controlling the bleed rate to the thickener should be calibrated every shift by using a simple Denver pulp tester or comparable device. This will permit control of the solids concentration in the spray liquor. The scrubber slurry also should be manually checked for pH once per shift to verify the inplace pH instruments. In addition, on limestone systems the available limestone in the slurry should be checked, because pH is a poor control indicator in such systems.

11. *Thickener underflow.* Check the solids concentration in the underflow. The sludge blanket depth should be maintained in the 3- to 7-ft range and blown down only at 4-ft intervals; continuous blowdown is to be avoided. The blanket depth can be checked from the top of the thickener with a dipstick. Too high or low a concentration can be controlled by adjusting blowdown frequency, the solids levels of incoming system bleed streams, amount of flocculant addition, and rake operation. In severe instances, the sludge blanket will float so that it cannot be pumped out by the underflow pumps. This may be due to too much flocculant

addition or too high a rake speed (both compounded by inattentive operation). The incoming bleed stream must be shut off, as well as the rake. Where possible, the rake should be moved up and operated in an on–off fashion to try to promote settling (e.g., 30 seconds on every 3 minutes). If it is possible to introduce the system bleed at another point, this may also help initiate settling. Proper thickener operation is critical for overall system operation and should be carefully monitored by a full-time operator.

12. *Thickener overflow.* The solids concentration in the thickener overflow should be measured periodically to determine system performance. Visual inspection of the overflow clarity is usually sufficient for assessing performance on a routine basis and should be done at least twice per shift.

Startup

The initial checkout and operation of a new FGD system is a critical phase in the life of the unit. An FGD system is a complex process unit, and no two are exactly alike. Unexpected problems develop during startup, and the available personnel are often unable to handle all the problems at once. Incidents of severe damage to FGD systems have occurred during the first days of operation.

Three considerations are of primary importance during the startup phase:

1. *Full and complete mechanical checkout of every component in the system.* There is usually a strong economic pressure to get the unit on line, but extended delays and greatly increased costs can result from attempts to operate the plant before construction and checkout are completed.

2. *Prevention of system overheating.* Loss of recycle slurry flow will cause hot flue gas to pass through the scrubber/absorber, which can severely damage rubber linings, plastic mist eliminators, and metal components not designed for high temperature. Loss of recycle flow can occur through pump or electrical failure, severe plugging from inadequately mixed slurry, pipeline failure, loss of liquid level in the recycle tank, and for many other reasons. At all times during operation, but especially during startup, flue gas must be immediately prevented from passing through the absorber vessel if recycle slurry flow stops. Various systems permit this to be accomplished in a number of ways, ranging from opening a bypass damper to shutting down the boiler.

3. *Prevention of scaling.* When process chemistry is badly out of balance, as it often is during startup, severe scaling can occur in a matter of a few hours. Speed in approaching normal pH and solids content during startup is of prime importance.

Actual startup is not difficult. It consists of first establishing slurry recycle, then starting the flow of flue gas. The slurry should be as close to optimum conditions as possible. If another FGD system is operating at the station, underflow from the second thickener can be used to prepare a slurry for the new unit. If a second unit is unavailable, best practice includes the preparation of a synthetic slurry using purchased gypsum. In either instance, rapid adjustment of pH is important, and the pH controller should be placed in the automatic mode as soon as the proper pH value can be attained.

Many FGD systems include a startup line to permit the return of thickener underflow slurry to the recycle tank. The objective is to allow a more rapid approach to operating slurry density. If the necessary recycle line has not been provided, use of a portable pump and a hose line may be indicated. If a sludge filter is used in the sludge disposal area, manual return of filter cake to the recycle tank is another option. Still another technique is the use of reagent slurry that is more concentrated than normal.

Especially for use prior to a planned shutdown in which the recycle tank will be drained, special sumps with the necessary piping are designed into some FGD systems to store slurry and thereby simplify restart. Accumulation of these solids in the thickener is also possible unless the thickener is common to several modules.

11.5 MAINTENANCE

In a large, complex, and continuously operating process unit, programs of maintenance have evolved almost into a science, which cannot be discussed fully in this chapter. Two general areas of maintenance are recognized: emergency and preventive. Emergency maintenance follows an unexpected failure of a plant component, which often disrupts plant operation, sometimes causes a major shutdown, and almost always results in an abnormally high repair expense. Preventive maintenance is scheduled so as not to disrupt operations, and it does not involve overtime use of maintenance personnel and the high costs of emergency procurement of parts or tools. The purpose of preventive maintenance is to minimize emergency maintenance, and a judicious program of preventive maintenance is vital to the continued satisfactory performance of an FGD system.

The program is based on an inventory of every component of the plant, categorized with respect to its potential for failure and its availability for service. Some components, such as rotating machinery, slurry nozzles, and control valves, will wear more rapidly than major process vessels or utility piping. Preventive maintenance programs must assign a frequency of attention to all these items; although the components least likely to fail will be examined less frequently, they must not be ignored entirely.

Availability for service defines when each component will be examined. Components will be classified into four groups. A program of routine renovation may not be applied to a few noncritical or expendable items, which can be quickly repaired or replaced by shift personnel. Electrodes for pH measurement may be in this category. Three schedules of availability will be prepared, however, to cover all other equipment items.

A program of regular renovation is practiced while the system continues to operate on those equipment items where redundant equipment was included as part of the original design. In some FGD systems, this may be limited to pumps and fans; in other systems, it may include entire scrubbing modules. A second program of renovation covers items that, through minor modifications to operation, can be removed temporarily from service for cleaning, replacement of parts, or total replacement with a spare unit. This program includes such items as ball mills, sludge filters, and most instruments. Finally, a third program of renovation covers those equipment items that cannot be removed from service without shutting down the plant. This may be a major part of some systems, whereas in others it may only involve low-maintenance devices such as clarifier vessels and major sections of ductwork and piping. An intensive effort must be made to refurbish this class of equipment during scheduled shutdowns, which also usually involve the boiler and the power-generation eqiupment.

A well-conceived maintenance program replete with appropriate spare equipment and trained personnel, coupled with a systematic approach to monitoring and control of process operations, will pay significant dividends: an on-line FGD system and continued plant operation.

11.6 IMPROVING OPERATION AND PERFORMANCE

Optimum conditions are expressed primarily in the chemical and physical characteristics of the scrubbing slurry. Of all the measurable characteristics, slurry pH is most important. If all other variables are constant, a consistent pH ensures that the amount of reagent fed to the scrubber/absorber is in the proper ratio to the amount of SO_2 being absorbed from the flue gas. If the pH varies, the variation will be reflected in other parameters throughout the system. If pH is too low, less SO_2 will be absorbed and more SO_2 will pass through into the stack. If pH is too high, an excess of reagent will be present in the slurry and formation of scale may result.

In general, most lime-based FGD systems operate best at a pH between 8.0 and 8.5, and limestone-based systems operate best at a pH between 5.5 and 6.0. The optimum pH will be different for different systems, however, depending on the location within the system where pH is measured, the reactivity of the reagent, and details of the unit design. The degree to which pH can be permitted to vary from optimum also differs for different systems. Units designed for low- and medium-sulfur coal can usually tolerate a rather wide variation in the percentage of total SO_2 that is absorbed from the flue gas and still remain within emission limits. Since high-sulfur coal produces a flue gas richer in SO_2, an equivalent percentage variation cannot usually be tolerated, and closer control of pH is necessary.

11.7 CONCLUSIONS

In the last decade, FGD has emerged as the most important technology for the control of SO_2 emissions from combustion sources. The selection of an FGD process for a specific plant generally includes preliminary screening of applicable processes and final evaluation of the selected processes. The preliminary screening would

consider performance requirements, site-specific constraints, raw material requirements, and desired end products. The final evaluation should consider operational
experience with similar systems, control efficiency, energy needs, and costs.

The most advanced FGD processes from a commercialization standpoint are
wet lime and limestone scrubbing. Proper system design is crucial to successful
operation of any FGD system. Most of the operational problems of FGD systems
are related to two primary design considerations: mechanical and chemical. The
mechanical design considerations consist of materials of construction, process control, scrubber and recycle tank sizing and configuration, component redundancy,
gas and liquid flow distribution, and accessibility for operation and maintenance.
The chemical design considerations include recycle slurry pH, liquid/gas ratio,
reagent stoichiometry, degree of oxidation, and homogeneous precipitation of solids
in the recycle tank. Inadequate mechanical design will lead to plugging and erosion
of mechanical components in the scrubber and auxiliary equipment. Inadequate
chemical design, or loss of chemical control, will lead to scaling (deposition of
calcium sulfite/sulfate/carbonate solids) and corrosion in the FGD system. Many
of the problems are interrelated, and it is essential to recognize the key variables
in the design phase.

Control of calcium sulfite scale formation is achieved by controlling slurry pH
and reagent stoichiometry below a certain level and operating at a high liquid/gas
ratio and recycle tank residence time. Control of calcium sulfate formation is
accomplished either by coprecipitation, when the oxidation level is below 16% and
the scrubbing liquor is subsaturated with gypsum, or by circulating gypsum seed
crystals for supersaturation, when the oxidation level is above 16%. Supersaturation
can also be controlled by operating in a forced-oxidized mode. Magnesium additives
(MgO or $MgSO_4$) or buffer additives (adipic or benzoic acid) have also been used
to control scaling and improve SO_2 removal efficiency and reagent utilization.

Specification of proper materials of construction is another important consideration. The materials generally preferred for the scrubber vessel are 316 and 316L
stainless steel, although rubber or resin lining is also widely used. The capital-cost
advantage of lining should be weighed against the potential failure due to improper
application or contact with hot flue gases during upset conditions. Higher-grade
alloys are used for special areas in the scrubber. The use of natural rubber, PVC,
and FRP predominates in the liquid circuits of the FGD system. The materials for
stack lining when high-sulfur, high-chloride coal is burned require careful consideration, and no single material has appeared as the predominantly successful choice.

The process control systems using integration of automatic and manual control
techniques have proven to be much more successful than fully automatic systems.
The primary variables that require process control include gas flow, recycle slurry
pH, reagent feed rate, recycle slurry density, and overall water balance.

An FGD system is essentially a chemical processing plant; thus, the installation
philosophy used in the chemical industry should be followed. The rotating machinery, such as fans, pumps, and agitators, should be installed with sufficient free
space for maintenance. Easy accessibility is also essential for high-maintenance

instruments such as SO_2 analyzers and pH electrodes. The initial checkout and startup is a critical phase for an FGD system. The most important considerations in system startup are complete mechanical checkout of every component, prevention of hot gas flow through the scrubber, and rapid attainment of stable system chemistry.

During initial normal operation, the optimum values of process parameters should be identified, and organized operating and maintenance procedures established to hold the system close to the optimum conditions. In general, the optimum pH at the scrubber inlet varies between 8.0 and 8.5 for lime, and 5.5 and 6.0 for limestone systems. The optimum recycle slurry solids content is 8 to 10%, and 12 to 15% for the lime and limestone systems, respectively. Typical manual diagnostic checks include pressure in the recycle header, pressure drop across the absorber and the mist eliminator, SO_2 removal efficiency, pH and composition of recycle slurry and mist-eliminator wash water, and slurry solids content for the reagent feed, recycle, and thickener underflow streams.

It is essential that personnel be assigned exclusively to the FGD operation and maintenance team. These operators should be well trained in emergency and preventive maintenance programs. It is only through proper training of operator and maintenance personnel that the careful attention paid in design, construction, and initial operation phases will yield the long-term dividends of successful system operation.

11.8 REFERENCES

[1] DEVITT, T., ET AL. *Flue Gas Desulfurization System Capabilities for Coal-Fired Steam Generators,* 2 vols. EPA-600/7-78-032, March 1978.

[2] ELECTRIC POWER RESEARCH INSTITUTE, *Lime FGD Systems Data Book.* Final report, FP-1030, Research Project 982-1. Palo Alto, Calif., May 1979.

[3] ELECTRIC POWER RESEARCH INSTITUTE, *Lime/Limestone Scrubber Operation and Control Study.* Final report, RP-630-2. Palo Alto, Calif., April 1978.

[4] HEAD, H., AND WANG, S., *EPA Alkali Scrubbing Test Facility: Advanced Program, Fourth Progress Report,* 2 vols. EPA-600/7-79-244, November 1979.

[5] LASEKE, B., AND DEVITT, T., "Status of Flue Gas Desulfurization," *Chem. Eng. Prog.,* 75(2), 37–50, February 1979.

[6] *Proceedings of the Symposium on Flue Gas Desulfurization, Las Vegas, Nevada, March 1979,* 2 vols. EPA-600/7-79-167, July 1979.

[7] SLACK, A., AND HOLLIDEN, G., *Sulfur Dioxide Removal from Waste Gases.* Noyes Data Corp., Park Ridge, N.J., 1975.

[8] SMITH, M., ET AL. *EPA Utility FGD Survey: December 1978–January 1979.* Updated through July–September 1979. EPA-600/7-79-022, May–October 1979.

12

Stacks

Eldon R. Dille

Chemical Engineer Consultant
Burns & McDonnell
Kansas City, Missouri

12.1 DESCRIPTION OF CONTROL DEVICE

After the 1972 Clean Air Act passed the U.S. Congress, the world of coal-fired power plant stack design and maintenance entered a period of dynamic change.

Use of precipitators and flue gas desulfurization (FGD) units drastically altered the environment inside stacks. Gas escape velocities were cut in half, stack temperatures went down, and the incidence of acid condensation on the stack liner increased markedly. Stack maintainers found themselves with an increased rate of liner disintegration and failure, while stack designers found themselves specifying new liners that had no proven track record.

Part of the problem came from the scrubbers. During the 1970s, there were 15 major and 20 minor scrubber suppliers in the United States [1]. Each one produced several process design combinations. Engineers specified different processes from different manufacturers, so the power plant owner usually ended up with a scrubber system that resembled a Chinese dinner—one from column A, one from column B, and so on.

Another part of the problem came from the power production equipment. Each power plant had a different combination of boiler design, precipitator type, and associated pieces of equipment. The amounts of sulfur trioxide, sulfur dioxide, and nitrogen oxide compounds found in the flue gas varied with the new boiler designs. Also, the quantities of fly ash allowed to enter the FGD system varied, depending on the design of the precipitators. The final result was a flue gas at each plant that was probably like no other in the world.

Each flue gas did share one common thread—the creation of a stack environment conducive to sulfuric acid condensation on the liner wall. To reduce this acid attack, many engineers recommended installation of reheaters to get the stack temperatures back up beyond the sulfur trioxide dew point, but not everyone agreed to their use. Some engineers argued that gas- or oil-fired reheaters were too expensive to operate and too difficult to maintain; besides, there was another way to get stack temperatures up—divert heat directly from the boiler to the stack. This could be done economically and the final gases would still meet clean air requirements. Of course, this bypass approach would mean increasing the volume of corrosive acids in the stack, but "the stack could take it."

Perhaps the one thing that has not changed since the 1972 Clean Air Act is this common attitude that "the stack can take it." This attitude has been around a lot longer than this latest debate that generated it. Stack care and maintenance have always been a low priority for plant owners. The stack is a passive structure. It is easily forgotten until a crisis occurs. But the acidic nature of the environment found in modern stacks mandates a reexamination of this negligent attitude. Some stacks built before and since 1972 may not be able "to take it."

Advances in scrubber technology, which are being made at an exponential rate, may or may not solve many of the problems associated with wet, low-temperature flue gases. Even with the best precipitator and the most efficient FGD system, some percentage of sulfur dioxide, sulfur trioxide, hydrochloric acid, hydrofluoric acid, nitric acid, and so on, that result from coal combustion will end up in the stack. What these acids do there will affect your inspection and maintenance programs and your need for costly repairs.

A good portion of this chapter on maintenance concentrates on stack design and the corrosive effects of sulfuric acid on the stack and related equipment. Of course, you will find procedures for inspection, maintenance, and repair, but you will not find any operational procedures here. That is simply because there is nothing on a stack to operate.

12.2 DESIGN PROCEDURES

Since care and maintenance of a stack can be only as good as the maintainer's knowledge of it, let us start at the beginning and see what stacks are all about.

The function of a modern stack is to lift plant flue gases to a suitable elevation above the plant so that prevailing horizontal winds can disperse and help dilute the effluent contents.

Reinforced-concrete stack design is one of two types: boiler stacks designed to operate from 120 to 600°F, and manufacturing-process stacks designed to operate up to 2000°F. This chapter concentrates on the first type of stack and the changing nature of designs caused by air quality requirements.

To do its job right, the average boiler stack has to rise about 600 ft above the general plant elevation. Some stacks, depending on plant location and elevation, may go up only 350 ft, whereas others may rise to 1000 ft.

The stack height of conventional coal-fired power plants is determined from 44 FR 2608, dated January 12, 1979. This is a proposed federal regulation that places the stack height at the building height plus one and a half times the lesser height of the building or width projection facing the prevailing wind. For example, a power plant building 250 ft high by 200 ft wide by 500 ft long would have a stack height of 250 ft plus 300 ft or a total height of 550 ft.

A typical stack consists of a concrete shell designed to withstand the stresses and loadings associated with a 500- to 600-ft vertical structure, plus a lining of independent brick, steel or concrete, corbel-supported brick, or gunned concrete. The type of lining used depends on the temperature tolerances and corrosive resistance required and the seismic zone in which the stack is located. On the stack top there is a cap made of metal, ceramic, or fiberglass.

Because of the stack environments caused by clean air regulations, most modern stacks were built in a period of design flux. Stacks built for power plants with flue gas desulfurization systems are all state-of-the-art designs. As such, some will perform better than others.

There are numerous new designs and construction materials used in state-of-the-art stacks. In theory, each design will give satisfactory service for the life of the power plant. However, until they actually have been in service for 30 years, the state-of-the-art designs will have to be considered "new" designs.

Three "new" design techniques used in power industry stacks will be considered here:

1. One method that has proven its ability to handle all types of acid condensation while permitting complete flexibility of operation with any type of fuel is a double-column construction using a concrete outer column and a completely separate inner brickwork column. Proper brick and the use of sodium-free potassium silicate mortars can make the inner column resistant to acid condensation. The space between the two columns can be narrow or, preferably, sufficiently wide to permit inspection and repair. In this design, the acid-resisting requirements of the inner column and the physical requirements of the outer column can be separated. In addition, several methods can be used to prevent acid condensation from reaching and damaging the foundation.

2. Another variation of the double-column design is a concrete outer column and a steel inner column. The carbon steel column is insulated on its outer surface (facing the concrete column) to keep the inside surface of the steel above the acid condensation temperature. When this method is used, the

temperatures at the inside steel surface must not be permitted to drop below the acid condensation point. Should the steel surface temperature drop below its acid dew point, corrosive attack of the carbon steel will occur at a rapid rate. This corrosive attack could occur at a rate of ¼ in. of steel per month of operation.

3. Still another variation is the fiberglass liner inside a concrete outer column. These fiberglass liners have to be joined to the concrete outer column with metals that are resistant to the corrosive liquids inside the liner. This assures the structural integrity of the liner if the acid should permeate the liner. Special precautions must also be taken to allow for the high thermal expansion of the fiberglass liner. Should these expansion characteristics not be taken into account, there will probably be structural damage to the liner. Also, since the fiberglass is flammable, special fire protection systems must be installed and maintained to assure fire protection. In addition, the sulfuric acid concentrations and flue gas temperatures that will be on the surface of the FRP liner must be considered.

Stack designers use the American Concrete Institute's *Specification for Design and Construction of Reinforced Concrete Chimneys (ACI 307-69)* to plan for material, construction, and design requirements.* This specification spells out the basic design standards needed to fit the right type of stack to a particular power plant. Some designers criticize the specification as being too conservative in some areas, but deviations from it are rare.

ACI 307-69 "sets forth recommended loadings for the design of reinforced concrete chimneys and recommended methods for determining the stresses in the concrete and reinforcement resulting from these loadings."

Although the specification is on firm ground about its structural standards, ACI 307-69 admits that the sections on lining materials "do not guarantee the adequacy of these materials to resist temperatures, abrasion or corrosion resulting from chimney gases." As you might expect, the area where engineering experience is sketchiest is where most maintenance problems occur—the liner is usually a maintainer's major concern.

When a maintenance problem occurs, it may be tempting for you, the maintainer, to point the finger of blame at the designer or the contractor. You may be right in doing so, but in the vast majority of cases stack maintenance problems stem from the flue gases the stack must handle and/or the natural forces that a 500- to 600-ft concrete spire must face. If you experience the same maintenance problem repeatedly, however, it could be caused by a construction or design flaw. Report it. Your company probably has policies established to initiate corrective action.

---

*Note that the words "chimney" and "stack" are used interchangeably in this chapter. Individuals involved in the construction of this unit usually refer to it as a chimney; individuals involved in its use, particularly those in air pollution, refer to the unit as a stack.

The Stack Foundation

Stacks are often supported on pile foundations that support a concrete mat footing, usually a circular slab. A lead baseplate is located between the footing and the concrete slab. In most stack foundations, the slab provides a wearing surface and a drain for water that might collect at the stack base.

ACI 370-69 (3.10) says

In chimneys conducting heated gases, consideration shall be given to the protection against the radiant heat of the gases for any part of the foundation which is exposed within the chimney and concrete floors either at the base of the chimneys or supported from the chimney column.

It has been demonstrated that radiated heat from gases causes high temperatures at top surface of foundation or floor if not insulated. Even in modern power plants with moderate flue gas temperatures, this radiant heat may be sufficient to crack the chimney foundation or floor although flue openings enter the chimney an appreciable distance above the foundation.

A satisfactory protection for the foundation floor is a ventilated insulation in the form of hollow tile covered with insulating brick or fire brick depending upon the temperature involved, with the voids of the hollow tile communicating directly to the atmosphere by means of air ducts when the chimney is lined or by openings in the shell when the chimney is unlined.

Lead pans at the bottom of the flue(s) can be eaten away by acid attack. Since these pans are located below 1 to 1½ ft of concrete, attack of the lead pan by the acid will occur after the concrete has been attached. When a flue begins to leak, a subverted lead pan beneath the concrete may be the cause.

In brick-lined stacks, some mortars in the lining are designed to contend with acid attack but not water. During the annual maintenance shutdown or whenever there is an extended boiler outage, the top of the stack should be covered if there is a possibility that rain may enter the stack. Water in the stack without flue gas present can result in a corrosive attack of the mortar. Water can change the acid-resistant mortar into a paste that has no structural or cohesive strength. This is especially true at the base of the stack if water is allowed to collect. This explanation of the way the mortar fails is being reevaluated by the industry.

The Shell

All concrete stack shells built with ACI 370-69 recommendations are reinforced with vertical and circumferential reinforcing steel. The size, spacing, and amount of reinforcing bars used in construction depends on loading factors and the wall thickness required. Shell walls that are 18 in. or thicker have inner-face reinforcement of No. 4 bars on 24-in. centers vertically, and No. 4 bars on 12-in. centers horizontally. Some older stacks may not have inner-face reinforcing, so they are more subject to thermal and wind-load cracking than reinforced inner faces.

ACI 370-69 says: "The minimum shell wall thickness for any chimney with an internal diameter of 20 feet or less shall be seven inches. When the internal diameter exceeds 20 feet, the minimum thickness shall be increased ¼-inch for each two feet increase in internal diameter."

Since the external shell is concrete, it is resistant to most forms of mechanical attack (i.e., precipitation and wind). But eventually, every concrete shell will experience spalling. Thermal changes from normal day and night exposures can cause small cracks in the concrete. These will gradually increase in size until spalling occurs. In colder climates, water driven into these cracks can freeze and cause physical damage more rapidly. With the advent of FGD systems, there can be acid downwash along the top of the shell. This can result in chemical attack of the concrete. (Consider coating the outside of the concrete shell with a coating that is resistant to sulfuric acid attack. A coating based on polyester or epoxy resins could prove adequate in this environment. Fluoroelastomeric coatings also have the necessary acid resistance plus the advantage of bridging small hairline cracks in the concrete.)

Sulfuric acid in hairline cracks from downwash or internal infiltration can form salts and accelerate spalling. Structural failure may occur if the acid and water are allowed to reach the reinforcing bars. If the attack is allowed to go this far without detection, a major sectional repair or total replacement may be required.

Wind can also cause a cracking problem, or make an existing problem worse, especially in unreinforced areas of concrete. A timely system of crack inspections and a policy of instant repair will help you avoid the difficulty and expense of repairs or reconstruction.

Another form of mechanical attack is seismic disturbance. The shell design in earthquake-prone areas is usually based on the minimum lateral seismic force data given in ACI 370-69. A dynamic analysis based on the local seismic-response spectrum may also be a part of the design planning. A dynamic analysis gives insights into the local geology which may directly affect design decisions [2].

The Liner

Unlike the concrete shell, the brick liner is not built to precise engineering specifications. The type of liner used in your stack was chosen according to seismic conditions, the type of service (constant or intermittent peaking), gas temperature and gas composition in the flue, and the degree of reliability required. Structural requirements were, of course, a part of the liner selection.

The function of a liner is to accept the thermal stress in the stack column and to help protect the shell from the corrosive acids of coal-fired combustion. There are six basic types of liners: independent brick, steel, gunned concrete (gunite), fiberglass-reinforced resin (more commonly known as FRP), alloys, and cast-in-place concrete. The use of high-alloy metal for liners is nascent; cast-in-place concrete liners are all but extinct. A liner that was popular in the past—sectional brick liners supported by corbels in the concrete shell—is rarely found today because acid penetration can subvert the support corbels.

Each modern-era liner will be discussed separately, but first let us look at the environments a stack may encounter.

The major concern in liner design is acid formation that leads to corrosion. Acids generated in the coal combustion process include sulfuric acid, sulfurous acid (an early stage of sulfuric acid), hydrofluoric acid, hydrochloric acid, and nitric acid. The dew-points of sulfurous, hydrochloric, and hydrofluoric acids are well below the dew-point of sulfuric acid. Prior to the advent of FGD systems, these did not pose a threat to the stack. Now, temperatures in the stack can be within or below the dew-points of sulfurous, hydrochloric, and hydrofluoric acids.

Hydrofluoric acid is the Achilles' heel of inorganic monolithic gunned linings, potassium silicate mortars, bricks, and fiberglass fillers. Concentrations of this acid are beginning to be detected in linings and are being monitored to determine if they are reaching problem concentration levels.

Nitric acid, an oxidizing acid, will readily attack most components of the stack. All the corrosion products of nitric acid are water-soluble, so corrosion will proceed until the nitric acid is used up. Only scattered reports have been made of nitric acid presence in power plant stacks. It is expected that new boiler designs that minimize the formation of nitrogen oxide compounds will result in minimizing the production of stack-bound nitric acid.

Brick liners Independent brick liners, the most widely used type, evolved from corbel brick liners, the standard design of the 1930s and 1940s. The corbel liner consisted of 30- to 40-ft lengths of brickwork, usually 4 in. thick, that were supported by corbels (ledges) built into the liner face of the concrete shell. A space called an annulus, 2 in. or more, separated the lining from the column. The annulus was often filled with insulation to offset thermal stresses within the shell.

This corbel design worked fine when stack temperatures were 400 to 600°F and the stack was self-drafting. But after World War II, power plant designs became more efficient, stack temperatures dropped to 300°F, and the positive pressure inside the stack forced acid through the brick to the corbels and to the shell. The high alumina-based mortar used in the corbel brick liners began to soften as the pH reached 4.5, then the mortar failed completely at 3.5 pH. Mortars containing sodium compounds mixed with the sulfuric acid to produce growth crystals that swelled to 300% of their original volume. Because of these failures, the days of the corbel brick liner were numbered.

Replacing the corbel liner was an independent brick liner that was actually a stack built within a stack. The minimum annular space between the liner and the shell was increased to about 2 ft and newly developed acid-proof potassium silicate mortars were used. All the inherent structural strength and high-temperature resistance of the independent brick design was retained while acid penetration of the shell was reduced by the larger annular space. And fewer mortar failures due to acid attack were reported.

Then came wet flue gas desulfurization systems. Large quantities of sulfuric acid at different concentrations formed on the liner face as flue temperatures dropped to 130°F and positive pressure in the stack increased. The acid worked its way through the brick and attacked the steel tension bands used to control thermal brick expansion. It also attacked the buckstays supporting these bands, the testing doors and ports, and the concrete shell itself. To solve this problem, the annular space was pressurized so that the flow of acid from the inside of the stack to the outside of the brickwork would be stopped. To get an effective pressure in the annular space, a cap had to be put on the stack top. After the cap was made as leaktight as possible, all the doors and openings in the stack were sealed. This allowed the annular space to be pressurized with a relatively minimum-sized blower.

Not all modern-day independent brick-lined stacks are pressurized. Some power companies have chosen to closely monitor the shell conditions and pressurize only if there is evidence of acid attack. There is mounting evidence that the positive pressure in the stack from a flue gas desulfurization system causes acid migration through the brickwork and that a pressurized annular space may become a universal design requirement on future brick-lined stacks.

Independent Brick Liners

Brick liners in plants with FGD units should be built with ASTM Specifications C-279 acid-resistant brick. If a wet scrubber is used, the mortar should be sulfuric-acid and sulfonation resistant. Expect stacks made of standard brick and nonacid resistant mortars to fail from acid attack of the mortar once an FGD system becomes operational.

Since a brick liner is subject to positive pressure from the FGD system, there has to be some type of protection for the metal portions of the stack shell. Annular space pressurization and metal coatings that are acid and sulfonation resistant help prevent this acid attack. The stack cap and the exterior of the shell can also be damaged by the corrosive acids. Condensed gases can be carried down the outside of the shell by a draft and attack the concrete shell, the protective system, the navigational lights, access ladders, and service platforms.

Variations on the general theme drawn so far may exist in your independent brick liner. There are no strict design standards here; most of the techniques used in the design and construction of independent brick liners are a function of experience. According to a paper prepared by Custodis Construction Company,

> Excepting for the design procedures which we have applied for seismic conditions (which allow us some design control for areas of moderate seismic activity), and for reviewing stresses between openings, if two openings are built close together, there are no rational design procedures or codes available to enable us to design the work to the "limits" of safety. Therefore, a great

deal of the "design" of brick linings is based upon our in-house standards and "rules of thumb," which have been proven over a long period of time to provide a very stable and conservative brick lining, often lasting the lifetime of the plant [3].

It is almost inevitable that cracks will appear in a brick lining. These cracks are controlled by steel tension bands spaced regularly up the liner height.

Here's how the bands work: consider a typical brick liner with a 12-inch-thick wall and an internal diameter of 20 feet. Assume that a temperature gradient of 150 degrees F exists between the hot and cool faces. For these conditions, calculations show that the inner face of the lining tends to "grow" by only a fraction of an inch. Neglecting its own "growth," the steel band must resist this strain, or a portion of it, to control crack formation.

You can see that, unless the bands are perfectly tight and in contact with the brick at all points around the circumference, they will not be effective in controlling cracks. Too often, bands are installed with gaps between steel and brick large enough to pass a finger. Such bands are of absolutely no value [4].

Around openings in the lining, where the buckstay is supposed to accept the band loadings, this steel beam may or may not be stiff enough to do its job. A shear key (stepped bricks from the buckstay, to the band, to the chimney shell), helps "relatively light structural sections to transfer uniform band reactions directly to the brickwork as compressive stresses [5].

In a radial brick chimney, the vertical open space in the bricks is sometimes filled with mortar during construction. If this mortar is conventional calcium-based mortar, salts will be formed when acid migrates into the brick. Since the brick is usually in compression, the inside brick face will fall off due to the internal pressure generated by the acid/mortar reaction [6]. This internal spalling is difficult to detect while the stack is in operation and may lead to liner failure if it goes undetected for an extended period of time.

Sodium silicate mortars are also susceptible to salt's growth. The stress caused by a water reaction with the sulfated mortar can cause failure of the brickwork. Evidence of this failure is bending brickwork and loss of strength in the mortar. Brickwork in this condition is subject to any stress caused by vibration, including earthquakes and pressure changes in the boiler.

Two types of mortar are acceptable for use in liner construction. These are Class A Type 4 potassium silicate mortars and Class A Type 5 silica mortars. Most of the brick stack liners built in the United States since 1955 use potassium silicate mortars. These mortars have to be exposed to an acidic environment of a pH below 5.5. If the pH of the liquid on the brickwork goes above 5.5, these mortars begin to lose their desirable properties. This is the mortar type that may suffer damage at the top of the flue when exposed to rainwater. Remember, potassium silicate mortars were designed to withstand acid, not water.

Class A Type 5 silica mortars are not affected by rainwater or sulfuric acid. However, they are more expensive to install than potassium silicate mortars and all bricks must be wire-cut or scratched before laying. This helps compensate for a lower bonding strength [7].

A third type of mortar is emerging from European designs. This is a sodium-free potassium silicate which decreases the effects of water on the still curing mortar.

The type of mortar used in a brickwork lining depends on numerous factors. Table 12-1 shows what kind of performance you can expect from your mortar.

Steel liners Steel liners were introduced around 1960 to help power plant owners cope with the corrosion problems then being experienced with corbel brick linings. The stated benefits of a steel liner were a lower first cost than brick liners, resistance to earthquakes, and high resistance to corrosive forces [8]. Prior to passage of the 1972 Clean Air Act, the performance of steel liners met most expectations. There were some cases of liner buckling, but steel liner failures were very rare.

Steel liners were able to cope with acid attack because they were exposed to sulfuric acid in its concentrated form. Also, the temperature of the flue gases passing through the steel liners was above the acid dew point of the sulfuric acid in the flue gas. Some of the steel liners were given additional protection by coating the liner with inorganic materials that formed a monolithic, protective layer on the inner face of the liner. This liner within a liner was either formed, troweled, or gunited into place (guniting is a process that pressure sprays refractory concrete or calcium aluminate cement onto a surface).

With the advent of FGD systems, the first protection given steel liners was to coat them with organic resins, polyesters, and combinations thereof, many of which contained glass flakes and fibers. Rolling, brushing, and spraying were the techniques used to apply these coatings 60 to 70 mils thick [9]. As power plants began to bypass the scrubber units and stack temperatures went up beyond the organic coating tolerance, the incidence of coating failures increased.

The present use of organic coatings on steel linings (in stacks that handle scrubbed and bypass gases) is restricted to use of an organic lining material equivalent to or surpassing the capabilities of a fluoroelastomeric coating. As of this writing, fluoroelastomeric coatings are the only organic coatings that can withstand both the relatively high temperatures and high concentrations of sulfuric acid found in bypass gases and the low-temperature, high-humidity, dilute acid conditions found in scrubbed gases. Conventional polyester and epoxy coatings can withstand the latter condition, but not the former. Fluoroelastomeric coatings are capable of resisting temperatures up to 400°F continuously and 85% sulfuric acid concentrations without any adverse damage to the coating. This coating is not expected to give indefinite service life, however. All organic coatings are sacrificial in nature and will have to be maintained at some interval. The precise interval required between repairs will depend on the service conditions experienced by the coating, quality

TABLE 12-1 Power industry chimneys brickwork lining performance.

Resistances by Type of Mortar	A-5 Silica	A-4 Pot. Sil.	A-2 Sod. Sil.	A-2A Excess	D-1 Fireclay	D-2 Cal. Alum.
Degree resistance to sulphation by free acid	Unlimited life	Unlimited life	2–10 years	2–5 years	3 months–5 years	2 months–5 years
Degree resistance to combustion gases	Unlimited life	Unlimited life	Unlimited life	Unlimited life	Unlimited life	Unlimited life
Degree resistance to weathering (before use)	Unlimited life	2 years; needs 3 acid washes	2 years; needs 1 acid wash	2–5 years	Unlimited life	Unlimited life
Degree resistance to weathering (after use)	Unlimited life	Unlimited life	Unlimited life	Unlimited life	Unlimited life	Unlimited life
Degree resistance to continuous acid exposure	Unlimited life	Unlimited life	2–10 years	2–5 years	1 week	1 month
Degree resistance to intermittent operation	Unlimited life	Unlimited life	1–5 years	6 months–3 years	1–5 years	1–5 years

Note: This guide chart is for the selection of mortars for use in power-plant chimneys (stacks), based on the North American and European experience of Pennwalt Corporation.

of the application of the coating, and the compatibility of the coating with the actual chemical conditions inside the liner.

If a separate liner is used for the scrubbed gases, a conventional polyester resin-based coating can be used to protect the liner. This choice is based on the low-temperature, low-acid concentration found under this operating condition. The upper section of this liner can be made of a high-alloy metal to resist the corrosion caused by thermal upset conditions.

For a polyester resin to protect carbon steel in a power plant liner, it now appears that the moist flue gases should not go over about 180°F. This means that bypass unscrubbed gases cannot be passed through a liner coated with polyester resins.

A new design being used to minimize the attack of sulfuric acid on the stack is dual flue liners in a single concrete shell. The design has one steel flue lined with an organic coating optimized for the expected chemical environment for the FGD system. The other steel liner is for the bypass gases. Since this dual flue liner concept is unproved, it will have to be followed closely to see how the design actually performs in service. The bypass liner and FGD liner are fabricated of high-alloy metal for the upper exit section of each liner. This was done to minimize the corrosive attack of the down gases into each liner at the top of the liner. Should failure occur in these liners, it would probably be in the uncoated liner. The mode of attack would come from sulfuric acid deposited on the steel that is then diluted with water absorbed from the operation of the FGD system. To minimize the chance of this failure, the high-alloy upper liner section was used. There is the possibility of operating the bypass liner at a low gas flow level to minimize the effect of the wet cool gases entering it from the FGD liner. A fluoroelastomeric coating over carbon steel is being tested in the bypass flue. This coating could possibly be used to protect the carbon steel bypass liner should a protective coating become necessary. Another technique for handling the corrosion problems is to keep both liners in service all the time. This could be considered if the total emissions for both liners would meet the EPA requirements. The use of the dual-liner concept is being tried and will be closely monitored to determine if this is a viable alternative to an alloy liner.

Another liner uses dual materials of construction. The bottom portion of the liner is fabricated of brick with acid-proof mortar and the upper section is made of nickel-chromium high alloy. The concept behind this design is that the upper portion of the stack liner has the most severe chemical exposure and thus needs maximum protection.

There are three types of inorganic coatings used in steel liners: inorganic silicate, modified silicate base, and modified inorganic silicate base.

The first is an *inorganic silicate* composition which resists all acids (except HF), as well as water, oil, most solvents, and temperatures to 955 degrees C. It is suitable for use over a pH range of 0.0 to 7.0. This acidproof concrete may be applied by guniting, pouring, or forming. During the more than 23 years it has been on the market, millions of pounds have been used successfully to withstand corrosion, abrasion, and elevated temperatures.

Now there is a new generation of corrosion resistant inorganic monolithic linings available. The first is a *modified silicate* base which is a single component system, a powder to be mixed with water. It hardens within 24 hours at 20 degrees C to produce a high strength corrosion resistant lining with extremely good adhesion to steel, concrete, and brick. It is not affected by exposure to acids (except HF), mild alkalies, water, oil, solvents, and temperatures to 955 degrees C and is suitable for use over a pH range of 0.0 to 9.0.

The second product is a *modified inorganic silicate* base, which is thermally insulating and light-weight. It is also a single component system, a powder to be mixed with water when used. This material develops a compressive strength of 3900–4200 psi and a tensile strength of 400–500 psi. It has a coefficient of thermal expansion of 7.1×10^{-6} which is very close to that of steel, thereby minimizing cracking due to temperature changes. It weighs approximately 98 pounds per cubic foot and is considerably lighter in weight than most other corrosion resistant linings, thereby reducing the dead load. It has a *"k"* factor of 2.25–2.50 and will provide a temperature drop of approximately 50 percent per inch of thickness. In other words, a 2-inch thickness with the hot face at 260 degrees C will drop the temperature on the cold face to 130 degrees C. It has excellent resistance to abrasion and a bond strength of 300 psi to steel. It withstands temperatures to 955 degrees C and is unaffected by water, moisture, steam, acids and alkalies in a pH range of 0.0 to 9.0 [10].

The acidproof concrete lining, while not completely impervious to the sulfuric acid, has a low initial effective porosity and acts as an efficient filter. Due to this low effective porosity the penetration rate of the corrodent to the surface of the base metal is greatly reduced. This slow penetration allows the acid reaction to proceed to completion with the alkaline concrete lining material to form sulfates or anodic inhibitors. The sulfates fill any available pores in the lining. At the same time a cementitious form of silicic acid is also precipitated from the matrix and the reaction reduces the porosity of the lining and makes it a stronger and denser physical barrier to the acid. When the acid eventually reaches the base metal there is no fluid flow and so no kinetic motion by which deposits may be carried away. Since the initial acid reaching the metal surface will be weaker, it will react more slowly and thus form a protective film of insoluble iron salts on the face of the metal. These salts further inhibit the corrosive action and form an integral cathodic inhibitor barrier held in place by the concrete lining. [11]

As long as the monolithic acidproof concrete lining maintains its physical integrity and remains intact, there is no corrosion at the interface of the base metal. To have corrosion it is necessary for oxygen, water, or an electrolyte to be present, so if the lining is broken or cracked by accident, oxygen can then reach the metal and cause localized cathodic corrosion and damage to the metal shell [12].

Another design aspect of interest to you is the use of insulation on the steel liner exterior. This insulation is installed to prevent heat from escaping through the liner. If more heat is retained in the stack, the chances of condensation on the liner's interior are reduced. The insulation also helps keep thermal stresses on the concrete shell within tolerable levels.

Most insulation is attached to the steel liner sections on the ground before the liner sections are assembled. The spun-glass insulation board is installed by centering stick pins into the boards no more than 16 in. apart and no more than 4 in. from the insulation board edges. A stud welding gun secures the boards to the liner. Wire mesh is placed over the insulation board, then stud-welded into the liner. The mesh ends are laced together where they overlap [13].

Sometimes, tape is used instead of mesh to secure the insulation boards and/or to seal joints. The selection of the tape or the wire mesh should be made after consulting with the plant engineer.

Fiberglass liners The lining materials of construction for power plant liners alternated between carbon steel and brick during the pre-FGD era. With the lowering of the flue gas to around 120°F without reheat and to 180°F with reheat, a new lining material was needed. In an attempt to find a new material in the 1960s, a coating of fiberglass-reinforced plastic coating was applied to steel liners. This composite material experienced some failures. However, the coating did withstand the corrosive liquids found in the liner.

Based on the chemical resistance of the plastic itself, a stack fabricator erected the first fiberglass-reinforced plastic stack liner in 1968. Installed at the Electric Reduction Company plant in Long Harbour, Newfoundland, the liner was designed to resist chemical attack from flue gases from an oil-fired boiler and hot gases from a phosphate production plant. The flue gases contained hydrochloric acid, sulfuric acid, hydrofluoric, and fluorosilic acids. Numerous FRP liners were then installed in various industries [14].

Finally, in 1974, the first fiberglass-reinforced plastic liner (FRP) was installed in the power industry. This first liner was 805 ft high and 15 ft in diameter. It was installed at East Kentucky Rural Electric Company's Spurlock Station, Unit No. 1, Maysville, Kentucky. This liner was completed in 1976 and went into service in 1977 [15].

The latest generation of fiberglass liners have helically wound fiberglass-reinforced plastic linings that have an angle of helical windings that change with the elevation of the fiberglass in the stack. This angle change provides the axial and hoop strengths required [16].

Fiberglass liners have several advantages over conventional brick liners. An outstanding feature is their chemical resistance to most liquids found in the flue gas. Another desirable feature is the resistance to high lateral forces from earthquakes.

FRP liners are not without problems, however. Usually, the FRP liner has a higher expansion due to heat than carbon steel. Since the FRP liner has a lower physical strength than steel, provisions must be made for additional structural support. Not to be overlooked are the sulfuric acid concentrations and temperatures expected on the surface of the liner.

Several layers are required in FRP liners to achieve the desired physical properties. The outer layer will contain an ultraviolet inhibitor and be essentially all resin. Next, the structural layer is fabricated of 60 to 65% fiberglass filament and 35 to 40% resin. The helix angle of the fiberglass filament will vary depending on the location of the section in the stack. (On one liner this angle varies from 8 to 30°.) Next comes the interior layer. This layer consists of 20 to 30% glass and 70 to 80% resin. Finally, the innermost layer is the chemical barrier layer. This layer is 100% resin, usually, but may contain some glass veil. Since this layer is the one that conducts the static charge from the liner, it is filled with conductive carbon pigment. It is important that a continuous electrical path be made between the inner carbon-filled layers. This continuous electrical path can be made by physically connecting each section on the inner layer with woven grounding cables. These cables should be designed to withstand the corrosive liquids expected in the liner.

One of the weak points in the design of an FRP liner is the fiberglass filament. This filament is an excellent structural material unless the flue gas that passes through the liner contains a significant amount of hydrofluoric acid. Hydrofluoric acid in high concentrations will readily dissolve glass and thus cause failure of the lining. Should the hydrofluoric acid concentration be excessive, a synthetic fiber such as Dacron should be considered for the main structural filament. Since the fiberglass liners are usually designed to operate after FGD systems, there is a high probability that the hydrofluoric acid will be removed by the FGD system and will not appear in the fiberglass liner in significant quantities. This should be determined on each individual FGD system before a design contract is issued.

Since a tall fiberglass liner cannot support its own weight, special design considerations must be made. Several techniques have been developed for compensating for the weakness of FRP liners. One of them is a counterweight and pulley system. This patented system allows the liner to expand and contract without incurring structural damage to the liner. Another support technique uses intermediate supports with expansion joints between the individual sections. This keeps the compressive loads on the liner below the design limits.

One of the major considerations in the design of the FRP liner is the temperature limitations on the liner. These limitations are determined by the resin used in the liner fabrication. Failure to observe the temperature restrictions of the resin used in the fabrication of the liner results in structural failure of the liner. Excessive temperatures in the liner can cause softening of the resin and result in bending of the liner.

A major concern in the fabrication of FRP liners is flammability. The addition of a fire-retardant chemical (antimony trioxide) to certain layers of the FRP liner will retard a fire initiation in the liner. And once the fire begins, the antimony trioxide continues to restrict the rate of burning. It is standard practice to install a water spray system in the ductwork ahead of the fiberglass liner. This emergency water spray system is activated by a present temperature excursion in the FRP liner. The precise temperature that activates the emergency water-quench system should be selected based on the physical properties of the resin used in the fabrication of the liner.

The precise design used for introducing the quenching water into the FRP liner is important. Care should be taken in the design to ensure that the fire-protection quenching water does not directly inpinge onto the FRP liner. Direct impingement can result in liner damage. Whenever the emergency fire protection system has been used, an inspection of the FRP liner should be made for possible thermal damage.

Fiberglass liners with flue gas passing through them act as a static electricity generator. This makes the electrical continuity of the grounding system very important. The inspection and maintenance of this grounding system cannot be overstressed. It would be helpful to have a static electricity meter to use when working around the FRP liner. This instrument allows the inspector and workers to determine if the grounding system is operating before they approach the liner.

One stack designer has developed a table for the visual inspection and repair requirements of FRP stack liners. This table is included in the section on inspection.

Alloy liners   An increasing interest is being shown in the use of high-corrosion-resistant alloys for the fabrication of stack liners. Since each coal and FGD system results in differing corrosive liner environments, the precise alloy being used varies. The most often used materials are nickel and chromium alloys with a 6% or higher molybdenum content.

These materials are known for their resistance to sulfuric acid corrosion and a varying resistance to chloride pitting. The precise chloride resistance of each alloy is dependent on the pH together with the temperature and chloride concentrations the alloy will experience.

The use of alloys is so new that there are no design techniques unique to alloys, and the designs being used have not settled on any fabrication techniques. Most designs have consisted of replacing the carbon steel used on non-FGD boilers with corrosion-resistant alloys.

A test that can help determine the proper alloy for corrosion resistance is the coupon test. This test involves a set of various alloys assembled on a corrosion coupon rack as per ASTM G4-68. Installation of these racks in the liner at various places of concern allows an in-place test of the corrosion effect on a variety of alloys. By checking the corrosive attack of each alloy sample, the most economical alloy for the liner can be selected. It is a practice to select a slightly higher corrosion-resistant alloy than the coupon tests indicate. The ultimate evaluation of the alloys and materials of construction is to fabricate the test material into the wall of the liner. The resulting tests should be identical to the results obtained when the new construction material is put into service.

The corrosion coupons are usually round in shape and are electrically insulated from each other by Teflon spacers. The rack used to hold the coupons is fabricated from an alloy considered to be sufficiently resistant to the chemical environment of the liner to remain intact for the duration of the test. The coupons are measured for thickness and weighted before and after the exposure test. The corrosion rates can be determined by use of the formula

$$p = \frac{W(365 \times 10^3)}{(2.54)^3 A d t}$$

where p = average penetration, mpy (mils per year)

           d = density of metal, g/cm^3

           t = time of exposure, days

          W = weight loss, g

          A = specimen surface area, in.^{2}

The corrosion coupons can be installed on the wall of the liner or on special racks installed in the liner ports. The precise location depends on the time of the test, probability of the unit being off line for removal of installed coupons, and the configuration of the lining. The tests can run from 1 to 6 months (the normal test is long enough for 10 to 15% weight loss of the coupons).

These test coupons will not corrode at the same rate as alloys installed in the lining. The coupon test results should be used for comparative purposes (i.e., comparing one coupon with another). Since the results obtained from these coupons are not available quickly, it is a good idea to install the test coupons when the lining is first put into service. Then, if any liner problem should occur, there are some data indicating what a suitable alternative might be.

At least one power plant has an entire stack lining of ASTM B443–72 alloy. This alloy was chosen after extensive stack testing of various alloys. Sections of several high alloys were welded onto the base of the liner and exposed to the various operation conditions of the boiler and FGD system. With the particular coal and operating parameters found in this stack, ASTM B443–72 alloy was the most corrosion-resistant.

Other power plants are being fabricated with high-nickel/chromium/molybdenum alloys as just a part of the stack. Some stacks have high alloy where the metal parts contact the brick liners; others have a partially or completely fabricated liner of high alloy.

To assure that the drains and various metal parts of any stack with FGD systems are not corroded by the environment of the scrubbed gases, some engineers specify that all parts of the stack base be fabricated of high nickel-chromium-molybdenum alloy. The baseplate of the stack has been fabricated of lead for some plants.

Inside many carbon steel liners coated with protective organic fluoroelastomeric coatings, the top section of the liner has been fabricated of high-alloy materials. The particular alloy used depends on the precise chemical conditions extant. Other carbon steel liners coated with the protective organic fluoroelastomeric coatings have the coating extended to the top of the liner with no alloy to protect the top section of the carbon steel metal liner. The fluoroelastomeric coating may also be used on the outside of the carbon steel liner above the cap.

The top platform of many new stacks uses high alloys. Ranging from 316L stainless to the nickel/chromium/molybdenum alloys. One of the difficulties experienced by designers has been finding standard fabricated alloy parts. Some substitutions are being made with the free knowledge that they will have to be replaced at frequent intervals due to corrosive attack.

Conventional materials used on some of the new FGD stacks have a short life span. In less than 1 year of operation, conventional galvanized steel at the top of the FGD stack may have the galvanized coating completely removed by corrosion. This results in rapid attack of the remaining steel and failure of the entire structure (working platform and operating warning lights). Each FGD system and its respective coal should be carefully evaluated to determine the chemical environment that will be found inside the stack liner and at the top of the stack.

There will be acid penetration through the brick liner to some degree even with a pressurized annular space. This migration of acid can occur when the seals on the doors in the chimney are not kept airtight or properly sealed. Any migration of acid into the bands (conventional carbon steel) around the brick liner will result in rapid attack of the carbon steel metal. Failure of the bands will allow the brick liner to expand when it cracks and can result in total failure of the brick liner. The use of the proper protective coating or lining in steel bands will assure continual retention of the brick liner.

One concept in alloy fabrication being examined for possible use in the stack is explosion-clad carbon steel. The explosion cladding process allows the primary structural strength of the liner to be supported by carbon steel. The carbon steel base metal in the cladding is on the nonwetted side of the liner. Thus, standard construction techniques can be used between the liner and the shell of the stack. However, on the wetted side of the cladding, high-nickel alloy is used. This high-nickel alloy is selected to have complete chemical resistance to the expected chemical environment in the liner. It is important that the cladding alloy have sufficient thickness to resist the corrosion of the chemical environment in the liner for the life of the power plant. Otherwise, corrosion will destroy the alloy and the carbon steel will rapidly disappear.

Special fabrication techniques are required when constructing a liner with the explosion-clad alloys. A technique that has been tested for corrosion resistance consists of welding the alloy side of the cladding with a three-pass weld of high-alloy welding wire. The use of cladding allows a given amount of high alloy to be used in the fabrication of a number of liners.

Other types of cladding can be used for the fabrication of clad liners. These primarily consist of rolling a corrosion-resistant alloy and carbon steel through roller mills to form a single-bonded metal. The welding techniques required to join these sheets of clad metal would basically be similar to those required for joining explosion-clad metal sheets.

The fabrication of stainless steel and nickel-based alloys presents a completely new problem for the power industry. The fabrication techniques previously used with carbon steel are not necessarily applicable to alloys. As a matter of fact, it can be assumed that welders will have to be retrained to assure a satisfactory fabrication of the high alloys. Nickel alloys require the use of special welding techniques.

The remainder of this section is a discussion of some of the characteristics and techniques associated with corrosion-resistant alloys, as noted by the Technology Division of High Technology Corporation [17].

Hot Cracking

Welding nickel-based alloys can be a major problem because of "hot cracking." Hot cracking is caused by stress and the presence of a "strain-intolerant microstructure." Since stress of the metal is a natural occurrence during welding, care must be taken to minimize the stress by proper weld design and proper welding technique. The minimization of the strain intolerant microstructure can be achieved by minimizing the heat generated during the welding process. This is done by keeping the current required on the welding machine as low as possible and using a proper-sized electrode to maintain the minimum current load. Excessive heat input will increase the amount of metal exposed to the crack-sensitive temperature range, and increase the duration of the elevated temperature exposure. A common source of excessive heat input is from "weaving" the electrode instead of using the preferred stringer bead welding technique. The weaving should not exceed three times the diameter of the electrode being used; however, the best precaution is to rely on sound welding practice and good common sense. A welder has described the proper welding current for alloys as the current required to weld so that it feels like you are taking the metal from the electrode and putting it on the metal.

Fusion-Zone Hot Cracking

Bend contour and shape are important factors in minimizing fusion-zone (weld metal) hot cracking, especially when using butt welding plate of pipe which requires multiple passes. The individual stringer passes or beads should be deposited with a distinct crown. Weld beads that are deposited with a convex or flat surface are more susceptible to centerline cracking. If coated electrodes are used to make the welds, all the slag should be removed between passes so that there will not be any slag inclusion in the weldment. All flaws in the weld should be removed before another pass is placed in the welding area.

Cracks in the alloy can also occur at the edge of the heat-affected zone. These cracks are caused by excessive welding heat input coupled with stresses associated with welding.

Some manufacturers recommend that no welding of any kind occur if the metal temperature is below 32°F. Ideal welding temperature is 72°F. Interpass temperatures should not exceed 200°F. No cooling techniques should be used between passes that can introduce oil, grease, dirt, or contamination into the weld. This includes shop air, oil-soaked rags, hard-water deposits, and so on. For best results, allow the metal to cool naturally.

The surfaces of all areas adjacent to the welding joint should be cleaned to bright metal prior to welding. This means that all grease, oil, crayon marks, dirt, magic marker marks, and so on, should be removed by wiping the area with the recommended solvent and lint-free rags. Failure to do this can result in porosity and increased susceptibility to corrosion at the weldment.

Coated electrodes should be stored in sealed moisture-resistant containers until they are used. Electrodes that are removed from the original containers should be heated at least 15°F above the ambient temperature to assure that the electrodes stay dry. If the electrodes get wet, they should be dried as recommended by the manufacturer. One manufacturer recommends that the electrodes be dried at 250 to 300°F for at least 2 hours before they are used.

Porosity can occur in shielded metal arc (SMA) welding when welded vertically. This probably occurs because gravitational forces prevent the molten slag from properly protecting the weld pool. Another cause may be the breaking of the molten slag layer by improper rod movements. As much as possible, the SMA welding should be done in the horizontal position. Check with the supplier of the alloy and the welding rod to determine the proper rod for vertical welding if it is required.

Two other techniques are also used in welding alloys. These are gas tungsten arc (GTA) and gas metal arc (GMA). In both processes, the gas shielding the molten pool should precede the arc initiation by several seconds. Also, the gas flow should not be excessive. Excessive gas flow can result in porosity of the resulting weld. This excessive gas flow causes air to be brought into the weld puddle region. Should the gas flow be too low, there is insufficient gas flow to properly protect the molten pool, and air is brought into the molten pool. When welding with either GTA or GMA, care should be taken to keep the diffuser screen clean and free of splatter.

Other Shielding-Related Problems

Another problem related to proper shielding is the inability to deposit a successful root pass, especially when butt welding pipe. The underside of a root pass should be adequately shielded with inert gas or else the weld bead will not "wet" and "flow" to produce the desired penetration effect. Proper shielding should also include the use of purge dams. When initiating the root pass, a temporary covering of tape over the remainder of the joint will minimize the loss of purge gas through the root opening.

The surface of weld beads deposited by the gas tungsten arc process and gas metal arc process will occasionally appear to have very thick oxide rafts, sometimes referred to as "slag spots." Although these artifacts may be questionable from a cosmetic viewpoint, they very rarely interfere with the subsequent weld passes or with overall integrity of the welded joint.

Repair of Welds

All rejected welds should be removed by a suitable grinding method. Remelting of the weld beads or depositing additional weld beads over the defect is definitely not recommended.

In joining a high alloy to carbon steel, an alloy welding filler material should be used that will give a final metal composition that is equal to or better than the parent metal.

One of the alloy manufacturers has provided a suggested list of fabricating tips when working with corrosion-resistant alloys. These tips are guides for maintaining the corrosion-resistant properties of the alloys [18].

DO:

Follow acceptable welding practices as set forth in the welding section of the ASME Boiler and Pressure Vessel Code.

Use filler metal of the same composition as the alloy base metal.

Achieve complete penetration of the weld to avoid crevices which might lead to accelerated corrosion and stress raisers in the fabrication.

Remove all traces of foreign material from the alloy before heating to above the black heat range when welding or for any other reason.

100% dye-check all welds to assure crack-free joints for corrosion applications as required by ASME Boiler and Pressure Vessel Code.

10% X-ray where required, to conform to code.

Solution-heat-treat after fabrication if the alloy and the corrosion application warrant it.

When heat treating, use fuels of low sulfur content to avoid sulfur penetration and subsequent detrimental effect on mechanical properties and corrosion resistance.

DO NOT:

Allow direct flame impingement on the alloys.

Overheat when solution-heat-treating or hot working. Overheating of some of the alloys by as little as 25 to 50°F above the recommended solution heat-treatment temperature can cause incipient melting.

Age the alloys for prolonged periods of time in the 1100 to 1800°F range. Aging causes carbide precipitation in the grain boundaries, which leads to loss of ductility and greater susceptibility to corrosive attack. If aging does occur, solution heat treatment at the recommended temperature will restore much of the original ductility and corrosion resistance to the alloy.

Use oxy-acetylene for welding the alloys unless recommended by the manufacturer.

Use brazing alloys that contain copper.

Make intermittent cuts when machining these alloys. They have a tendency to work-harden, which makes subsequent cuts much more difficult.

Use shot blasting to clean the alloys. Iron particles that are embedded in the surface of the alloys can cause serious product contamination and/or severe corrosion problems in some media.

12.3 SULFURIC ACID ATTACK

Sulfuric acid is primarily responsible for the corrosion attack on any material used in the stack liners. It is not the only acid you might find eating away at your liner, but it is the most plentiful and potentially the most destructive. This acid comes from the sulfur trioxide and sulfur dioxide produced in the boiler during fuel combustion. Other acids that can increase the corrosion rate of sulfuric acid are hydrochloric and hydrofluoric acids.

The Stationary Source Performance Standard of 1979 allows 10% of the sulfur dioxide emissions from solid and solid-derived fuels combustion to pass through the FGD system into the stack. Since this sulfur dioxide is saturated with water when it leaves the wet FGD system, there will be sufficient moisture to make an acidic solution in the stack.

There is also sulfur trioxide present in the flue gas when it comes from the boiler. Depending on the sulfur content of the coal, oil, or other fuel, the boiler design, vanadium content of the ignitor oil, vanadium pentoxide deposits on the boiler tubes, pattern of varying loads on the boiler, and so on, various amounts of sulfur trioxide can be produced. The Shawnee Pilot Plant tests have shown that about 58% of the sulfur trioxide present in the unsaturated flue gas is removed by the FGD system. Some sulfur trioxide is also removed in the cold-side electrostatic precipitator. At some plants, if the fly ash being collected in the electrostatic precipitator has a very high electrical resistivity, it is sometimes the practice to add sulfur trioxide before the precipitator to improve the removal efficiency of the unit.

Sulfur trioxide that passes through the precipitator and FGD system goes through the demisters and into the stack, where it is available to make sulfuric acid. The quantity of sulfuric acid will depend on the precise amount of sulfur trioxide present in the treated flue gas.

Sulfur trioxide has a high affinity for water. When the sulfur trioxide concentration goes below its dew point, sulfur trioxide reacts with water vapor in less than 1 second to form minute droplets of sulfuric acid. These droplets will be very concentrated, approximately 98% sulfuric acid, but they will continue to react with water until the concentration on the liner is lowered to its boiling point equilibrium concentration. This medium-concentrated acid will either react with components of the flue lining or continue to absorb water and form more dilute sulfuric acid. The pH value of dilute sulfuric acid that finally forms after a wet FGD system is 0.5 to 2.2.

Sulfuric acid can be found whenever the flue gas temperature drops below 400°F. In the past, flue gases from the preheater were around 350°F. Even though this temperature is below the maximum acid condensation temperature, there were only minimal corrosion problems. In the contemporary stack environment of about 130°F, copious amounts of sulfuric acid are formed when sulfur trioxide gas goes below its dew point.

There has been considerable discussion about what the dew point of sulfuric acid is. It can be defined as the temperature at which liquid sulfuric acid forms in equilibrium with the sulfuric acid in the vapor state. Research to develop some instrument to accurately determine the sulfuric acid dew point has produced several mechanical techniques. These instruments generally determine the sulfuric acid dew point by the use of electrical conductivity probes. These probes are put into the flue gas stream and then slowly cooled. At the point that sulfuric acid and water condense on the probe, a sudden change will be seen on the meter, indicating the conductivity change across the probe. The temperature at which this sudden change in conductivity occurs is the dew point of the flue gas. Since sulfuric acid is an unusual compound, this dew-point temperature does not mean that liquid sulfuric acid will be seen on the walls of the stack.

Care must be taken when using these instruments to assure that a value is obtained indicative of what is happening in the liner. Gibson and Kufta suggest that the corrosion rate with sulfuric acid increases from the theoretical dew point of the acid to a temperature about 85°F below the sulfuric acid dew point [18]. From this temperature to the dew point of water, the corrosion rate decreases. Below the water dew point, the corrosion rate rapidly increases. This lower condition is the primary area of interest for scrubbed flue gases.

To illustrate the effect of trace amounts of sulfur trioxide on the dew point of sulfuric acid, here is an example taken from a paper by Robert R. Pierce:

Thus, if a combustion gas containing as much as 10 percent water vapor by volume also contains as little as 40 parts per million sulfur trioxide by volume (0.004 percent), it has a weight percentage of sulfuric acid (based on sulfur trioxide and water alone) of only 0.22 percent. The vapor phase in equilibrium with the liquid phase has a concentration of 82.5 percent by weight sulfuric acid at its boiling point at 0.1 atmosphere pressure which is 148 degrees C (298 degrees F). Thus, as soon as the temperature of such a mixture of 10 percent water vapor and 40 parts per million sulfur trioxide by volume drops below 148 degrees C, liquid must form because the gas becomes super-saturated with 82.5 percent sulfuric acid. This powerfully acidic liquid will condense on any solid surface which is cool enough, which is why it is called the dew point or sometimes called the acid dew point. If the gas is cooled below the dew point temperature by radiation, etc., a mist of acid droplets will be formed. It may seem strange that a gas containing a mixture of sulfur trioxide and water corresponding to only 0.22 percent by weight of sulfuric acid will form a droplet or a film on a surface containing 82.5 percent by weight of sulfuric acid but such is the nature of this acid. Of course, only a very small amount of the water in the gas is condensed out with the acid [19].

As was mentioned earlier, considerable work has been done to develop some type of instrument to accurately determine the sulfuric acid dew point. But in the literature reviewed, there seems to be no agreement about the accuracy of the available mechanical techniques. There is, however, a method for determining the sulfur trioxide concentration in the flue gas. Method 8 from the *Federal Register,* Appendix A-32, can be used to determine the sulfur trioxide concentration. The concentration determined by the EPA Method 8 is in grams per dry standard cubic meter or pounds per dry standard cubic foot. This value can be converted to parts per million (ppm) by use of the equation

$$SO_3 \text{ (ppm)} = (4.82 \times 10^6) \, C_{SO_3}(\text{lb/dscf})$$

To use the data presented in Figure 12-1, a moisture determination needs to be made. This can be done from EPA Method 8, paragraph 4.2. This moisture determination will be used to determine the mm Hg water for use in Figure 12-1. The conversion of the volume of water collected in Method 8 converts to mm Hg by the equation

$$\text{water (mm Hg)} = 760 \frac{V_{w \text{ std}}}{V_{m \text{ std}} + V_{w \text{ std}}}$$

where $V_{w \text{ std}} = 0.0474 \text{ ft}^3/\text{ml } (V_{lo})$

$V_{lo} = $ ml of condensate collected

$V_{w \text{ std}} = $ equivalent cubic feet of moisture in vapor state at STP

$V_{m \text{ std}} = $ dry gas measured in cubic feet at STP

Not only is sulfuric acid present in a dilute concentration in the lining, but at the low temperatures after the scrubber, hydrochloric acid can also be present. If there is some bypass of the unscrubbed flue gas and mixing of this unscrubbed flue

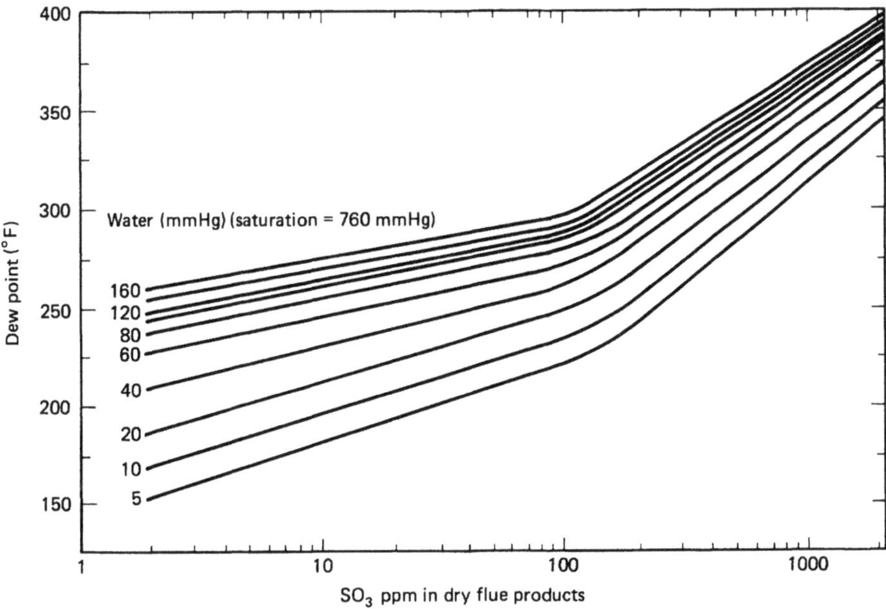

FIGURE 12-1 Determination of SO_3 concentration in flue gas.

gas with scrubbed flue gas, hydrochloric acid will be present in the liner of the stack. A set of conditions found in one stack that burned pulverized coal with 2.67% sulfur and 0.066% chlorides with no scrubber were discussed by Robert R. Pierce and D. J. Kossler. Their data are presented in Table 12-2 [20].

From these data we can determine that hydrochloric acid can exist after a wet scrubber. The precise concentration of the hydrochloric acid will have to be determined and will probably not reach the concentrations shown in this table.

In 1962, Pennsalt Chemical Corporation began a program in 11 chimneys to determine the sulfate content, free sulfuric acid, and pH. Their data are shown in Table 12-3 [21].

The data from Table 12-3 shows that the same acidic condition exists throughout the entire height of the lining. In the case of brick liners, it seems that the damage

TABLE 12-2 Conditions in stack burning pulverized coal.

Temperature (°F)	Sulfuric Acid (%)	Hydrochloric Acid (%)
242	64.6	—
233	48.8	—
244	43.8	—
124	38.8	8.8
106	12.8	17.2
90	0.6	0.4

TABLE 12-3   Location of acidity in chimney linings.

Height	Sulfate in Mortar (%)	Free Sulfuric Acid in Mortar (%)	Acidity of Mortar (pH)
Top level	31.8	10.0	2.3
75% level	31.6	9.7	2.0
50% level	28.9	14.7	1.8
25% level	32.1	12.9	1.9
Bottom	21.3	10.1	2.0

from the acid attack is more evident at the top of the lining. This is probably due to the fact that there is less weight on the bricks at the top of the lining. Also, there is more bending stress at the top of the lining. Another consideration is that the top of the lining is more likely to be exposed to rain from the environment, with resulting sulfation and hydration of the mortar.

One of the operating conditions that has become more apparent with the advent of flue gas desulfurization systems is the formation of acid smut in the stacks (chimneys). Gills and Lees reported:

Based on many years' study of this problem, including examinations of the inside of chimneys, microscopic examination and analysis of the deposits, and examination of the smuts, it has been concluded that acidic smuts consist of large agglomerates of black spheres bonded together with sulfuric acid or acidic salts.

Smuts form where chimney surfaces are below 250 degrees F and occur chiefly with metal stacks, but they have also been observed with concrete and brick stacks. It is possible that they can also form in ducts at the base of stacks.

It is believed that the four stages in the formation of smuts are:

1. Sulfuric-acid solution is first deposited from the flue gas stream on to the cool surface where it reacts with the metal and forms an acidic salt.

2. This is a sticky layer and consequently soot particles adhere to it. Careful control of combustion conditions and use of good burners will reduce the concentration of these carbonaceous particles, but they will not be entirely eliminated. Hence smut emission will be reduced, but not eliminated by these measures.

3. Part of the acidic layer containing carbonaceous matter then flakes off due to a change in load or temperature.

4. The detached flakes break up into agglomerates or smuts. Some fall to the bottom of the chimney (where they form a thick acidic sooty layer) and some are carried out by the gases into the atmosphere. Owing to their large size and weight they fall near to the stack. These appear as large black

smuts, often shaped like a cone up to one-half-inch diameter. Because of their acidic nature they will:

a. Damage the cellulose finish on cars.

b. Damage fabrics, particularly synthetic fibers, such as nylon.

c. Cause a brown stain on fabrics and asbestos roofs due to the acidic iron salts present in the smuts.

d. Damage leaves of plants.

e. Appear as sticky black smudges on surfaces on which they fall. (This is often mistakenly described as an oil-smudge. It is not oily, but acidic.)[22]

Up until the advent of flue gas desulfurization systems, the basic technique to minimize the formation of acid smut was to operate the stack above the sulfuric acid dew point. Now that the flue gas is cooled by the flue gas desulfurization system, there is a good possibility that acid smut will be formed in the stacks and ducts. Fortunately, as previously mentioned, a flue gas desulfurization system does remove part of the sulfur trioxide from the flue gas.

Another technique for the control of the acid smut mentioned by Gills and Lees is the neutralization of the sulfuric acid by an additive. The injection of finely divided magnesium hydroxide gives effective results in some cases. In this country, a number of companies offer additives for reduction of the sulfur trioxide. Most help decrease sulfuric acid formation in the stack.

Another cause of acid smut is air infiltration at the top of the stack. This effect can occur with any stack liner and should be considered in the initial design. Because of this effect, some liners are now designed with a high-corrosion-resistant upper liner section. You should consider installing a highly corrosion resistant material inside your existing liner or in place of the existing liner (depending on the rec-ommendations of your engineer), if you find an accelerated corrosion rate at the top of your stack.

To get an indication of the corrosion rate in the stack, conduct a coupon corrosion test. Discussed earlier in the chapter, these tests are described in detail by ASTM Standards G4 and D2688.

A device to determine the actual corrosion rate experienced by the metal liner is an ultrasonic thickness gauge. With a gauge capable of reading 0.001-in. dif-ferences of metal thickness, you can take initial readings at various locations on the liner. Then, at selected intervals of plant operation, additional readings can be taken at the same locations and the corrosion rate of your liner determined. This measurement technique has the advantage of allowing you to make measurements from the dry side of the liner while the liner is in operation. The main limitation is that the liner must be fabricated of solid metal.

There are a number of methods in addition to corrosion coupon tests and sulfur trioxide EPA Method 8 sampling for determining the corrosive conditions in the

liner. These include an air-cooled constant temperature corrosion probe, the deposition rate probe, and the electrical conductivity dew point meter. In Table 12-4
a comparison is made of the various methods and the results obtained from each
one. One of the problems with sulfuric acid is that the maximum rate of deposition
of the acid occurs 20 to 90°F below the dew point of the acid. It also appears to
be a function of the flue gas sulfur trioxide concentration. In a paper by Gibson and
Kufta [23] a comparison is made between the various cold-end corrosion monitoring
methods. If any of these methods are being considered for determining the corrosive
environment in your liner, it is suggested that this reference be reviewed.

12.4 INSPECTION AND REPAIR OF LINERS

With the advent of FGD systems, the inspection of the stack liner has taken on
increased importance. It is suggested that you contact a stack inspection firm to
make a formal inspection of your stack liner after its first year of operation. The
frequency of additional inspections should be determined after consultations with
the stack inspectors and the engineer.

Inspection and repair information in this section will concentrate on five liner
types:

1. Independent brick
2. Steel
3. Gunned concrete
4. Fiberglass
5. Alloy

Independent Brick Liners

Some of the inspection criteria for brick liners has already been discussed in
the section on design of brick liners.

One of the more frequent defects in the brick liner is cracking. There are
basically three types of cracks. The first is the crack caused by a settling foundation.
These cracks generally run upwards or zigzag through mortar joints. Next is the
crack caused by excessive wind loads. These cracks usually go through the mortar
joints at 45° angles. The last is the thermal crack that extends upward through the
brick and mortar. If any of these cracks are observed, special attention should be
given to the metal tension bands holding the bricks together. It is possible that acid
has migrated through the cracks in the bricks and attacked the metal bands. Repair
all cracks based on the recommendations of the stack designer; replace all damaged
bands.

Should brick liners be out of service for an extended period of time, place a
cap over the top of the liner. If rain or dew is allowed to wet acid-saturated

TABLE 12-4 Comparison of cold-end corrosion monitoring methods.

Method	Variable Measured	Time Required	Cost	Advantages	Limitations
Corrosion coupon	Test specimen	1–3 months	Low	Simple	Data after the fact; does not indicate effect on corrosion from changing variables
Air-cooled constant-temperature corrosion probe	Test specimen corrosion	1 day–1 week	Low	Can determine corrosion rate as a function of temperature	Temperature can be difficult to control; not good for long-term test
Recirculating fluid constant temperature corrosion probe	Test specimen corrosion	1 day–1 month	Low to moderate	Can determine corrosion rate as a function of temperature	Data after the fact; not good for long-term test
Correlating flue gas sulfur trioxide with theoretical acid dew point	Flue gas sulfur trioxide	3–5 hours, including setup time	Moderate to high	Flue gas SO_3 can be estimated to give quick estimation of problems	Only indicates approximate acid dew point; does not indicate if corrosion will occur or its severity
Deposition rate probe	Rate of acid deposition	3–5 hours including setup and lab work	Low to moderate	Can determine acid dew point and rate of acid depositions; chemical analysis of probe sample can provide added information	Data sometimes scattered and therefore hard to interpret; distinct acid dew point sometimes hard to find
Electrical conductivity dew point meter	Temperature at which acid condenses on probe; rate of acid deposition	30 min to several hours	Moderate	Quick and easy; measures both acid dew point and rate of buildup	Very low sulfur fuels (0.3%) can give inconsistent readings or require extensive time to find dew point; high dust loadings may interfere with RBU measurements; repeatability sometimes poor

brickwork, sulfation of the mortar joints can occur. This will lead to cracking and spalling when the stack is put back into service.

Inspection of the liner should occur during the annual outage. Failure to make adequate preventative inspections can result in major failure of the liner. During the inspection look for:

Stack - Loose, missing or spalled brick, weathered joints, and cracked or weathered corbels caused by water or acid draining from the exterior stack wall projections.

Cap - Tightness of the seal to the shell, loose or missing lightning rod points, loose or broken cable point connections, corrosion in the entire system.

Ladders and platforms - Corrosion of the structure, proper attachment to the shell.

Lining - The metal bands—are they structurally sound and not being corroded? Is the packing intact at the breeching entry?

As part of the annual outage, remove soot from the lining. On old pre-FGD stacks, use a long-handled hoe to remove the soot through the cleanout door. Do not use water for soot removal—water can cause rapid chilling of the hot masonry walls, which can lead to excessive thermal stress and possible wall cracking. On new FGD system linings, water on the lining can result in mortar attack and failure of the structure.

Stacks that were previously used for nonscrubbed gases are sometimes subjected to scrubbed flue gas due to retrofit FGD systems. These linings will probably have to be inspected more frequently for corrosion attack. This is true whether there is reheat or not. The addition of the FGD system will add moisture to the stack and lower the temperature of the gases in the stack. One firm has suggested that a solution to the acid attack of these retrofitted flues is to shotcrete the existing lining with an acid-resistant material. Another solution is to line the existing lining with a fiberglass liner. This liner will restrict the maximum operating temperature of the flue gases to 350°F or the upper operating limit of the resin used in the fiberglass liner. Lining the brickwork with an alloy liner is an alternative to totally replacing the brick liner. Installing any type of liner inside a brick liner should only be considered if the existing brickwork is in excellent structural condition or can be restored to excellent structural condition.

It is best to consult the engineer and stack contractor before any repairs are made to the stack. By doing this you can ensure that all design considerations are taken into account before any changes are made to the stack.

Steel Liners

Steel liners are found in two basic forms in the power industry. The first is the bare carbon steel liner that is used on a conventional power plant that does not have a flue gas desulfurization system. The second is the carbon steel liner coated with some organic or inorganic material used after a flue gas desulfurization system.

Inspection of the bare carbon steel liner has been a standard procedure for a number of years. The main areas of concern are the places where the temperature is allowed to drop below the acid dew point for sulfuric acid. At these locations there will be corrosion of the steel. Areas susceptible to this form of corrosion are the entry doors into the liner, sample ports, inlet breechings, and the liner that extends beyond the cap. The basic repair procedure is to remove the corroded steel and replace it with new metal.

An inspection should be made of the insulation around the steel liner. Any missing insulation should be replaced to maintain the steel temperature above the acid dew point.

Failure of the gunited inorganic coating in a steel liner will be either mechanical, chemical, or both. Mechanical failure produces extensive cracking, spalling, and chucks falling from the coating. Chemical failure appears as a soft and mushy texture of the lining.

The first repair step is to contact the lining manufacturer. Depending on the particular coating used, it may or may not be possible to join new lining patches to existing lining. In general, defective portions of the existing inorganic coating can be removed from the steel liner with water lances. Repairs should be made to the remaining coating according to the manufacturer's recommendations.

Since repairs of organically coated steel liners require the services of the coating applicator, the recognition of coating failures will be discussed here.

NACE Publication 6D170, *The Causes and Prevention of Coating Failures* [24], describes some types of coating failures:

1. Chalking
2. Color fading or color change
3. Cracking, checking, and alligatoring
4. Peeling, flaking, and delamination
5. Rusting
6. Blistering
7. Lifting and wrinkling
8. Failures due to chemical attack or solvent attack
9. Failures around weld areas
10. Edge failures

This list does not include mechanical damage, erosion, temperature excursions, and so on.

A detailed explanation of each of these modes of failure is beyond the scope of this chapter, but here is a brief description of each mode of failure:

1. *Chalking.* Term used to describe the formation of a powdery layer on a coating surface exposed to weather. This powdery layer is usually white.
2. *Color facing or color change.* Is usually caused by reaction of the pigment and/or binder with sunlight or the chemical environment.

3. *Cracking, checking, and alligatoring.* Cracking and checking are caused by shrinkage within the film during aging. Alligatoring is a film rupture which is usually caused by application of a hard, brittle film over a more flexible film.

4. *Peeling, flaking, and delamination.* These are caused by poor adhesion. Flaking is a failure where the coating comes off in small flakes. Peeling is a failure where the coating comes off in large sections or sheets. When flaking or peeling occurs between coats, this failure is called delamination.

5. *Rusting.* Usually appears as (1) spot rusting at damaged areas, (2) pinhole rusting at minute areas, (3) rust nodules breaking through the coating, and (4) underfilm rusting, which eventually causes peeling or flaking of the coating. When rusting reaches 25% of the total coated surface area, some companies remove the entire coating and replace it.

6. *Blistering.* This describes small or large round projections or pimples that appear on the coating surface.

7. *Lifting and wrinkling.* Lifting describes a phenomenon that occurs when the solvent of a coat of paint softens the previous coat of paint too readily. Wrinkling describes the appearance of a paint surface that has not dried to a smooth layer, but rather into a roughly ridged surface.

8. *Failure due to chemical attack or solvent attack.* This will be evident by the apparent disintegration or dissolution of the film.

9. *Failures around weld areas.* Weld defects result in underfilm rusting, formation of rust nodules, and film peeling.

10. *Edge failures.* Usually appear in the form of rusting through the film and eventual rust creepage under the film.

Gunned Concrete

Problems encountered with gunned concrete liners are the same as those experienced with gunned coatings on steel liners. The repair and inspection procedures are also the same.

FRP Liners

These inspection procedures are covered quite adequately by the information furnished by Pullman Power Products in Table 12-5.

Specifications used as part of the specifications for fiberglass liners include:

ASTM Standard D3299

NBS Voluntary Product Standard PS15-69

ASTM Standard D2393-1972

ASTM Standard D2471-1971

ASTM Standard D2583-1975

ASTM Standard D2584-1972

ASTM Standard B84-1975

All resins used in fiberglass liners should contain a fire suppressant (antimony trioxide—less than 5% concentration) and an ultraviolet absorber to increase resistance to the destructive effect of the sun on the resin. Where a coating that is resistant to the chemical environment in the absorber is applied over the fiberglass liner, the life of the fiberglass is increased. (These coatings should contain a high-ultraviolet-reflecting pigment such as titanium dioxide.)

If your FRP liner experiences a thermal excursion or fire, it should be inspected. Look for delamination of the resin-rich layers from the fiberglass-rich layers, and blistering, charring, and fracturing of the resin-rich layer.

On some liners, a coating repair may be all that is required. Remember that the application conditions for the repair should be the same as those specified by the manufacturer for the coating application. Two forms are included that will help ensure a good coating application or repair. Before the repair work begins, a prejob conference should be held (Figure 12-2). The purpose of this conference is to establish details of how the coating work is to be done. At this meeting, establish mutual agreement on all parts of the job. Then, when the work actually begins, everyone will know how the repairs are to be made. The agreements reached at this meeting are the basis for the inspection criteria during the application of the coating.

During coating application, daily coating reports (Figure 12-3) should be issued. By making a continuing inspection, the final performance of the coating should be satisfactory.

Alloy Liners

Alloy liners are constructed of various metals. The precise alloy selected for your liner depends on the chemical environment it will face. To get maximum corrosion resistance from alloys, it is important that they be kept clean. That is true for all types of alloys. There is a saying in the alloy field that "stainless steel to be stainless must be clean." To a certain extent this applies to the nickel alloys as well as to the stainless steels.

To examine the alloys, it is necessary to remove the dust and scale deposits from the metal surface. The engineer should recommend the type of tools and cleaning devices needed for your alloy type.

Because there are a number of alloys being used in the fabrication of liners, it is hard to generalize about the precise modes of failure that might occur in your alloy. There are several generalizations that can be made, however. For example, the use of steel shovels to clean an austenitic stainless steel can leave iron on the surface of the stainless steel. These iron scrapings will rust when the liner is put back into service and possibly leave pits in the stainless steel. These pits could then

TABLE 12-5 Visual inspection criteria and repair requirements for FRP chimney liners [25].

	Discrepancy	Description	Size—Description	No./Ft.^{2}	Repair Requirements
A. Pipe ID surface					
1.	Air bubble (voids)	Air entrapment within and between the plies of liner reinforcement usually spherical in shape	a. $^1/_{16}$ in.	See Note 1	None required
			b. In excess of a	—	See Note 2
	Porosity	Numerous very fine air bubbles trapped in the interior surface			
2.	Crack, surface	Cracks existing only on the surface of the laminate	a. Hairline cracks × 10 in. long (less than ¼ in. not considered)	6	Overlay with equivalent liner material
	Crazing	Fine cracks at or under the laminate	b. In excess of a	—	See Note 3
3.	Crack	An actual separation of laminate, visible on opposite surfaces, and extending through the thickness	No cracks allowed	—	Not allowed
4.	Dry-spot	Area where veil has not been thoroughly wetted with resin	a. 6 × 3 in. long	1	Repair with equivalent liner material
			b. In excess of a	—	See Note 4
5.	Lack of fillout	An area, occurring usually in a radius where the resin has drained from the reinforcement	a. Gasket sealing surface 1½ × ⅜ × ¼ in. deep	2	Fill void with filled resin
			b. Nonscaling area 1½ × ⅜ × ¼ in. deep	6	None required
			c. In excess of a or b	—	See Note 4
6.	Pits (pinholes)	Small crater in the surface of the laminate	a. ⅛ × $^1/_{32}$ in. deep	10	None required
	Chip	A small piece broken off an edge of surface	b. In excess of a	—	Fill with filled resin

7.	Scratch	Shallow mark, groove, furrow, or channel caused by improper handling, storage, or extraction of part from form	a. $1/32$ in. deep b. In excess of $1/32$ in. but not exceeding $1/16$ in. c. In excess of b	—	Coat with liner resin /Cover with resin impregnated veil See Note 4
8.	Wrinkles	In a laminate, an imperfection that has the appearance of a wave molded into one or more plies of fabric or other reinforcement material		See Note 3	Build up liner to required thickness
B. Pipe OD surface					
1.	Burned	Showing evidence of thermal decomposition, through some discoloration, distortion, or destruction of the surface of the laminate; discoloration is usually white and gradually becomes darker brown as it becomes more severe; usually caused by excessive exothermic temperature	a. 4 in across b. In excess of a	1 —	None required See Note 4
2.	Crack, surface	Cracks existing only on the surface of the laminate	a. $1/16$ in. deep	—	None required
	Crazing, surface crack	Fine cracks at or under the laminate	b. In excess of a	—	Consider as a fracture as in B-5
	Scratch	Shallow mark, groove, furrow, or channel caused by improper handling, storage, or extraction of part from form			
3.	Crack	An actual separation of laminate visible on opposite surfaces, and extending through the thickness	No cracks allowed	—	Not allowed

TABLE 12-5 Continued

	Discrepancy	Description	Size—Description	No./Ft.^{2}	Repair Requirements
4.	Dry-spot	Area where reinforcement has not been thoroughly wetted with resin	a. 4 in. across	2	Coat with resin
	Lack of fillout	An area, occurring usually in a radius area where the resin has drained from the reinforcement	b. In excess of a	1	See Note 4
5.	Fracture	Rupture of laminate surface without complete penetration	No fractures allowed (see B-2)	—	See Note 4
6.	Orange-peel	An uneven surface somewhat complete resembling an orange-peel	a. ¼ in. max. height	—	None required
	Pimple	Small, sharp, or conical elevation on exterior surface of laminate	b. In excess of a	—	Grind excess
	Drip, resin	Resin protrusions			
	Fish-eye	Small globular mass that has not blended completely into the surrounding material and is particularly evident in a transparent or translucent material			
7.	Pits (pinholes)	Small crater in the surface of the laminate	a. $^3/_{16} \times {}^1/_{16}$ in. deep	20	None required
	Chip	A small piece broken off an edge	b. In excess of $^1/_{16}$ up to ⅛ in. deep	10	Fill with filled resin
			c. Chips in excess of b	—	See Note 4
C. Laminate wall					
1.	Air bubble (blister, porosity)	Air entrapment within and between the plies of liner reinforcement usually spherical in shape	a. Air bubbles ⅛ in. dia. ($^1/_{16}$ in. or less not considered)	6	None required

	Voids (bridging)	Small air inclusions entrapped between the rovings of the filament winding	b. Voids and wormholes $1/8 \times 1\frac{1}{2}$ in. long	6	None required
	Wormhole	Elongated air entrapment which is either in or near the surface of a laminate and maybe covered with a thin film of cured resin	c. In excess of a or b above	—	See Note 4
2.	Burned	Same as in B-1	Consider as in B-1		
3.	Dry-spot	Area where reinforcement has not been thoroughly wetted with resin	a. 2 in. across	1	None required
	Delamination	Separation between layers of material. Usually appears as a thin light-shaded (whitish) layer	b. In excess of a	—	See Note 4
4.	Foreign inclusion	Particles of a substance included in a laminate which seems foreign to its composition	a. $\frac{1}{2}$ in. b. In excess of a	2	None required Overlay required if within $1/8$ in. of surface; repair with liner material if near ID and fab/mat if near OD

General note regarding all discrepancies: In some cases, repairs may not be permitted if the discrepancy or number of discrepancies are deemed excessive for repair.

Note 1: No limitation on number provided that surface is resin-rich.

Note 2: Where larger bubbles than specified are pressed with a blunt instrument, and the inner surface fractures, the fractured surface will be removed and the void filled with a resin putty. Bubbles that cannot be opened should be treated as a laminate air bubble as C-1.

Note 3: No limitations provided that full liner thickness or minimum ID is not impaired by a shallow void or fragile crest.

Note 4: Proposed repair procedure must be approved by the buyer.

411

Project _________________________ Location _________________________

Meeting Place _________________ Date ________________ Time _________

Owner ______________________ General Contractor ________________

Owner Representative ____

Engineer Representative ____

General Contractor Representative _______________________________

Paint Manufacturer Representative _______________________________

Paint Contractor Representative _________________________________

All parties have read and understand specifications?

Coating System					
Surface Preparation					
Field Demonstration					
Anchor Pattern					
Prime-Time and Date					
Mils Dry Film					
Required Field Prep.					
Field Touch-Up					
Mils Dry Film					
Dry or Cure Time					
1st Finish Coat					
Mils Dry Film					
Recoat Time					
2nd Finish Coat					
Mils Dry Film					
Recoat Time					
3rd Finish Coat					
Mils Dry Film					
System Min. Dry Film					

FIGURE 12-2 Prejob conference.

COATING REPORT

Project _________________ Location _____________________________ Date _________

Contractor ___________________ Foreman _______________________ Surface Prep _______

Profile: Spec ____________ ;   Obtained _________ ;   Quality ______ ;   Time Completed ____

Primer: Spec ___________ ;   Used ________ ;   Air Temp ____ °F;   Surface ___________ °F

Dew Point _________ °F Dry Film: Spec _______________ mils ;   Obtained _______ mils

Relative Humidity _____________________ % Time Completed _________________________

Field Touch-Up Contractor ____

Foreman _________________ Date ___________ Time _________ Air Temp _________ °F

Surface Temp __________ °F;   Relative Humidity _________ % ;   Dew Point _________ °F

Spec Paint _______________ ;   Item Used __________ ;   Dry Film:Spec ________ mils

Obtained _______________ mils

Finish Coats: Contractor ____

Foreman _________________ Date ___________ Time __________ Air Temp ________ °F

Surface Temp __________ °F;   Relative Humidity _________ % ;   Dew Point _________ °F

Paint Spec ________________ ; Item Used ___________ ; Dry Time Before Recoat _________

Cure Time Allowed ___________________ Dry Film: Spec ________ mils Obtained ______ mils

Call contractors' attention at once to any deviation from the specification.
Unless corrective action is taken at once (same day), send owner a letter of
exception to such procedure, citing the error and remedy requested. Copies
of such letter got to engineer and contractors.

FIGURE 12-3 Coating report.

become points where chloride pitting might occur. Therefore, it is standard practice
to use stainless steel tools when working with stainless steel.

In an austenitic stainless steel, such as 316, 326L, 317, 317L, 317LM, and
904L, a loss of passivity of the metal will appear as a light layer of rust on the
alloy surface. Removal of this rust will probably reveal pitting corrosion underneath.
These alloys will first show corrosion in areas of the liner that do not have rapid
gas flow, for example where sample pipes go through the wall of the liner, at the
surface of the wall of the liner, and recessed flanges out of the main flow of gas
through the vessel.

In nickel-based alloys, there should not be any iron-colored corrosion products on the metal surface. Should a rustlike corrosion be found on the surface, it probably came from damage caused by iron tool scraping of the alloy.

Should scale be allowed to form and remain on the surface of the alloy, a form of crevice corrosion can begin. The concentration of corrosive salts under the scale can exceed the concentration found in the main liquid flow in the liners. Some recent corrosion studies by one of the major alloy suppliers has verified this increased corrosion effect in FGD systems and thus in liners of stacks. Scale that forms on the alloys should not be removed with carbon steel equipment—stainless tools have to be used.

Handling of the nickel alloys requires special techniques in order to maintain the corrosion resistance for which they are noted. The engineer should be involved in the inspection and advise the owner about repairs. Each alloy supplier has an excellent selection of instructions for the welding and repair of their alloys. These should be followed in detail.

Alloy welding rods of MIG or TIG welding techniques are needed for all alloy liner repairs. This may require the purchase of new pulse arc welding machines and the retraining of plant personnel. Corrosion-resistant alloys cannot be treated as carbon steel and retain their corrosion-resistant properties.

Some FGD system owners have already found that they must clean the stainless steel in their FGD systems to keep the corrosion to a minimum. This is also true of the stainless steels in the stack area. Since it is not as feasible to clean the stack liner as it is to clean the FGD system, it seems logical to construct the liners of more-corrosion-resistant metals than are used in the FGD system.

When welding and repairing the alloy metals, be sure that the correct welding rods or wires are used. If there is any chance that carbon steel can be brought into the molten weld pool, the alloy rod or wire must have sufficient metallurgical composition not to lose the desired corrosion resistance when the iron is brought into the weld. It is usually estimated that about 25% of the molten weld pool, when welding an alloy to carbon steel, can come from the carbon steel. The alloy manufacturer can recommend the proper welding rod or wire for its alloy to minimize the effect of this iron dilution.

It is important that good fabrication techniques be used in working with the alloys in the stack. For example, the alloy should be wiped on both sides of the area to be welded with a solvent-soaked rag. This is to remove any organic material that might be present. During the welding process, this organic material can be burnt and brought into the molten weld metal. This will increase the carbon content of the final weld and lower the corrosion resistance of the weld.

12.5 CONCLUSIONS

From the day a stack is built, both external and internal forces are at work to destroy it.

External villains are much easier to design for than internal ones, so exterior inspection and maintenance procedures can usually be based on known design parameters and realistic life expectations of construction materials. Despite this relative stability, however, internal corrosion may affect external performance. It cannot be overstressed that programmed inspections, timely crack repairs, and scheduled maintenance procedures are the keys to continuing stack shell integrity.

Inside the stack, knowing the strengths and weaknesses of your particular type of liner is just as important as knowing the kind of environment your liner must face. Independent brick liners and the emerging family of alloy liners have the best record and highest potential for success in the battle against low-temperature, wet stack environments, but even with these design Goliaths there may be a David hiding in the post-FGD or bypass effluent. You *must* determine effluent temperature at various stack heights, dew point, condensation, and a myriad of other stack environment data. Only then can you tell what your liner can tolerate and for how long.

Continuing research and development portend optimism for increased stack longevity and performance reliability. Someday there may be a stack that "can take it," but that day is not here yet.

I would like to thank Mr. Robert R. Pierce of Pennwalt Corp. for his extensive review of this chapter. Many of his helpful, personal comments have been incorporated into this chapter.

12.6 REFERENCES

[1] SNOOK, R. W., "The Impact of Flue Gas Desulfurization on Chimney Design in the United States," Third International Chimney Design Symposium, Munich, Germany, October 25, 1978.

[2] COOLEY, B. L., "Consider Dynamic Wind and Earthquake Loads When Designing Reinforced-Concrete Stacks," *Power*, p. 24, October 1977.

[3] CUSTODIS CONSTRUCTION COMPANY, "Design and Construction of Independent Brick Chimney Linings," revised paper, Chicago, Ill., April 12, 1979.

[4] COOLEY, "Consider Dynamic Wind," p. 27.

[5] Ibid.

[6] KOSSLER, D. J., AND PIERCE, R. R., "Power Industry Chimneys: Brickwork Lining Performance," Canadian Electrical Association Thermal Power Section Eastern Zone Meeting, Mont Gabriel, Quebec, October 6, 1964.

[7] KOSSLER AND PIERCE, Ibid.

[8] SNOOK, R. W., chairman, "Interim Report of the ASCE Task Committee on Steel Chimney Liners," American Power Conference, Chicago, Ill., April 20, 1972, p. 1.

[9] READ, G. W., JR., "Corrosion-Resistant Linings for Chimneys, Scrubbers and Ducts of Flue Gas Systems," American Power Conference, 40th Annual Meeting, Chicago, April 24–26, 1978, p. 2.

[10] Ibid., p. 4.

[11] SNOOK, "Interim Report of the ASCE Task Committee," p. 8.

[12] READ, G. W., JR., "Acidproof Concrete Lining for Steel Stacks of Power Plants," Sauereisen Cements Company, Pittsburgh, Pa., p. 4.

[13] "Detroit Edison First to Insulate Stack Liner with New High Temperature Fiber Glass Board," *Combustion*, p. 45, September 1977.

[14] CUSTODIS CONSTRUCTION COMPANY, "Sample Specification for Fiberglass Reinforced Plastic Chimney Liners," Chicago, Ill., p. 2.

[15] Ibid., p. 3.

[16] Ibid., p. 5.

[17] *Weldability Characteristics of Stellite Division High Performance Alloys*, Technology Division, Cabot Corporation Stellite R&D Department, September 12, 1977.

[18] GIBSON, G., AND KUFTA, R., "Cold-End Corrosion Monitoring Methods," *Power Eng.*, p. 71, April 1977.

[19] PIERCE, R., "Acid Dew Point Estimating Method," 1976 Chimney Design Symposium and Exhibition, TH 2281.C55, 1976, Quarto Vol. 3, p. 143.

[20] KOSSLER AND PIERCE, "Power Industry Chimneys."

[21] Ibid.

[22] GILLS, B. G., AND LEES, B., "Insulated Steel Chimneys Avoid Corrosion and Smut Emission," American Society of Mechanical Engineers Winter Annual Meeting, Chicago, November 7–11, 1965, p. 2.

[23] GIBSON AND KUFTA, "Cold-End Corrosion." p. 73.

[24] *The Causes and Prevention of Coating Failures*, NACE Publication 6D170 prepared by Task Group T-6D-22 (A. K. Long, Chairman) and Unit Committee T-6D (R. R. Rosenthal, Jr.), pp. 33–36.

[25] PULLMAN POWER PRODUCTS.

Appendix

SI Units

Louis Theodore, D. Eng. Sc.

Professor of Chemical Engineering
Manhattan College
Bronx, New York

A.1 THE METRIC SYSTEM

The need for a single worldwide coordinated measurement system was recognized over 300 years ago. Gabriel Mouton, Vicar of St. Paul in Lyons, proposed in 1670 a comprehensive decimal measurement system based on the length of 1 minute of arc of a great circle of the earth. In 1671, Jean Picard, a French astronomer, proposed the length of a pendulum beating seconds as the unit of length. (Such a pendulum would have been fairly easily reproducible, thus facilitating the widespread distribution of uniform standards.) Other proposals were made, but over a century elapsed before any action was taken.

In 1790, in the midst of the French Revolution, the National Assembly of France requested the French Academy of Sciences to "deduce an invariable standard for all the measures and all the weights." The commission appointed by the academy created a system that was, at once, simple and scientific. The unit of length was to be a portion of the earth's circumference. Measures for capacity (volume) and mass (weight) were to be derived from the unit of length, thus relating the basic units of the system to each other and to nature. Furthermore, the larger and smaller versions of each unit were to be created by multiplying or dividing the basic units by 10 and its multiples. This feature provided a great convenience to users of the system, by eliminating the need for such calculations as dividing by 16 (to convert ounces to pounds) or by 12 (to convert inches to feet). Similar calculations in the metric system could be performed simply by shifting the decimal point. Thus, the metric system is a "base 10" or "decimal" system.

The commission assigned the name "metre" (which we now spell meter) to the unit of length. This name was derived from the Greek word "metron," meaning "a measure." The physical standard representing the meter was to be constructed so that it would equal one ten-millionth of the distance from the North Pole to the equator along the meridian of the earth running near Dunkirk in France and Barcelona in Spain.

The metric unit of mass, called the "gram," was defined as the mass of one cubic centimeter (a cube that is 1/100 of a meter on each side) of water at its temperature of maximum density. The cubic diameter (a cube 1/10 of a meter on each side) was chosen as the unit of fluid capacity. This measure was given the name "liter."

Although the metric system was not accepted with enthusiasm at first, adoption by other nations occurred steadily after France made its use compulsory in 1840. The standardized character and decimal features of the metric system made it well suited to scientific and engineering work. Consequently, it is not surprising that the rapid spread of the system coincided with an age of rapid technological development. In the United States, by act of Congress in 1866, it was made "lawful throughout the United States of America to employ the weights and measures of the metric system in all contracts, dealings or court proceedings."

By the late 1860s, even better metric standards were needed to keep pace with scientific advances. In 1875, an international treaty, the "Treaty of the Meter," set up well-defined metric standards for length and mass, and established permanent machinery to recommend and adopt further refinements in the metric system. This treaty, known as the Metric Convention, was signed by 17 countries, including the United States.

As a result of the treaty, metric standards were constructed and distributed to each nation that ratified the Convention. Since 1893, the internationally agreed-to metric standards have served as the fundamental weights and measures standards of the United States.

By 1900 a total of 35 nations, including the major nations of continental Europe and most of South America, had officially accepted the metric system. Today, with the exception of the United States and a few small countries, the entire world is using the metric system predominantly or is committed to such use. In 1971, the Secretary of Commerce, in transmitting to Congress the results of a three-year study authorized by the Metric Study Act of 1968, recommended that the United States change to predominant use of the metric system through a coordinated national program.

The International Bureau of Weights and Measures located at Sèvres, France, serves as a permanent secretariat for the Metric Convention, coordinating the exchange of information about the use and refinement of the metric system. As measurement science develops more precise and easily reproducible ways of defining the measurement units, the General Conference of Weights and Measures—the diplomatic organization made up of adherents to the Convention—meets periodically to ratify improvements in the system and the standards.

A.2 THE SI SYSTEM

In 1960, the General Conference adopted an extensive revision and simplification of the system. The name "Le Système Internationale d'Unités" (International System of Units), with the international abbreviation SI, was adopted for this modernized metric system. Further improvements in and additions to SI were made by the General Conference in 1964, 1968, and 1971.

The basic units in the SI system are the kilogram (mass), meter (length), second (time), kelvin (temperature), ampere (electric current), candela (the unit of luminous intensity), and radian (angular measure). All are commonly used by the engineer. The Celsius scale of temperature $0°C = 273.15$ K) is commonly used with the absolute Kelvin scale. The important derived units are the newton (SI unit of force), the joule (SI unit of energy), the watt (SI unit of power), the pascal (SI unit of pressure), and the hertz (unit of frequency). There are a number of electrical units: coulomb (charge), farad (capacitance), henry (inductance), volt (potential), and weber (magnetic flux). One of the major advantages of the metric system is that larger and smaller units are given in powers of 10. In the SI system a further simplification is introduced by recommending only those units with multipliers of 10^3. Thus, for lengths in engineering, the micrometer (previously micron), millimeter, and kilometer are recommended and the centimeter is generally avoided. A further simplification is that the decimal point may be substituted by a comma (as in France, Germany, and South Africa), while the other numbers, before and after the comma, will be separated by spaces between groups of three; for example, one million dollars will be "$1 000 000,00." More details are provided below.

Seven Base Units

1. *Length—meter (m)*. The meter (common international spelling, metre) is defined as 1 650 763.73 wavelengths in vacuum of the orange-red line of the spectrum of krypton-86. The SI unit of area is the square meter (m^2). The SI unit of volume is the cubic meter (m^3). The liter (0.001 cubic meter), although not an SI unit, is commonly used to measure fluid volume.

2. *Mass—kilogram (kg).* The standard for the unit of mass, the kilogram, is a cylinder of platinum–iridium alloy kept by the International Bureau of Weights and Measures at Sèvres. A duplicate in the custody of the National Bureau of Standards serves as the mass standard for the United States. This is the only base unit still defined by an artifact. The SI unit of force is the newton (N). One newton is the force which, when applied to a 1-kilogram mass, will give the kilogram mass an acceleration of 1 (meter per second) per second. $1 \text{ N} = 1 \text{ kg·m/s}^2$. The SI unit for pressure is the pascal (Pa). $1 \text{ Pa} = 1 \text{ N/m}^2$. The SI unit for work and energy of any kind is the joule (J). $1 \text{ J} = 1 \text{ N·m}$. The SI unit for power of any kind is the watt (W). $1 \text{ W} = 1 \text{ J/s}$.

3. *Time—second (s).* The second is defined as the duration of 9 192 631 770 cycles of radiation associated with a specified transition of the cesium-133 atom. It is realized by tuning an oscillator to the resonance frequency of cesium-133 atoms as they pass through a system of magnets and a number of periods or cycles per second is called frequency. The SI unit for frequency is the hertz (Hz). One hertz equals one cycle per second. The SI unit for speed is the meter per second (m/s). The SI unit for acceleration is the (meter per second) per second (m/s^2).

4. *Electric current—ampere (A).* The ampere is defined as that current which, if maintained in each of two long parallel wires separated by one meter in free space, would produce a force between the two wires (due to their magnetic fields) of 2×10^{-7} Newton for each meter of length. The SI unit of voltage is the volt (V). $1 \text{ V} = 1 \text{ W/A}$. The SI unit of electrical resistance is the ohm (Ω). $1 \text{ } \Omega = 1 \text{ V/A}$.

TABLE A-1 Multiples and prefixes.

Multiples and Submultiples		*Prefixes*[1]	*Symbols*
1 000 000 000 000	10^{12}	tera (ter ´ a)	T
1 000 000 000	10^{9}	giga (ji ´ ga)	G
1 000 000	10^{6}	mega (meg ´ a)	M
1 000	10^{3}	kilo (kil ´ o)	k
100	10^{2}	hecto (hek ´ to)	h
10	10^{1}	deka (dek ´ a)	da
Base Unit 1	10^{0}		
0.1	10^{-1}	deci (des ´ i)	d
0.01	10^{-2}	centi (sen ´ ti)	c
0.001	10^{-3}	milli (mil ´ i)	m
0.000 001	10^{-6}	micro (mi ´ kro)	u
0.000 000 001	10^{-9}	nano (nan ´ o)	n
0.000 000 000 001	10^{-12}	pico (pe ´ ko)	p
0.000 000 000 000 001	10^{-15}	femto (fem ´ to)	f
0.000 000 000 000 000 001	10^{-18}	atto (at ´ to)	a

[1] These prefixes may be applied to all SI units.

TABLE A-2 Selected common abbreviations.

A°, A	angstrom unit of length	kg	kilogram
abs.	absolute	km	kilometer
amb.	ambient	liq.	liquid
app. mol. wt.	apparent molecular weight	L	liter
atm.	atmospheric	log	logarithm (common)
at. wt.	atomic weight	ln	logarithm (natural)
b.p.	boiling point	m.p.	melting point
bbl	barrel	m, M	meter
Btu	British thermal unit	μm	micrometer (micron)
cal	calorie	mks system	meter–kilogram–second
cg	centigram		system
cm	centimeter	mph	miles per hour
cgs system	centimeter-gram-second	mg	milligram
	system	ml	milliliter
conc.	concentrated, concentration	mm	millimeter
cc, cm^3	cubic centimeter	mμ	millimicron
cu ft, ft^3	cubic feet	min	minute
cfh	cubic feet per hour	mol. wt.	molecular weight
cfm	cubic feet per minute	oz	ounce
cfs	cubic feet per second	ppb	parts per billion
m^3, M^3	cubic meter	pphm	parts per hundred million
°	degree	ppm	parts per million
°C	degree Celsius, degree Centigrade	lb	pound
		psi	pounds per square inch
°F	degree Fahrenheit	psia	pounds per square inch absolute
K	degree Kelvin		
°R	degree Reaumur, degree Rankine	psig	pounds per square inch gage
		rpm	revolutions per minute
ft	foot	sec	second
ft lb	foot pound	sp. gr.	specific gravity
fpm	feet per minute	sp. ht.	specific heat
fps	feet per second	sp. wt.	specific weight
fps system	foot–pound–second system	sq	square
f.p.	freezing point	scf	standard cubic foot
gr	grain	STP	standard temperature and pressure
g, gm	gram		
hr	hour	temp.	temperature
in.	inch	wt	weight
kcal	kilocalorie		

5. *Temperature—kelvin (K).* The kelvin is defined as the fraction 1/273.16 of the thermodynamic temperature of the triple point of water. The temperature 0 K is called "absolute zero." On the commonly used Celsius temperature scale, water freezes at about 0°C and boils at about 100°C. The °C is defined as an interval of 1 K, and the Celsius temperature 0°C is defined as 273.15 K. 1.8 Fahrenheit degrees are equal to 1.0°C or 1.0 K; the Fahrenheit scale uses 32°F as a temperature corresponding to 0°C.

TABLE A-3 Common conversion factors.[1]

To Convert from:	To:	Multiply by:
milligrams/m^3	micrograms/m^3	1,000
	micrograms/liter	1.0
	ppm by volume (20°C)	(24.04/M)
	ppm by weight	0.8347
	lb/ft^3	62.43×10^{-9}
micrograms/m^3	milligrams/m^3	0.001
	micrograms/liter	0.001
	ppm by volume (20°C)	(0.02404/M)
	ppm by weight	834.7×10^{-6}
	lb/ft^3	62.43×10^{-12}
micrograms/liter	milligrams/m^3	1.0
	micrograms/m^3	1,000
	ppm by volume (20°C)	(24.04/M)
	ppm by weight	0.8347
	lb/ft^3	62.43×10^{-9}
ppm by volume (20°C)	milligrams/m^3	(M/24.04)
	micrograms/m^3	(M/0.02404)
	micrograms/liter	(M/24.04)
	ppm by weight	(M/28.8)
	lb/ft^3	$(M/385.1 \times 10^6)$
ppm by weight	milligrams/m^3	1.198
	micrograms/m^3	1.198×10^{-3}
	micrograms/liter	1.198
	ppm by volume (20°C)	(28.8/M)
	lb/ft^3	7.48×10^{-6}
lb/ft^3	milligrams/m^3	16.018×10^6
	micrograms/m^3	16.018×10^9
	micrograms/liter	16.018×10^6
	ppm by volume (20°C)	$(385.1 \times 10^6/M)$
	ppm by weight	133.7×10^3
grams	milligrams/m^3	35.3145×10^3
	grams/m^3	35.314
	micrograms/m^3	35.314×10^6
	micrograms/ft^3	1.0×10^6
	lb/1000 ft^3	2.2046
grams/m^3	milligrams/m^3	1000.0
	grams/ft^3	0.02832
	micrograms/m^3	1.0×10^6
	micrograms/ft^3	28.317×10^3
	lb/1000 ft^3	0.06243
micrograms/ft^3	milligrams/m^3	35.314×10^{-3}
	grams/ft^3	1.0×10^{-6}
	grams/m^3	35.314×10^{-6}
	micrograms/m^3	35.314
	lb/1000 ft^3	2.2046×10^{-6}
no. of particles/ft^3	no./m^3	35.314
	no./L	35.314×10^{-3}
	no./cm^3	35.314×10^{-6}

422

TABLE A-3 Continued

To Convert from:	To:	Multiply by:
tons/mi^2	lb/acre	3.125
	lb/1000 ft^2	0.07174
	grams/m^2	0.3503
	kg/km^2	350.3
	milligrams/m^2	350.3
	milligrams/cm^2	0.03503
	grams/ft^2	0.03254
lb	gr	7,000.0
micrometer	in.	3.937×10^{-5}
	mm	1.0×10^{-3}

[1]*Note:*
cm   = 0.0328 ft
gal (U.S.) = 0.1337 ft^3
liter   = 0.03532 ft^3 = 0.001 m^3
microgram   = 0.000001 g
micrometer = 0.0000394 in. = 0.001 mm
milligram   = 0.001 g
lb   = 7000 grains = 453.6 g
M   = molecular weight

6. *Amount of substance—mole (mol).* The mole is the amount of substance of a system that contains as many elementary entities as there are atoms in 0.012 kilogram of carbon-12. When the mole is used, the elementary entities must be specified and may be atoms, molecules, ions, electrons, other particles, or specified groups of such particles. The SI unit of concentration (of amount of substance) is the mole per cubic meter (mol/m^3).

7. *Luminous intensity—candela (cd).* The candela is defined as the luminous intensity of 1/600 000 of a square meter of a blackbody at the temperature of freezing platinum (2045 K). The SI unit of light flux is the lumen (lm). A source having an intensity of 1 candela in all directions radiates a light flux of 4π lumens.

Two Supplementary Units

1. *Plane angle—radian (rad).* The radian is the plane angle with its vertex at the center of a circle that is subtended by an arc equal in length to the radius.

2. *Solid angle—steradian (sr).* The steradian is the solid angle with its vertex at the center of a sphere that is subtended by an area of the spherical surface equal to that of a square with sides equal in length to the radius.

Multiples and prefixes assigned to SI units are given in Table A-1. Common abbreviations employed in engineering practice, including those for SI units, are given in Table A-2. Common conversion factors are presented in Table A-3.

Index

Springer-Verlag
and the Environment

We at Springer-Verlag firmly believe that an international science publisher has a special obligation to the environment, and our corporate policies consistently reflect this conviction.

We also expect our business partners – paper mills, printers, packaging manufacturers, etc. – to commit themselves to using environmentally friendly materials and production processes.

The paper in this book is made from low- or no-chlorine pulp and is acid free, in conformance with international standards for paper permanency.

MIX
Papier aus verantwortungsvollen Quellen
Paper from responsible sources
FSC® C105338
FSC
www.fsc.org

If you have any concerns about our products,
you can contact us on
ProductSafety@springernature.com

In case Publisher is established outside the EU,
the EU authorized representative is:
Springer Nature Customer Service Center GmbH
Europaplatz 3, 69115 Heidelberg, Germany

Printed by Libri Plureos GmbH
in Hamburg, Germany